Advances in Industrial Control

Advances in Industrial Control is a series of monographs and contributed titles focusing on the applications of advanced and novel control methods within applied settings. This series has worldwide distribution to engineers, researchers and libraries.

The series promotes the exchange of information between academia and industry, to which end the books all demonstrate some theoretical aspect of an advanced or new control method and show how it can be applied either in a pilot plant or in some real industrial situation. The books are distinguished by the combination of the type of theory used and the type of application exemplified. Note that "industrial" here has a very broad interpretation; it applies not merely to the processes employed in industrial plants but to systems such as avionics and automotive brakes and drivetrain. This series complements the theoretical and more mathematical approach of Communications and Control Engineering.

Indexed by SCOPUS and Engineering Index.

Proposals for this series, composed of a proposal form (please ask the in-house editor below), a draft Contents, at least two sample chapters and an author cv (with a synopsis of the whole project, if possible) can be submitted to either of the:

Series Editor

Professor **Michael J. Grimble**:
Department of Electronic and Electrical Engineering, Royal College Building, 204 George Street, Glasgow G1 1XW, United Kingdom;
e-mail: m.j.grimble@strath.ac.uk

or the

In-house Editor

Mr. **Oliver Jackson**:
Springer London, 4 Crinan Street, London, N1 9XW, United Kingdom;
e-mail: oliver.jackson@springer.com

Proposals are peer-reviewed.

Publishing Ethics

Researchers should conduct their research from research proposal to publication in line with best practices and codes of conduct of relevant professional bodies and/or national and international regulatory bodies. For more details on individual ethics matters please see: https://www.springer.com/gp/authors-editors/journal-author/journal-author-helpdesk/publishing-ethics/14214

Jinna Li · Frank L. Lewis · Jialu Fan

Reinforcement Learning

Optimal Feedback Control with Industrial Applications

Springer

Jinna Li
School of Information and Control
Engineering
Liaoning Petrochemical University
Wanghua District, Fushun, Liaoning, China

Frank L. Lewis
UTA Research Institute
University of Texas at Arlington
Fort Worth, TX, USA

Jialu Fan
State Key Lab of Synthetical Automation
for Process Industries
Northeastern University
Heping District, Shenyang, Liaoning, China

ISSN 1430-9491 ISSN 2193-1577 (electronic)
Advances in Industrial Control
ISBN 978-3-031-28393-2 ISBN 978-3-031-28394-9 (eBook)
https://doi.org/10.1007/978-3-031-28394-9

Mathematics Subject Classification: 15-02, 15Axx

This Springer imprint is published by the registered company Springer Nature Switzerland AG
The registered company address is: Gewerbestrasse 11, 6330 Cham, Switzerland

Series Editor's Foreword

Control engineering is viewed rather differently by researchers and those that must design, calibrate, implement, and maintain control systems. Researchers often develop algorithms for general control problems with a well-defined mathematical basis; engineers have more immediate concerns over the limitations of equipment, quality of control, safety, security, and plant downtime. The series *Advances in Industrial Control* attempts to bridge this divide by encouraging the adoption of advanced control techniques when they offer benefits such as higher performance.

The rapid development of new control theory, techniques, and technology has an impact on all areas of engineering applications. This monograph series focuses on applications that often stimulate the development of new more general control paradigms. This is desirable if the different aspects of the 'control design' problem are to be explored with the same dedication that 'control synthesis' problems have received. The series provides an opportunity for researchers to introduce new ideas possibly motivated by the applications of interest. It raises awareness of the various benefits that advanced control can provide whilst explaining the challenges that can arise.

This *Advances in Industrial Control* series monograph is concerned with the application of reinforcement learning (RL) for use in industrial applications. The use of artificial intelligence (AI) in business has attracted great attention and is being applied in many industries. The use of machine learning in control applications has generated similar interest but the techniques are still not well established and proven. This text plays a valuable role by providing a theoretical basis for RL, which is one of the most important topics with good application potential.

Chapter 1 introduces the topic, the optimal control results required, and the architecture of networked control systems. Chapter 2 covers the use of robust H_∞ control that is normally associated with model-based control design methods, where the models are based on a reasonable understanding of the plant and uncertainties. The chapter introduces game problems and reveals how reinforcement learning can add a new dimension to the solution of systems where the uncertainties are difficult to categorise and to some extent define. A Q-learning algorithm is introduced that is data-driven and does not require model knowledge. A winding machine example and results are presented.

Chapter 3 is on robust tracking control and output regulation—common control problems. The Q-learning algorithm is justified theoretically, and stability and convergence issues are addressed. Chapter 4 considers 'adaptive' robust controller design using adaptive dynamic programming. In fact, there is the potential of breathing new life into the adaptive systems area through developments in learning techniques.

Chapter 5 considers the existence of network-induced delays and interestingly uses a Smith predictor which has proven itself over the years in less exalted and more mundane company. An F-16 aircraft autopilot example is included in this chapter.

Chapter 6, on Interleaved Q-Learning, is for affine nonlinear systems. One of the main strengths of Q-learning is that it can update the control action without requiring knowledge of the models of the environment. Neural networks are used, and novel results are described for this difficult problem of uncertain nonlinear systems. Chapter 7 considers off-policy game reinforcement learning for solving the synchronization problem of multi-agent systems.

Chapter 8 is very substantial and represents the main contribution from the series viewpoint. It concerns industrial applications of reinforcement learning. Operational indices must be defined if the system is to be optimized. This requires plant-wide control to be used and both an architecture and an optimal 'operational control problem' for nonlinear systems are discussed. The H_∞ tracking-control problem for operational processes is solved assuming local linearization can be employed, an assumption which is often valid for process control applications. Rougher flotation processes are usually placed at the beginning of a flotation circuit of a mineral concentration process and are one of many industrial processes that are difficult to model and can benefit from a reinforcement learning approach. This problem is considered in the very interesting main design example, which utilizes the data-based techniques introduced in previous chapters. The comparison in the example with classical methods reveals the possible benefits of the learning-based algorithms.

The authors have strong publication records and have provided a valuable text in an important and growing area of data-driven AI-based algorithms and applications. It is sometimes the case that useful ad hoc algorithms are suggested in this area but with a little underpinning theoretical basis. This will certainly deter some potential users, but it is a problem the authors have clearly avoided. The monograph should be useful to control design engineers and to students interested in the theoretical methods that can be applied. It may even motivate members of the wider AI community to put greater effort into industrial control problems research.

Glasgow, UK
December 2022

Michael J. Grimble

Preface

Reinforcement learning (RL) aims at solving decision-making and optimal control problems for complex known and unknown dynamic systems, and its powerful and potential value has been proven not only from theoretical analysis, but also from a variety of applications, such as iron ore processing, power grid systems, unmanned automatic vehicles, and communication networks.

This book provides an in-depth development of RL for optimal feedback control problems of nonlinear systems, multi-player systems, multi-agent systems, and networked control systems with illustrative numerical examples and industrial applications. A description of classical RL and a brief introduction to typical RL algorithms (Chap. 1) build the foundation for the remainder of the book. This is followed by the research of data-driven robust control for unknown dynamics of nonlinear systems and multi-player systems (Chaps. 2–4) in the framework of RL and dynamic programming. Chapters 5 and 6 respectively dedicate to the data-driven optimal control of networked single and multiple-player systems, in which novel RL algorithms with network-induced delay or packet dropout compensation and for improving learning efficiency are developed. Chapter 7 studies the multi-agent RL for finding the synchronization policies using measured data. In Chap. 8, the data-driven game RL methods are presented for optimal operational control (OOC) and the plant-wide operational optimization of industrial processes.

The rigorous theoretical analysis and proofs of stability and performance optimality of systems are provided in this book. For graduate students, scholars, and engineers, this book offers a thorough introduction to both the scientific research and technological innovation of RL-based feedback control. The combination of practical applications, theoretical analysis, and comprehensive examples presented in this book will also interest researchers and practitioners who have been working in the fields of optimal and adaptive control, machine learning, artificial intelligence, and operations research, such that the advanced treatment to control will be applied to the practical industrial control systems.

Fushun, China Jinna Li
Fort Worth, USA Frank L. Lewis
Shenyang, China Jialu Fan

Acknowledgements

We wish to thank the people who have been very enthusiastic about helping us develop the overall view of this book: Tianyou Chai; Jinliang Ding. We thank the people who have made contributions in umpteen ways: Sarangapani Jagannathan, Zhengtao Ding, Zhenfei Xiao, Hao Nie.

This work was supported by the National Natural Science Foundation of China (Grant Nos. 62073158), the Liaoning Revitalization Talents Program (Grant No. XLYC2007135) and the Basic Research Project of Education Department of Liaoning Province (Grant No. LJKZ0401).

Contents

1 Background on Reinforcement Learning and Optimal Control 1
1.1 Fundamentals of Reinforcement Learning and Recall 1
1.2 Fundamentals of Optimal Control with Dynamic Programming 3
1.3 Architecture and Performance of Networked Control System 4
1.4 The State of the Art and Contributions 5
References 7

2 H_∞ Control Using Reinforcement Learning 11
2.1 H_∞ State Feedback Control of Multi-player Systems 11
2.1.1 Problem Statement 13
2.1.2 Solving the Multi-player Zero-Sum Game 17
2.1.3 Off-Policy Game Q-Learning Technique 21
2.1.4 Simulation Results 27
2.2 H_∞ Output Feedback Control of Multi-player Systems 41
2.2.1 Problem Statement 42
2.2.2 Solving the Multi-player Zero-Sum Game 45
2.2.3 Off-Policy Game Q-Learning Technique 49
2.2.4 Simulation Results 56
2.3 Conclusion 60
References 61

3 Robust Tracking Control and Output Regulation 65
3.1 Optimal Robust Control Problem Statement 67
3.2 Theoretical Solutions 70
3.2.1 Solving the Regulator Equations with Known Dynamics 71
3.2.2 Solving Problem 3.2 with Known Dynamics 71
3.3 Data-Driven Solutions 73
3.3.1 Data-Driven OPCG Q-Learning 73
3.3.2 No Bias Analysis of the Solution for the Proposed Algorithm 79

3.4 Illustrative Examples 81
3.5 Conclusion 88
References 88

4 Interleaved Robust Reinforcement Learning 91
4.1 Robust Controller Design and Simplified HJB Equation 93
4.2 Interleaved RL for Approximating Robust Control 98
4.2.1 Theoretical Analysis 101
4.3 Illustrative Examples 104
4.4 Conclusion 110
References 110

5 Optimal Networked Controller and Observer Design 113
5.1 Off-Policy Q-Learning for Single-Player Networked Control Systems 113
5.1.1 Problem Formulation 115
5.1.2 Optimal Observer Design 118
5.1.3 Optimal Controller Design 125
5.1.4 Simulation Results 133
5.2 Off-Policy Q-Learning for Multi-player Networked Control Systems 139
5.2.1 Problem Formulation 140
5.2.2 Main Results 142
5.2.3 Illustrative Example 148
5.3 Conclusion 150
References 150

6 Interleaved Q-Learning 155
6.1 Optimal Control for Affine Nonlinear Systems 155
6.1.1 Problem Statement 157
6.1.2 On-Policy Q-Learning Formulation 158
6.2 Off-Policy Q-Learning Technique 163
6.2.1 Off-Policy and Q-Learning 163
6.2.2 Derivation of Off-Policy Q-Learning Algorithm 163
6.2.3 No Bias of Off-Policy Q-Learning Algorithm 165
6.3 Neural Network-Based Off-Policy Interleaved Q-Learning 166
6.3.1 Model Neural Network 166
6.3.2 Actor Neural Network 167
6.3.3 Critic Neural Network 168
6.3.4 Interleaved Q-Learning 169
6.3.5 Optimal Control for Linear Systems 172
6.4 Illustrative Examples 175
6.5 Conclusion 181
References 182

7 Off-Policy Game Reinforcement Learning 185
7.1 Graphical Game for Optimal Synchronization 185
7.1.1 Preliminaries 187
7.1.2 Multi-agent Graphical Games 188
7.1.3 Off-Policy Reinforcement Learning Algorithm 193
7.1.4 Simulation Examples 204
7.2 Nonzero-Sum Game for Cooperative Optimization 207
7.2.1 Problem Statement 208
7.2.2 Solving the Nonzero-Sum Game Problems 211
7.2.3 Finding K_i^* $(i = 1, 2, \ldots, n)$ by the On-Policy Approach 212
7.2.4 Finding K_i^* $(i = 1, 2, \ldots, n)$ by the Off-Policy Approach 215
7.2.5 Simulation Results 219
7.3 Conclusion 230
References 230

8 Industrial Applications of Game Reinforcement Learning 233
8.1 Rougher Flotation Processes and Plant-Wide Optimization Control 233
8.2 Optimal Operational Control for Industrial Process 237
8.2.1 Problem Statement 238
8.2.2 H_∞ Tracking Control for Operational Processes 240
8.2.3 Solving the H_∞ Tracking Control Problem 243
8.2.4 Off-Policy Reinforcement Learning Algorithm 246
8.2.5 Simulation Results 251
8.3 Optimal Set-Point Design for Rougher Flotation Processes with Multiple Cells 256
8.3.1 Problem Statement 258
8.3.2 H_∞ Tracking Control for Rougher Flotation Processes 259
8.3.3 On-Policy Q-Learning Based on Zero-Sum Game 262
8.3.4 Off-Policy Q-Learning Algorithm 265
8.3.5 Optimum Set-Point Selector 269
8.3.6 Simulation Results 270
8.4 Plant-Wide Optimization Using Game Reinforcement Learning 274
8.4.1 Problem Statement 275
8.4.2 Nonzero-Sum Graphical Game for Solving Multi-objective Optimization Problem 279
8.4.3 Model Free Solution to Nonzero-Sum Graphical Game 285
8.4.4 Industrial Application in Iron Ore Processing 289
8.5 Conclusion 298
References 299

Appendix A 303

Index 309

Abbreviations

ACL	Adaptive critic learning
ADHDP	Action-dependent heuristic dynamic programming
ADP	Adaptive dynamic programming
ADPRL	Adaptive dynamic programming combined with reinforcement learning
AI	Artificial intelligence
ARE	Algebraic Riccati equation
BLS	Batch least squares
CBR	Case-based reasoning
CT	Continuous time
DDP	Differential dynamic programming
DHDP	Dual heuristic dynamic programming
DP	Dynamic programming
DT	Discrete time
EMPC	Economic model predictive control
GARE	Game algebraic Riccati equation
HDP	Heuristic dynamic programming
HJB	Hamilton–Jacobi–Bellman
IC	Intersection coordination
LD	Lagrangean decomposition
LMI	Linear matrix inequality
LMS	Least mean squares
LQT	Linear quadratic tracker
MASs	Multi-agent systems
MCC	Mixed competitive and cooperative
MCS	Multilevel coordinate search
MDPs	Markov decision processes
MPC	Model predictive control
NCSs	Networked control systems
NNs	Neural networks
OOC	Optimal operational control
OPCG	Off-policy cooperative game

OTC	Optimal tracking control
PDE	Partial differential equations
PE	Persistent excitation
PI	Policy iteration
RADP	Robust adaptive dynamic programming
RL	Reinforcement learning
RLS	Recursive least squares
RTO	Real-time optimization
TCP	Transmission control protocol
TE	Trial-and-error
UDP	User datagram protocol
UUB	Uniformly ultimately bounded
VI	Value iteration

Chapter 1
Background on Reinforcement Learning and Optimal Control

In this chapter, we discuss the fundamentals of RL, optimal control with dynamic programming, the basic idea of network control as well as the motivation and contributions of this book. The discussion is preparatory to well handle optimal feedback control problems using the RL technique in subsequent chapters, with strong potentials and benefits for future practical applications, particularly industrial intelligent optimization and control. In addition, some typical literature are provided to help understand well.

1.1 Fundamentals of Reinforcement Learning and Recall

RL is the reward-guided behavior of an agent that learns by trial-and-error (TE) procedure and interacts with the environment with the goal of maximizing the reward for the agent (Sutton and Barto 2018). It is widely applied to the various fields of modern life, such as computer science, neuroscience, psychology, engineering, mathematics and economic (Littman 1996).

A Markov decision process (MDP) (Sutton and Barto 2018), generally composed by the sets of states, actions and rewards, is denoted as (S, A, R). The dynamics of the MDPs is well defined by a discrete probability distribution:

$$p(s^{'}, r|s, a) = P[S_t = s^{'}, R_t = r|S_{t-1} = s, A_{t-1} = a] \tag{1.1}$$

where $s^{'}, s \in S, r \in R, a \in A(s)$. The function p defines the dynamics of the MDP, t is a sequence of discrete-time steps, S_t represents the state perceived by the agent at each time step t and A_t denotes the selected action at time step t. The dynamics function $p : S \times R \times S \times A \longrightarrow [0, 1]$ is an ordinary deterministic function of four arguments. Since p specifies a probability distribution for each choice of s and a, that is, the following conclusion naturally holds.

J. Li et al., *Reinforcement Learning*, Advances in Industrial Control,
https://doi.org/10.1007/978-3-031-28394-9_1

$$\sum_{s' \in S} \sum_{r \in R} p(s', r|s, a) = 1 \quad \forall s \in S \quad a \in A(s). \tag{1.2}$$

Based on the above discrete probability distribution p with four parameters, we can also express the state-transition probability p but taking only three parameters (Sutton and Barto 2018):

$$p(s'|s, a) = P[S_t = s'|S_{t-1} = s, A_{t-1} = a] = \sum_{r \in R} p(s', r|s, a) \tag{1.3}$$

where a three-argument function $p : S \times S \times A \longrightarrow [0, 1]$. Next, we denote the expected reward in terms of state–action pairs:

$$r(s, a) = E[R_t|S_{t-1} = s, A_{t-1} = a] = \sum_{r \in R} r \sum_{s' \in S} p(s', r|s, a). \tag{1.4}$$

Furthermore, the expected reward for state–action-next-state triples is given by

$$r(s, a, s') = E[R_t|S_{t-1} = s, A_{t-1} = a, S_t = s'] = \sum_{r \in R} r \frac{p(s', r|s, a)}{p(s'|s, a)}. \tag{1.5}$$

RL technique is employed into MDPs with the critical task of finding an optimal decision-making policy for maximizing a reward or accumulative rewards. Based on RL methods, the MDPs problems can be formulated as the value function of the state s under the policy π:

$$V_\pi(s) = \mathbb{E}_\pi[R_{t+1} + \gamma R_{t+2} + \ldots |S_t = s]. \tag{1.6}$$

Similarly, the Q function of state–action pair (s, a) is formalized:

$$Q_\pi(s, a) = \mathbb{E}_\pi[R_{t+1} + \gamma R_{t+2} + \ldots |S_t = s, A_t = a]. \tag{1.7}$$

With the widespread use of RL, its techniques are beginning to be applied to multi-agent systems. Vamvoudakis et al. (2012) and Abouheaf et al. (2014) introduced a model-based RL algorithm to solve a multi-player differential game problem. Zhang et al. (2014, 2017) developed an adaptive dynamics programming (ADP) method for the optimal cooperative control problem with the outcome of finding the solution to the coupled Hamilton–Jacobi–Bellman (HJB) equations. Notice that Vamvoudakis et al. (2012), Abouheaf et al. (2014), Zhang et al. (2014, 2017) proposed RL algorithms for solving the optimal consensus of multi-agent systems, in which the system models are required to be completely known or partially known. The current challenges faced by multi-agent systems are the inter-coupling relationships among the agents and the unknown dynamics, which put great pressure on the research.

1.2 Fundamentals of Optimal Control with Dynamic Programming

Optimal control theory, classified into the modern control theory, devotes to the study of theories and methods for optimizing the performance indexes of the control systems. The essence of the optimal control problem is to design controllers for minimizing the time-dependent behavior of dynamic systems (Kirk 2004).

Dynamic programming (DP) is an important approach in optimization theory, proposed by the American mathematician Bellman in the mid-1950s (Bellman 1966). Initially, DP was used to solve the problem of multi-level decision process optimization. Later, this method was extended to the optimal control problem, becoming one of the important methods for the optimal control problem (Bertsekas 2012). The theoretical basis of DP is the principle of Bellman optimality, and it is originally applied to discrete systems in optimal control.

Let the dynamics of the discrete system be

$$x(k+1) = f(x(k), u(k), k) \quad k = 0, 1, \ldots, N-1 \tag{1.8}$$

where $x \in \mathbb{R}^n$ is the state, $u \in \mathbb{R}^m$ is the control input. The objective of optimal control is to design a control policy $u^*(k)$ for the systems (1.8), so as to minimize or maximize the cost function below

$$J(x(k)) = \Sigma_{k=0}^{N-1} L(x(k), u(k)) \tag{1.9}$$

where $L(x(k), u(k)) = x(k)^T Qx(k) + u(k)^T Ru(k)$. Based on the DP theory, one has the Bellman Equation below

$$J(x(k)) = L(x(k), u(k)) + J(x(k+1)). \tag{1.10}$$

The optimal u^* is obtained from

$$u^*(k) = \arg\min_{u(k)}[L(x(k), u(k)) + J^*(x(k+1))]. \tag{1.11}$$

Replacing $u(k)$ in (1.10) by $u^*(k)$, the HJB equation can be derived

$$J^*(x(k)) = L(x(k), u^*(k)) + J^*(x(k+1)). \tag{1.12}$$

The analytical solutions to the HJB equation cannot be derived due to the nonlinearity of the system. So, lots of theoretical results have been achieved for solving optimal control problems by using adaptive dynamic programming combined with reinforcement learning (ADPRL). For discrete-time (DT) systems, in Al-Tamimi et al. (2007b), adaptive critic ADP designs were derived to solve the DT zero-sum game in which the state and action spaces are continuous. Al-Tamimi et al. (2007a) was concerned with the application of ADP techniques to the DT linear quadratic zero-sum game with H_∞ optimal control. Q-learning algorithm was proposed to solve the infinite-horizon linear quadratic tracker (LQT) for unknown DT systems in a causal manner in Kiumarsi et al. (2014). Lewis and Vamvoudakis (2010) proposed the implementation of ADP using only the measured input/output data from a dynamical system. For continuous-time (CT) systems, Vrabie and Lewis (2009) presented a CT adaptive controller, based on policy iteration (PI), for learning the CT optimal control policy without the knowledge of the dynamics of nonlinear systems. In Modares and Lewis (2014), an online learning algorithm based on reinforcement learning was presented to find the solution to the LQT problem for unknown system drift dynamics and the command generator dynamics.

1.3 Architecture and Performance of Networked Control System

Networked control systems (NCS) is a feedback control system that forms a closed loop through a network, characterized by the fact that information can be exchanged between system components (sensors, controllers, actuators) through the network (Li et al. 2010). The introduction of communication networks into control systems has indeed brought the advantages, such as low cost, simple installation and maintenance as well as high reliability. Therefore, NCSs have been extensively applied to variety of practical applications, including robot control, industrial manufacture, aircraft and spacecraft, etc. (Liao et al. 2002; Wang et al. 2015, 2016; Jiang et al. 2015; Chai et al. 2012; Fan et al. 2016; Qu et al. 1991). Despite their advantages and wide applications, the negative impact caused by unreliable transmission, such as network-induced dropout (Chai et al. 2012; Fan et al. 2016; Qu et al. 1991; Gao and Chen 2008), network-induced time delay (Gao and Chen 2008; Zhang et al. 2005; Liu et al. 2008), network noise (Fan et al. 2017; Wu et al. 2017) and quantization problem (Gao et al. 2008; Liu et al. 2012; Wei et al. 2015; Chen et al. 2016; Shi et al. 2015), are still inevitable. There is no doubt that the aforementioned negative factors would definitely degrade the control performance of the closed-loop NCSs, even cause the unstability of systems. Increasing attention has been paid to control policy design for stablizing and optimizing the performance of NCSs.

The simple architecture of a NCS is shown in Fig. 1.1. From Fig. 1.1, one can find that the measured state $y(k)$ passes via a network to a controller, and the computed control input $u(k)$ will arrive at the plant via the network.

Hespanha et al. (2007) and Schenato et al. (2007) discussed the transmission control protocol (TCP) and the user datagram protocol (UDP). Sinopoli et al. (2004) and Imer et al. (2006) analyzed the relationship between the stability and the probability of the dropout. Yang et al. (2013) proposed the networked-predictive-control scheme to compensate for the network-induced delay. Notice that the actuate models need to be known a prior when using the aforementioned methods.

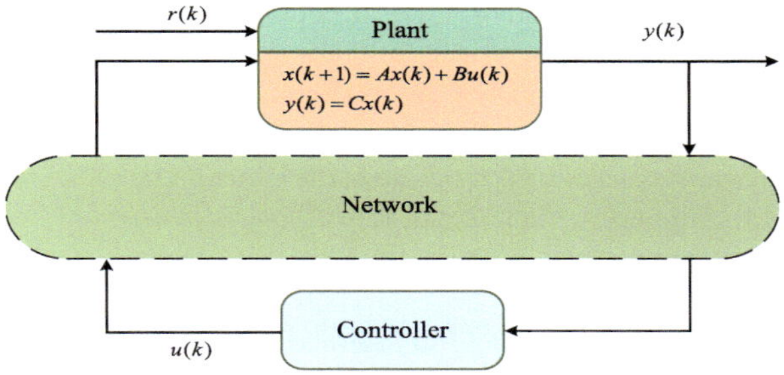

Fig. 1.1 The architecture of NCSs

To address the optimal control problem for NCSs with unknown system dynamics, many effors have been made. Xu et al. (2014) developed a stochastic Q-learning method for designing the optimal controller in the event-sampled environment. Xu et al. (2012) proposed a stochastic Q-learning technique, such that the optimal regulation was achieved. This is a data-driven approach for unknown NCSs with time-varying system matrices. Then, this approach was extended to the nonlinear case (Xu and Jagannathan 2013). Jiang et al. (2017) developed a new Q-learning approach for the purpose of finding the optimal tracking controller, such that the performance of NCSs subject to dropout was approximately optimized without the information of system dynamics. A novel methodology with a state observer and a virtual Smith predictor was presented, such that the nonzero-sum game of networked multi-player systems can be solved using only data (Li et al. 2022).

1.4 The State of the Art and Contributions

Regarding applications of RL methods to optimal control problems, there indeed exist several challenging problems, e.g., the robust control problems with external disturbances, the network control problems for systems with network-induced time delay and dropout, the multi-agent consensus issues with coupled relationship, and unknown and inaccurate models of practical industrial control systems.

With regard to uncertainties or disturbances existed in the unknown control systems, the H_∞ control and some variants of robust control methods under the framework of RL have been reported. In Al-Tamimi et al. (2007b), zero-sum game theory was employed to solve a game Bellman equation resulting in getting the H_∞ control policies. ADPRL and output regulation with game theory-based optimal tracking control was extended to robust tracking control, in which the dynamics of uncertainty is generally assumed to be linear only for CT systems (Zhang et al. 2015; Gao et al. 2019; Odekunle et al. 2020). Notice that robust tracking control problem for DT systems has been rarely studied. Besides, ADP methods were employed to design robust controller for CT nonlinear systems, in which the robust control problem was converted into the optimal control problem of nominal nonlinear systems (Wu et al.

2014; Wang et al. 2014, 2017; Liu et al. 2015; Zhao et al. 2020). The above research methods indeed lay a solid foundation for data-driven robust control using RL methods, while there still exist challenges of solving robust control problem for unknown DT nominal systems with matched or unmatched uncertainties. The challenges are basically caused by the natural difference between DT systems and CT systems and the nonlinearity characteristic.

As for the NCSs, the traditional approaches of solving them generally require the dynamics of systems to be known a priori. However, the NCSs with unknown dynamics have not been fully studied, especially using RL methods to design network controllers, even though researchers have made a few attempts.

In addition, the optimal consensus problem of multi-agent systems (MASs), also called the optimal synchronization, concerns not only consensus, but also performance optimization of MASs. There is a need to apply RL methods to effectively solve the optimal consensus problem of MASs. Chapter 7 is devoted to well addressing the model-free optimal consensus of MASs, as well as the multi-player game problem.

Conventional approaches to solving practical problems in industries include real-time optimization (RTO) and model predictive control (MPC), which all require that the system models are known. Whereas, two intractable challenges have to be faced, i.e., unknown dynamics of unit processes and operational indices and multiple time-scale industrial processes. The novel RL methods used for solving these challenges will be given in Chap. 8. In order to solve all these issues mentioned above, a series of RL methods integrated with game theory, on-policy learning and off-policy learning are developed. The contributions of this book are as follows.

Chapter 1 reviews the RL methods for solving the MDPs problems, optimal control problems with dynamics programming and the network control problems. Moreover, the state of the art of RL methods for optimal control problems and industrial applications of them are clearly discussed.

Chapter 2 addresses the H_∞ problem using ADPRL and game theory. Firstly, a model-free off-policy game Q-learning algorithm is proposed to solve the H_∞ state feedback control problem for linear DT multi-player systems with a single source of external disturbance. Secondly, the H_∞ output feedback control problem for linear DT systems with multiple players subject to multi-source disturbances is solved based on ADPRL and game theory.

Chapter 3 primarily concentrates on the robust tracking control problem for linear DT multi-player systems using ADPRL combined with output regulation. A novel off-policy cooperative game Q-learning algorithm is proposed for achieving optimal tracking control (OTC) of linear DT multi-player systems, suffering from exogenous dynamic disturbances without the information of multi-player systems, exogenous disturbances and command generator.

Chapter 4 tackles robust control problems for DT affine nonlinear systems. A novel adaptive interleaved RL algorithm is developed for finding a robust controller of DT affine nonlinear systems subject to matched or unmatched uncertainties, such that the uniformly ultimately bounded (UUB) stability of DT affine nonlinear systems can be guaranteed, allowing for all realization of unknown bounded uncertainties.

Chapter 5 mainly discusses the optimal control of linear DT networked controller systems and networked multi-player systems in the presence of network-induced delays. In the view of unknown system model parameters, unmeasurable system states, and the existence of network-induced delays, a model-free off-policy Q-learning algorithm with the delay compensation is designed. Thus, the Nash equilibrium of networked multi-player games can be achieved under the designed controllers only using data.

Chapter 6 settles optimal control problems of affine nonlinear DT systems by presenting a novel off-policy interleaved Q-learning algorithm using only the measured data along the system trajectories.

Chapter 7 deals with the cooperative optimal control problem for multi-agent and multi-player systems based on game RL. To this end, we respectively develop an off-policy RL algorithm to solve optimal synchronization of CT MASs and an off-policy game Q-learning algorithm to solve linear DT multi-player game problems, such that the Nash equilibrium is reached under the learned control policies.

Chapter 8 discusses how to apply game RL theory to practical industrial applications, mainly involving the optimal set-point design for industrial process operation, particularly dual-rate rougher flotation operation, and performance optimization problems for large-scale industrial processes. To this end, game off-policy RL algorithms are developed to find the optimal solution to industrial operational optimization using measured data in real time. Moreover, a RL framework was proposed for achieving plant-wide performance optimization for large-scale unknown industrial processes by integrating RL with multi-agent game theory.

References

Abouheaf MI, Lewis FL, Vamvoudakis KG, Haesaert S, Babuska R (2014) Multi-agent discrete-time graphical games and reinforcement learning solutions. Automatica 50(12):3038–3053

Al-Tamimi A, Lewis FL, Abu-Khalaf M (2007a) Model-free Q-learning designs for linear discrete-time zero-sum games with application to H-infinity control. Automatica 43(3):473–481

Al-Tamimi A, Abu-Khalaf M, Lewis FL (2007b) Adaptive critic designs for discrete-time zero-sum games with application to H_∞ control. IEEE Trans Syst Man Cybern Part B (Cybern) 37(1):240–247

Bellman R (1966) Dynamic programming. Science 153(3731):34–37

Bertsekas D (2012) Dynamic programming and optimal control: volume I, vol 1. Athena Scientific

Chai T, Zhao L, Qiu J, Liu F, Fan J (2012) Integrated network-based model predictive control for setpoints compensation in industrial processes. IEEE Trans Ind Inform 9(1):417–426

Chen C, Wen C, Liu Z, Xie K, Zhang Y, Chen CP (2016) Adaptive consensus of nonlinear multi-agent systems with non-identical partially unknown control directions and bounded modelling errors. IEEE Trans Autom Control 62(9):4654–4659

Fan J, Jiang Y, Chai T (2016) Operational feedback control of industrial processes in a wireless network environment. Acta Automatica Sinica 42(8):1166–1174

Fan J, Jiang Y, Chai T (2017) MPC-based setpoint compensation with unreliable wireless communications and constrained operational conditions. Neurocomputing 270:110–121

Gao H, Chen T (2008) Network-based H_∞ output tracking control. IEEE Trans Autom Control 53(3):655–667

Gao H, Chen T, Lam J (2008) A new delay system approach to network-based control. Automatica 44(1):39–52
Gao W, Jiang Y, Davari M (2019) Data-driven cooperative output regulation of multi-agent systems via robust adaptive dynamic programming. IEEE Trans Circuits Syst II: Express Briefs 66(3):447–451
Hespanha JP, Naghshtabrizi P, Xu Y (2007) A survey of recent results in networked control systems. Proc IEEE 95(1):138–162
Imer OC, Yüksel S, Başar T (2006) Optimal control of LTI systems over unreliable communication links. Automatica 42(9):1429–1439
Jiang Y, Fan J, Chai T, Lewis FL, Li J (2017) Tracking control for linear discrete-time networked control systems with unknown dynamics and dropout. IEEE Trans Neural Netw Learn Syst 29(10):4607–4620
Jiang Y, Fan J, Chai T, Chen T (2015) Setpoint dynamic compensation via output feedback control with network induced time delays. In: 2015 American control conference (ACC), IEEE, pp 5384–5389
Kirk DE (2004) Optimal control theory: an introduction. Courier Corporation
Kiumarsi B, Lewis FL, Modares H, Karimpour A, Naghibi-Sistani MB (2014) Reinforcement Q-learning for optimal tracking control of linear discrete-time systems with unknown dynamics. Automatica 50(4):1167–1175
Lewis FL, Vamvoudakis KG (2010) Reinforcement learning for partially observable dynamic processes: adaptive dynamic programming using measured output data. IEEE Trans Syst Man Cybern Part B (Cybern) 41(1):14–25
Li H, Sun Z, Chow MY, Sun F (2010) Gain-scheduling-based state feedback integral control for networked control systems. IEEE Trans Ind Electron 58(6):2465–2472
Li J, Xiao Z, Fan J, Chai T, Lewis FL (2022) Off-policy Q-learning: solving Nash equilibrium of multi-player games with network-induced delay and unmeasured state. Automatica 136:110076
Liao F, Wang JL, Yang GH (2002) Reliable robust flight tracking control: an LMI approach. IEEE Trans Control Syst Technol 10(1):76–89
Littman ML (1996) Reinforcement learning: a survey. J Artif Intell Res 4
Liu T, Jiang ZP, Hill DJ (2012) A sector bound approach to feedback control of nonlinear systems with state quantization. Automatica 44(1):4654–4659
Liu D, Yang X, Wang D, Wei Q (2015) Reinforcement-learning-based robust controller design for continuous-time uncertain nonlinear systems subject to input constraints. IEEE Trans Cybern 45(7):1372–1385
Liu H, Zhu Q, Jiang J, Wang Y, Yang H (2008) Guaranteed cost control of networked control systems with long time delay. In: 2008 IEEE Pacific-Asia workshop on computational intelligence and industrial application, IEEE, vol 1, pp 175–179
Modares H, Lewis FL (2014) Linear quadratic tracking control of partially-unknown continuous-time systems using reinforcement learning. IEEE Trans Autom Control 59(11):3051–3056
Odekunle A, Gao W, Davari M, Jiang ZP (2020) Reinforcement learning and non-zero-sum game output regulation for multi-player linear uncertain systems. Automatica 112:108672
Qu Z, Dorsey J et al (1991) Robust tracking control of robots by a linear feedback law. IEEE Trans Autom Control 36(9):1081–1084
Schenato L, Sinopoli B, Franceschetti M, Poolla K, Sastry SS (2007) Foundations of control and estimation over lossy networks. Proc IEEE 95(1):163–187
Shi P, Wang H, Lim CC (2015) Network-based event-triggered control for singular systems with quantizations. IEEE Trans Ind Electron 63(2):1230–1238
Sinopoli B, Schenato L, Franceschetti M, Poolla K, Jordan MI, Sastry SS (2004) Kalman filtering with intermittent observations. IEEE Trans Autom Control 49(9):1453–1464
Sutton RS, Barto AG (2018) Reinforcement learning: an introduction. MIT Press
Vamvoudakis KG, Lewis FL, Hudas GR (2012) Multi-agent differential graphical games: online adaptive learning solution for synchronization with optimality. Automatica 48(8):1598–1611

Vrabie D, Lewis F (2009) Neural network approach to continuous-time direct adaptive optimal control for partially unknown nonlinear systems. Neural Netw 22(3):237–246
Wang D, Liu D, Li H (2014) Policy iteration algorithm for online design of robust control for a class of continuous-time nonlinear systems. IEEE Trans Autom Sci Eng 11(2):627–632
Wang T, Gao H, Qiu J (2015) A combined adaptive neural network and nonlinear model predictive control for multirate networked industrial process control. IEEE Trans Neural Netw Learn Syst 27(2):416–425
Wang T, Gao H, Qiu J (2016) A combined fault-tolerant and predictive control for network-based industrial processes. IEEE Trans Ind Electron 63(4):2529–2536
Wang D, Liu D, Mu C, Zhang Y (2017) Neural network learning and robust stabilization of nonlinear systems with dynamic uncertainties. IEEE Trans Neural Netw Learn Syst 29(4):1342–1351
Wei L, Fu M, Zhang H (2015) Quantized output feedback control with multiplicative measurement noises. Int J Robust Nonlinear Control 25(9):1338–1351
Wu HN, Li MM, Guo L (2014) Finite-horizon approximate optimal guaranteed cost control of uncertain nonlinear systems with application to mars entry guidance. IEEE Trans Neural Netw Learn Syst 26(7):1456–1467
Wu Q, Fan J, Jiang Y (2017) Dual-rate operational optimal control for linear flotation process. In: 2017 IEEE international conference on information and automation (ICIA), IEEE, pp 448–453
Xu H, Jagannathan S (2013) Stochastic optimal controller design for uncertain nonlinear networked control system via neuro dynamic programming. IEEE Trans Neural Netw Learn Syst 24(3):471–484
Xu H, Jagannathan S, Lewis FL (2012) Stochastic optimal control of unknown linear networked control system in the presence of random delays and packet losses. Automatica 48(6):1017–1030
Xu H, Sahoo A, Jagannathan S (2014) Stochastic adaptive event-triggered control and network scheduling protocol co-design for distributed networked systems. IET Control Theory Appl 8(18):2253–2265
Yang R, Liu GP, Shi P, Thomas C, Basin MV (2013) Predictive output feedback control for networked control systems. IEEE Trans Ind Electron 61(1):512–520
Zhang L, Shi Y, Chen T, Huang B (2005) A new method for stabilization of networked control systems with random delays. IEEE Trans Autom Control 50(8):1177–1181
Zhang H, Zhang J, Yang GH, Luo Y (2014) Leader-based optimal coordination control for the consensus problem of multiagent differential games via fuzzy adaptive dynamic programming. IEEE Trans Fuzzy Syst 23(1):152–163
Zhang H, Liang H, Wang Z, Feng T (2015) Optimal output regulation for heterogeneous multiagent systems via adaptive dynamic programming. IEEE Trans Neural Netw Learn Syst 28(1):18–29
Zhang H, Yue D, Zhao W, Hu S, Dou C (2017) Distributed optimal consensus control for multiagent systems with input delay. IEEE Trans Cybern 48(6):1747–1759
Zhao J, Na J, Gao G (2020) Adaptive dynamic programming based robust control of nonlinear systems with unmatched uncertainties. Neurocomputing 395:56–65

Chapter 2
H_∞ Control Using Reinforcement Learning

In this chapter, we first present a model-free off-policy game Q-learning algorithm to solve the H_∞ control problem for linear discrete-time multi-player systems with a single source of external disturbances. The primary contribution lies in that the Q-learning strategy employed in the proposed algorithm is implemented in an off-policy policy iteration approach other than the on-policy learning. Then, we present a data-driven adaptive dynamic programming for solving the H_∞ output feedback control problem with multiple players subject to multi-source disturbances. Considering the advantages of off-policy RL over on-policy RL, a novel off-policy game Q-learning algorithm dealing with mixed competition and cooperation among players is developed, such that the H_∞ control problem can be finally solved for linear multi-player systems without the knowledge of system dynamics. Besides, rigorous proofs of algorithm convergence and unbiasedness of solutions are presented. Simulation results demonstrate the effectiveness of the proposed method.

2.1 H_∞ State Feedback Control of Multi-player Systems

The H_∞ control is a robust control method that aims at designing the controllers to attenuate the negative effects in the performance of dynamical systems caused by external disturbances, meanwhile to guarantee the stability of systems if no disturbance exists (Van Der Schaft 1992; Isidori 1994; Wang et al. 2018). This issue can be handled using the zero-sum game theory, that is, solving a game Bellman equation in the zero-sum game framework results for getting the H_∞ controller policies (Basar 2008; Al-Tamimi et al. 2007a). The truth of more complex and large-scale systems with multiple subsystems and multiple controllers in practical engineering applications makes anti-interference control of multi-player systems valuable and more complicated, thereby attracting increasing attention of researchers to H_∞ control for

J. Li et al., *Reinforcement Learning*, Advances in Industrial Control,
https://doi.org/10.1007/978-3-031-28394-9_2

multi-player or multi-agent systems (Jiang et al. 2018; Jiao et al. 2016; Lv and Ren 2019; Song et al. 2017).

By reviewing the existing results on H_∞ control for dynamical systems, it is not difficult to find that most of the researchers are concerned about model-based H_∞ controller design using the variety of methods, such as linear matrix inequality (LMI) (El Ghaoui et al. 1997; Geromel et al. 1998; Cao et al. 1998), zero-sum game (Jiang and Jiang 2012; Vamvoudakis and Lewis 2012; Werbos 1992; Vrabie 2010; Vamvoudakis and Lewis 2010), pole assignment (Kautsky et al. 1985; Amini and Samani 2014; Byers and Nash 1989), etc. Notice that the above-mentioned results require accurate system models to be known a priori, which prevents them from applications for systems with inaccurate or even completely unknown models. RL that can deal with controller design or decision-making problems in an uncertain or unknown environment is an alternative tool to solve H_∞ control for systems without the information of system dynamics, including DT systems (Al-Tamimi et al. 2007b; Kiumarsi et al. 2017; Kim and Lewis 2010; Rizvi and Lin 2018) and CT systems (Modares et al. 2015; Jiang et al. 2017; Luo et al. 2015, 2014; Kiumarsi et al. 2016; Fu and Chai 2016). It is worth pointing out that the designed H_∞ controllers in the aforementioned results are only for single-player systems. On the other hand, the existing reports on multi-player games (Li et al. 2017; Liu et al. 2014; Vamvoudakis and Lewis 2011; Vamvoudakis et al. 2012, 2017; Vamvoudakis 2017a) usually ignore the negative effects caused by disturbances on performance of systems. What we focus on here is how to attenuate the influence of disturbances for multi-player systems using only measured data even in completely unknown environments.

In Jiao et al. (2016), agents have their individual dynamics, and the anti-interference problem has been investigated for CT multi-player systems in Jiang et al. (2018); Lv and Ren (2019); Song et al. (2017). The main difference of multi-player systems from multi-agent systems is that all players in multi-player systems are capable of accessing the state of the overall systems, which makes us try to find a different manipulation from multi-agent systems to design H_∞ controllers. Like (Jiang et al. 2018; Lv and Ren 2019; Song et al. 2017), the model-free H_∞ controller design will be taken into account for multi-player systems in this paper, while the difference of nature of discrete-time sampling from CT processes makes it more complicated to solve H_∞ control problem from the DT system perspective, and multiple players and completely unknown dynamics of players increase this difficulty. In view of the advantages of off-policy learning over on-policy learning shown in our previous result (Li et al. 2019b) (Sect. 7.2), our target is to develop an off-policy game Q-learning algorithm to solve H_∞ control problem for DT linear multi-player systems by using the measured data.

Therefore, in the ADP architecture, this section presents a novel off-policy game Q-learning algorithm for finding multiple controllers that not only guarantee the stability of multi-player systems, but also the disturbance is limited within a disturbance attenuation bound. Moreover, under the premise of ensuring the persistent excitation (PE) condition, probing noises have to be added to the behavior control policy for each player when learning the target policies. The unbiasedness of solution to Nash equilibrium for zero-sum multi-player games using the off-policy game Q-learning algorithm is rigorously proven.

The structure of this section is shown below. In Sect. 2.1.1, the H_∞ control problem of multi-player linear DT systems is proposed and the conversion of H_∞ control into zero-sum game is presented. Section 2.1.2 focuses on solving the problem of multi-player zero-sum games with external disturbance. Moreover, the on-policy game Q-learning algorithm is proposed and the solution to the Nash equilibrium using this kind of algorithm is proven to be biased. In Sect. 2.1.3, an off-policy game Q-learning algorithm is developed together with the proof that the Nash equilibrium solution learned by the proposed algorithm is unbiased. The Sect. 2.1.4 shows the simulation experiments that are used to verify the effectiveness and contributions of the proposed method.

2.1.1 Problem Statement

In this subsection, the H_∞ control problem for linear DT multi-player systems is proposed first, and then it is converted into a zero-sum game problem where all players work together for minimizing the specific performance index, while the disturbance makes the performance worst on the opposite. The value function and the Q-function defined in terms of the performance index are proven to be quadratic finally.

2.1.1.1 H_∞ Control Problem

Consider the following linear DT multi-player system subject to exogenous disturbance:

$$x_{k+1} = Ax_k + \sum_{i=1}^{n} B_i u_{ik} + Ed_k, \tag{2.1}$$

where $x_k = x(k) \in \mathbb{R}^p$ is the system state with initial state x_0, $u_{ik} = u_i(k) \in \mathbb{R}^{m_i}$ $(i = 1, \ldots, n)$ are the control inputs and $d_k = d(x_k) \in \mathbb{R}^q$ is the external disturbance input. $A \in \mathbb{R}^{p\times p}$, $B_i \in \mathbb{R}^{p\times m_i}$, $E \in \mathbb{R}^{p\times q}$ and k is the sampling time instant.

Definition 2.1 (Kiumarsi et al. 2017; Modares et al. 2014) System (2.1) has L_2-gain less than or equal to γ if

$$\sum_{k=0}^{\infty}(x_k^T Qx_k + \sum_{i=1}^{n} u_{ik}^T R_i u_{ik}) \le \gamma^2 \sum_{k=0}^{\infty} \|d_k\|^2 \tag{2.2}$$

for all $d_k \in L_2[0, \infty)$. $Q \ge 0$, $R_i > 0$ and $\gamma \ge 0$ is a prescribed constant disturbance attenuation level.

The H_∞ control is to find a feedback control policy u such that the system (2.1) with $d_k = 0$ is asymptotically stable and it satisfies the disturbance attenuation condition (2.2). As claimed in Van Der Schaft (1992), the H_∞ control problem can be equivalently expressed as

$$J(x_0, U, d_k) = \min_U \max_{d_k} \sum_{k=0}^{\infty} (x_k^T Q x_k + \sum_{i=1}^{n} u_{ik}^T R_i u_{ik} - \gamma^2 \|d_k\|^2), \tag{2.3}$$

where $U = \{u_{1k}, u_{2k}, \ldots, u_{nk}\}$, which means the set U is composed of n players with each of them is a controller. As one can know from (2.3), the objective of these players U is to fight with the disturbance for minimizing the performance in (2.3), while the disturbance d_k could also be viewed as a player that tries to maximize (2.3). This is a typical zero-sum game problem.

Definition 2.2 (Saddle Point Solution) (Jiang et al. 2018; Modares and Lewis 2014b) A set of policies $(u_{1k}^*, u_{2k}^*, \ldots, u_{nk}^*, d_k^*)$ is called the game theoretical saddle point if it satisfies the form

$$\begin{aligned} J(x_0, u_{1k}^*, u_{2k}^*, \ldots, u_{nk}^*, d_k^*) &= \min_U \max_{d_k} J(x_0, U, d_k) \\ &= \max_{d_k} \min_U J(x_0, U, d_k) \end{aligned} \tag{2.4}$$

which indicates the Nash equilibrium condition holds, that is

$$J(x_0, U^*, d_k) \le J(x_0, U^*, d_k^*) \le J(x_0, U, d_k^*),$$

where $U^* = \left\{u_{1k}^*, u_{2k}^*, \ldots, u_{nk}^*\right\}$.

In such zero-sum game (2.3), the saddle-point solution exists if and only if there exists a function $V(x_k)$ satisfying the following HJB equation (Vamvoudakis et al. 2017; Al-Tamimi et al. 2008):

$$\begin{aligned} V^*(x_k) &= \min_U \max_{d_k} \sum_{l=k}^{\infty} \left(x_l^T Q x_l + \sum_{i=1}^{n} u_{il}^T R_i u_{il} - \gamma^2 \|d_l\|^2 \right) \\ &= \min_U \max_{d_k} \left(x_k^T Q x_k + \sum_{i=1}^{n} u_{ik}^T R_i u_{ik} - \gamma^2 \|d_k\|^2 + V^*(x_{k+1}) \right) \\ &= x_k^T Q x_k + \sum_{i=1}^{n} u_{ik}^{*T} R_i u_{ik}^* - \gamma^2 \left\|d_k^*\right\|^2 + V^* \left(A x_k + \sum_{i=1}^{n} B_i u_{ik}^* + E d_k^* \right), \end{aligned} \tag{2.5}$$

where $V^*(x_k)$ is viewed as the optimal value function.

The arguments provided above have illustrated that H_∞ control is closely related to zero-sum game (2.3). The ultimate target of designing H_∞ control policies in this paper can be achieved by seeking the saddle-point solution to the zero-sum game.

Similar to Al-Tamimi et al. (2007b); Li et al. (2017, 2019a), the optimal Q-function referring to (2.3) can be defined as

$$Q^*(x_k, U, d_k) = x_k^T Q x_k + \sum_{i=1}^{n} u_{ik}^T R_i u_{ik} - \gamma^2 \|d_k\|^2 + V^*(x_{k+1}). \tag{2.6}$$

Thus, the following relation holds:

$$V^*(x_k) = \min_U \max_{d_k} Q^*(x_k, U, d_k) = Q^*(x_k, U^*, d_k^*). \tag{2.7}$$

2.1.1.2 Quadratic Form Proof of Value Function and Q-Function

Definition 2.3 (Admissible Control Policies) (Al-Tamimi et al. 2007b; Vamvoudakis and Lewis 2011) Suppose $d_k = 0$, the control policies $u_1(x_k)$,$u_2(x_k)$,. . . ,$u_n(x_k)$ are defined as admissible with respect to (2.3) on $\Omega \in \mathbb{R}^p$, denoted by $u_i(x_k) \in \Psi(\Omega)$ if $u_1(x_k), u_2(x_k), \ldots, u_n(x_k)$ are continuous on Ω, $u_1(0) = 0$, $u_2(0) = 0, \ldots, u_n(0) = 0$, $u_1(x_k), u_2(x_k), \ldots, u_n(x_k)$ stabilize (2.1) on Ω and (2.3) is finite $\forall x_0 \in \Omega$.

Lemma 2.1 is given to show the quadratic forms of the optimal value function and the optimal Q-function associated with performance index (2.3).

Lemma 2.1 *Assume that there are admissible control policies $u_i = -K_i x_k$ and disturbance policy $d_k = -K x_k$, the quadratic forms of the optimal value function and the optimal Q-function can be expressed as*

$$V^*(x_k) = x_k^T P^* x_k \tag{2.8}$$

and

$$Q^*(x_k, U, d_k) = z_k^T H^* z_k, \tag{2.9}$$

where P^* and H^* are positive definite matrices, and

$$z_k = \begin{bmatrix} x_k^T & u_{1k}^T & u_{2k}^T & \cdots & u_{nk}^T & d_k^T \end{bmatrix}^T. \tag{2.10}$$

Proof

$$\begin{aligned} V^*(x_k) &= \min_U \max_{d_l} \sum_{l=k}^{\infty} (x_l^T Q x_l + \sum_{i=1}^{n} u_{il}^T R_i u_{il} - \gamma^2 d_l^T d_l) \\ &= \min_L \max_K \sum_{l=k}^{\infty} \left[x_l^T Q x_l + \sum_{i=1}^{n} (-K_i x_l)^T R_i (-K_i x_l) - \gamma^2 (-K x_l)^T (-K x_l) \right] \end{aligned}$$

$$= \min_{L} \max_{K} \sum_{l=k}^{\infty} x_l^T \left[Q + \sum_{i=1}^{n} (K_i)^T R_i (K_i) - \gamma^2 K^T K \right] x_l$$

$$= \min_{L} \max_{K} \sum_{l=0}^{\infty} x_{l+k}^T \left[Q + \sum_{i=1}^{n} (K_i)^T R_i (K_i) - \gamma^2 K^T K \right] x_{l+k}, \tag{2.11}$$

where $L = \{K_1, K_2, \ldots, K_n\}$ and $x_{l+k} = (A - \sum_{i=1}^{n} B_i K_i - EK)^l x_k = G^l x_k$. Further, one has

$$V^*(x_k) = \min_{L} \max_{K} \sum_{l=0}^{\infty} x_k^T (G^l)^T \left[Q + \sum_{i=1}^{n} (K_i)^T R_i (K_i) - \gamma^2 K^T K \right] (G^l) x_k, \tag{2.12}$$

then one has

$$V^*(x_k) = x_k^T P^* x_k, \tag{2.13}$$

where

$$P^* = \min_{L} \max_{K} \sum_{l=0}^{\infty} (G^l)^T \left[Q + \sum_{i=1}^{n} (K_i)^T R_i (K_i) - \gamma^2 K^T K \right] (G^l).$$

Then, one has

$$\begin{aligned}
Q^*(x_k, U, d_k) &= x_k^T Q x_k + \sum_{i=1}^{n} u_{ik}^T R_i u_{ik} - \gamma^2 d_k^T d_k + V^*(x_{k+1}) \\
&= x_k^T Q x_k + \sum_{i=1}^{n} u_{ik}^T R_i u_{ik} - \gamma^2 d_k^T d_k + x_{k+1}^T P^* x_{k+1} \\
&= x_k^T Q x_k + \sum_{i=1}^{n} u_{ik}^T R_i u_{ik} - \gamma^2 d_k^T d_k \\
&\quad + (A x_k + \sum_{i=1}^{n} B_i u_{ik} - E d_k)^T P^* (A x_k + \sum_{i=1}^{n} B_i u_{ik} - E d_k) \\
&= z_k^T H^* z_k,
\end{aligned} \tag{2.14}$$

where

$$
H^* = \begin{bmatrix} H^*_{xx} & H^*_{xu_1} & H^*_{xu_2} & \cdots & H^*_{xu_n} & H^*_{xd} \\ H^{*,T}_{xu_1} & H^*_{u_1u_1} & H^*_{u_1u_2} & \cdots & H^*_{u_1u_n} & H^*_{u_1d} \\ H^{*,T}_{xu_2} & H^{*,T}_{u_1u_2} & H^*_{u_2u_2} & \cdots & H^*_{u_2u_n} & H^*_{u_2d} \\ \vdots & \vdots & \vdots & \ddots & \vdots & \vdots \\ H^{*,T}_{xu_n} & H^{*,T}_{u_1u_n} & H^{*,T}_{u_2u_n} & \cdots & H^*_{u_nu_n} & H^*_{u_nd} \\ H^{*,T}_{xd} & H^{*,T}_{u_1d} & H^{*,T}_{u_2d} & \cdots & H^{*,T}_{u_nd} & H^*_{dd} \end{bmatrix}
$$

$$
= \begin{bmatrix} A^TP^*A+Q & \cdots & A^TP^*B_n & A^TP^*E \\ (A^TP^*B_1)^T & \cdots & B_1^TP^*B_n & B_1^TP^*E \\ (A^TP^*B_2)^T & \cdots & B_2^TP^*B_n & B_2^TP^*E \\ \vdots & \ddots & \vdots & \vdots \\ (A^TP^*B_n)^T & \cdots & B_n^TP^*B_n+R_n & B_n^TP^*E \\ (A^TP^*E)^T & \cdots & (B_n^TP^*E)^T & -\gamma^2 I + E^TP^*E \end{bmatrix}. \tag{2.15}
$$

By (2.13) and (2.14), one can get

$$
P^* = M^T H^* M, \tag{2.16}
$$

where

$$
M = \begin{bmatrix} I & -K_1^T & \cdots & -K_n^T & -K^T \end{bmatrix}^T.
$$

This completes the proof. □

Remark 2.1 Following the idea in Al-Tamimi et al. (2007b), where the quadratic forms of value function and Q-function for linear single-player systems are proven, the rigorous proof that the value function and Q-function defined for zero-sum game of multi-player systems are quadratic is presented in here.

2.1.2 Solving the Multi-player Zero-Sum Game

In this subsection, the theoretical solution of the zero-sum game for multi-player systems is first obtained according to the Bellman equation of Q-function. Then the on-policy game Q-learning algorithm is provided to solve this problem. Finally, it is proved that the Nash equilibrium solution obtained by the on-policy game Q-learning algorithm is biased.

2.1.2.1 Theoretical Solution

Now we are in a position to solve HJB equation based on game theory. By Lemma 2.1, referring to HJB equation (2.5) yields the optimal Q-function-based Bellman equation as follows:

$$z_k^T H z_k = x_k^T Q x_k + \sum_{i=1}^{n} u_{ik}^T R_i u_{ik} - \gamma^2 \|d_k\|^2 + z_{k+1}^T H z_{k+1}. \tag{2.17}$$

The optimal control policy u_i^* of each player i and the worst-case disturbance d_k^* should satisfy $\frac{\partial Q^*(x_k,U,d_k)}{\partial u_i} = 0$ and $\frac{\partial Q^*(x_k,U,d_k)}{\partial d_k} = 0$. Therefore, one has

$$u_i^*(k) = -K_i^* x_k \tag{2.18}$$

$$d_k^* = -K^* x_k, \tag{2.19}$$

where

$$K_i^* = H_{u_i u_i}^{-1} \left[H_{x u_i}^T - \left(\sum_{r=1, r\neq i}^{n} H_{u_i u_r} K_r + H_{u_i d} K \right) \right] \tag{2.20}$$

$$K^* = H_{dd}^{-1} \left[H_{xd}^T - \sum_{i=1}^{n} H_{d u_i} K_i \right]. \tag{2.21}$$

Substituting K_i^* in (2.20) and K^* in (2.21) into (2.17) yields the optimal Q-function-based game algebraic Riccati equation (GARE)

$$z_k^T H^* z_k = x_k^T Q x_k + \sum_{i=1}^{n} (u_i^*)^T R_i u_i^* - \gamma^2 \left\| d_k^* \right\|^2 + z_{k+1}^T H^* z_{k+1}. \tag{2.22}$$

One thing to note is that Al-Tamimi et al. (2007b) and Vamvoudakis et al. (2017) have proved that the following $K_i^* (i = 1, 2, ...n)$ and K^* can keep system (2.1) stable when $d_k = 0$ and achieve Nash equilibrium:

$$K_i^* = (H_{u_i u_i}^*)^{-1} \left[(H_{x u_i}^*)^T - \left(\sum_{r=1, r\neq i}^{n} H_{u_i u_r}^* K_r^* + H_{u_i d}^* K^* \right) \right] \tag{2.23}$$

$$K^* = (H_{dd}^*)^{-1} \left[(H_{xd}^*)^{-1} - \sum_{i=1}^{n} H_{d u_i}^* K_i^* \right]. \tag{2.24}$$

Note that solving (2.23) and (2.24), that is, solving the zero-sum game problem defined by (2.6), is to find the optimal control policies satisfying the disturbance attenuation condition (2.2).

Remark 2.2 Since the optimal control policies and disturbance policy in (2.23) and (2.24) are coupled, they are difficult to be solved. Therefore, an on-policy game Q-learning algorithm is going to be presented to overcome this difficulty resulting in obtaining the control laws $u_i^*(k) = -K_i^* x_k$ and disturbance policy $d_k^* = -K^* x_k$.

2.1.2.2 On-Policy Game Q-Learning Algorithm

The on-policy RL algorithms in Al-Tamimi et al. (2007a); Wei et al. (2017); Lee et al. (2012); Li et al. (2014) are extended to solve H_∞ control problem for multi-player systems, and thus the on-policy game Q-learning Algorithm 2.1 is proposed.

Algorithm 2.1 On-policy game Q-learning for the zero-sum game

1: Given an n-tuple initial admissible controller gains for $K_1^0, K_2^0, \ldots, K_n^0$ and disturbance policy gain K^0. Let $j = 0$ and j represents the number of iterations, and i denotes the player i ($i = 1, 2, \ldots, n$). Set $i = 1$;

2: Evaluate policies by solving H^{j+1}:

$$z_k^T H^{j+1} z_k = x_k^T Q x_k + \sum_{i=1}^{n} (u_{ik}^j)^T R_i u_{ik}^j - \gamma^2 \left\| d_k^j \right\|^2 + z_{k+1}^T H^{j+1} z_{k+1} \tag{2.25}$$

where $z_{k+1} = \left[x_{k+1}^T \quad (u_{i,k+1}^j)^T (i = 1, 2, \ldots, n) \quad (d_{k+1}^j)^T \right]^T$;

3: Update the control and disturbance policy:

$$u_{ik}^{j+1} = -K_i^{j+1} x_k \tag{2.26}$$

$$d_k^{j+1} = -K^{j+1} x_k \tag{2.27}$$

where

$$K_i^{j+1} = (H_{u_i u_i}^{j+1})^{-1} \left[(H_{x u_i}^{j+1})^T - \left(\sum_{r=1, r \neq i}^{n} H_{u_i u_r}^{j+1} K_r^j + H_{u_i d}^{j+1} K^j \right) \right] \tag{2.28}$$

$$K^{j+1} = (H_{dd}^{j+1})^{-1} \left[(H_{xd}^{j+1})^T - \sum_{i=1}^{n} H_{d u_i}^{j+1} K_i^j \right]; \tag{2.29}$$

4: If $i < n + 1$, then $i = i + 1$ and go back to Step 3. Otherwise, go to Step 5;

5: Stop when $\left\| H^{j-1} - H^j \right\| \leq \varepsilon$ with a small constant $\varepsilon (\varepsilon > 0)$. Otherwise $j = j + 1, i = 1$ and go back to Step 2.

Remark 2.3 In Algorithm 2.1, the training data z_k are generated by the iterative control policies (2.26) and iterative disturbance policies (2.27). In this sense, Algorithm 2.1 is indeed on-policy learning (Li et al. 2020; Vamvoudakis 2017b; Kiumarsi et al. 2014). Moreover, the control policy gains K_i^{j+1} are updated by solving matrix H^{j+1} in Q-function. As $j \to \infty$, H^{j+1} converges to the optimal value, resulting in K_i^{j+1} converges to the optimal value. This conclusion can be made and proven using a similar way to Al-Tamimi et al. (2007b); Li et al. (2019a).

2.1.2.3 Bias Analysis of Solution Learned by the On-Policy Game Q-Learning Algorithm

For systems with a single player, the existing results have proven that biased solutions to iterative Bellman equation can be generated by adding probing noise. The sequel is going to prove that this kind of biasedness of solutions still exists when solving the optimal Q-function based Bellman equation using the on-policy game Q-learning algorithm for multi-player systems.

To satisfy the PE condition in Algorithm 2.1, probing noises are added to u_{ik}^j. Thus, the actual control inputs applied to the system for collecting data are

$$\hat{u}_{ik}^j = u_{ik}^j + e_{ik} \tag{2.30}$$

with $e_{ik} = e_i(k)$ being probing noises and u_{ik}^j given by (2.26). Theorem 2.1 will prove that there exists the bias of solutions to (2.25).

Theorem 2.1 *Rewrite iterative Q-function-based Bellman Eq. (2.25) as*

$$\begin{aligned} x_k^T (M^j)^T H^{j+1} M^j x_k = x_k^T Q x_k + \sum_{i=1}^{n} (u_{ik}^j)^T R_i u_{ik}^j - \gamma^2 d_k^T d_k \\ + x_{k+1}^T (M^j)^T H^{j+1} M^j x_{k+1}, \end{aligned} \tag{2.31}$$

where

$$M^j = \begin{bmatrix} I & -(K_1^j)^T & \cdots & -(K_n^j)^T & -(K^j)^T \end{bmatrix}^T.$$

Let H^{j+1} be the solution (2.31) with $e_{ik} = 0$ and $\hat{H}^{j+1}$ be the solution to (2.31) with $e_{ik} \neq 0$. Then, $H^{j+1} \neq \hat{H}^{j+1}$.

Proof Using (2.30) with $e_i \neq 0$ in (2.31), the Bellman equation becomes the following:

$$\begin{aligned} x_k^T (M^j)^T \hat{H}^{j+1} M^j x_k = x_k^T Q x_k + \sum_{i=1}^{n} (u_{ik}^j + e_{ik})^T R_i (u_{ik}^j + e_{ik}) - \gamma^2 d_k^T d_k \\ + x_{k+1}^T (M^j)^T \hat{H}^{j+1} M^j x_{k+1}, \end{aligned} \tag{2.32}$$

where

$$x_{k+1} = A x_k + \sum_{i=1}^{n} B_i (u_{ik}^j + e_{ik}) + E d_k. \tag{2.33}$$

Further, (2.32) is rewritten as

$$
\begin{aligned}
x_k^T (M^j)^T \hat{H}^{j+1} M^j x_k &= x_k^T Q x_k + \sum_{i=1}^{n} (u_{ik}^j + e_{ik})^T R_i (u_{ik}^j + e_{ik}) - \gamma^2 d_k^T d_k \\
&\quad + \left(A x_k + \sum_{i=1}^{n} B_i (u_{ik}^j + e_{ik}) - E d_k \right)^T (M^j)^T \hat{H}^{j+1} \\
&\quad \times M^j \left(A x_k + \sum_{i=1}^{n} B_i (u_{ik}^j + e_{ik}) - E d_k \right) \\
&= x_k^T Q x_k + \sum_{i=1}^{n} (u_{ik}^j)^T R_i u_{ik}^j - \gamma^2 d_k^T d_k + x_{k+1}^T (M^j)^T \hat{H}^{j+1} M^j x_{k+1} \\
&\quad + 2 \sum_{i=1}^{n} e_{ik}^T R_i u_{ik}^j + \sum_{i=1}^{n} e_{ik}^T (B_i^T (M^j)^T \hat{H}^{j+1} M^j B_i + R_i) e_{ik} \\
&\quad + 2 \sum_{i=1}^{n} e_{ik}^T B_i^T (M^j)^T \hat{H}^{j+1} M^j x_{k+1}.
\end{aligned}
\tag{2.34}
$$

It can be seen that what we learned in Algorithm 2.1 needs to satisfy (2.34) other than (2.25). By comparing (2.25) with (2.34), one can find $H^{j+1} \neq \hat{H}^{j+1}$, which indicates that the control policies updated by (2.28) and (2.29) may be inaccurate and produce bias if probing noise is added when implementing Algorithm 2.1. This completes the proof. □

Remark 2.4 It can be noted that in Algorithm 2.1, the system must update $u_{ik}^{j+1} = -K_i^{j+1} x_k$ to generate data, which is a typical feature of the on-policy RL algorithm, which has to produce deviation of solution during the learning process.

Remark 2.5 One can also notice that the disturbance input must be updated in the prescribed manner (2.27) and applied to the system. However, this is not possible in practical applications, because the disturbance is generally independent and random.

In order to avoid these shortcomings of the on-policy game Q-learning algorithm mentioned in Remarks 2.3 and 2.4, behavior inputs and behavior disturbance policy are going to be introduced when learning the saddle point for multi-player systems subject to disturbance, which indicates we shall investigate the off-policy game Q-learning to solve the zero-sum game problem.

2.1.3 Off-Policy Game Q-Learning Technique

In this subsection, we propose an off-policy game Q-learning algorithm to solve the zero-sum game problem, such that the H_∞ controllers of multi-player systems can be found even when no information of system dynamics is available. Moreover, it is proved that this algorithm will not produce deviation of solution even though probing

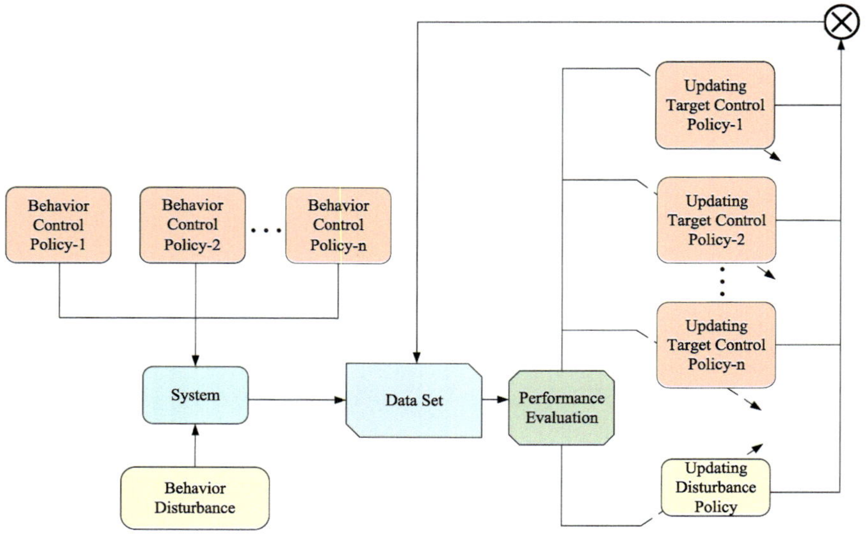

Fig. 2.1 The structure of the off-policy game Q-learning for H_∞ control

noises are added to the behavior control policies. The structure of the off-policy game Q-learning for the multi-player H_∞ control problem is shown in Fig. 2.1.

2.1.3.1 Derivation of Off-Policy Game Q-Learning Algorithm

From (2.25), one has

$$(M^j)^T H^{j+1} M^j = (M^j)^T \Lambda M^j + \left(A - \sum_{i=1}^{n} B_i K_i^j - E K^j\right)^T (M^j)^T H^{j+1} M^j \left(A - \sum_{i=1}^{n} B_i K_i^j - E K^j\right) \tag{2.35}$$

where

$$\Lambda = \mathrm{diag}(Q, R_1, R_2, \ldots, R_n, -\gamma^2 I).$$

Introducing auxiliary variables $u_i^j = -K_i^j x_k$ and $d_k^j = -K^j x_k$ to system (2.1) yields

$$x_{k+1} = A_c x_k + \sum_{i=1}^{n} B_i (u_{ik} - u_{ik}^j) + E(d_k - d_k^j), \tag{2.36}$$

where $A_c = A - \sum_{i=1}^{n} B_i K_i^j - E K^j$, u_i and d_k are called the behavior control policies and the behavior disturbance policy which are used to generate data, while u_{ik}^j and d_k^j are called the target control policies and the target disturbance policy which need to be learned. Along the system trajectory (2.36), one has

$$\begin{aligned}
&Q^{j+1}(x_k, U, d_k) - x_k^T A_c^T (M^j)^T H^{j+1} M^j A_c x_k \\
&\quad = x_k^T (M^j)^T H^{j+1} M^j x_k - \left(x_{k+1} - \sum_{i=1}^{n} B_i (u_{ik} - u_{ik}^j) - E(d_k - d_k^j) \right)^T \\
&\qquad \times (M^j)^T H^{j+1} M^j \left(x_{k+1} - \sum_{i=1}^{n} B_i (u_{ik} - u_{ik}^j) - E(d_k - d_k^j) \right) \\
&\quad = x_k^T (M^j)^T \Lambda M^j x_k.
\end{aligned} \tag{2.37}$$

In view that P^{j+1} and H^{j+1} are related as shown in (2.15) and (2.16), the following holds:

$$\begin{aligned}
&x_k^T (M^j)^T H^{j+1} M^j x_k - x_{k+1}^T (M^j)^T H^{j+1} M^j x_{k+1} + 2 \left(A x_k + \sum_{i=1}^{n} B_i u_{ik} + E d_k \right)^T \\
&\quad \times P^{j+1} \sum_{i=1}^{n} B_i (u_{ik} - u_{ik}^j) + 2 \left(A x_k + \sum_{i=1}^{n} B_i u_{ik} + E d_k \right)^T P^{j+1} E(d_k - d_k^j) \\
&\quad - \sum_{i=1}^{n} (u_{ik} - u_{ik}^j)^T B_i^T P^{j+1} \sum_{i=1}^{n} B_i (u_{ik} - u_{ik}^j) - 2 \sum_{i=1}^{n} (u_{ik} - u_{ik}^j)^T \\
&\quad \times B_i^T P^{j+1} E(d_k - d_k^j) - (d_k - d_k^j)^T E^T P^{j+1} E(d_k - d_k^j) \\
&\quad = x_k^T (M^j)^T \Lambda M^j x_k.
\end{aligned} \tag{2.38}$$

Further, one has

$$\begin{aligned}
&x_k^T (M^j)^T H^{j+1} M^j x_k - x_{k+1}^T (M^j)^T H^{j+1} M^j x_{k+1} \\
&\quad + 2 x_k^T \left[H_{xu_1}^{j+1} \quad H_{xu_2}^{j+1} \ldots H_{xu_n}^{j+1} \right] \sum_{i=1}^{n} (u_{ik} + K_i^j x_k) \\
&\quad + 2 \sum_{i=1}^{n} u_{ik}^T G^{j+1} \sum_{i=1}^{n} (u_{ik} + K_i^j x_k) + 2 d_k^T (H_{u_i d}^{j+1})^T \sum_{i=1}^{n} (u_{ik} + K_i^j x_k) \\
&\quad + 2 x_k^T (H_{xd}^{j+1})(d_k + K^j x_k) + 2 \sum_{i=1}^{n} u_{ik}^T H_{u_i d}^{j+1} (d_k + K^j x_k)
\end{aligned}$$

$$
\begin{aligned}
&-\sum_{i=1}^{n}\left(u_{ik}+K_i^j x_k\right)^T G^{j+1}\sum_{i=1}^{n}(u_{ik}+K_i^j x_k)-2\sum_{i=1}^{n}\left(u_{ik}+K_i^j x_k\right)^T H_{u_i d}^{j+1}\\
&\times(d_k+K^j x_k)-(d_k+K^j x_k)^T(H_{dd}^{j+1}+\gamma^2 I)(d_k+K^j x_k)\\
&=x_k^T(M^j)^T\Lambda M^j x_k,
\end{aligned}
\tag{2.39}
$$

where

$$
G^{j+1}=\begin{bmatrix}
H_{u_1u_1}^{j+1}-R_1 & H_{u_1u_2}^{j+1} & \dots & H_{u_1u_n}^{j+1}\\
(H_{u_1u_2}^{j+1})^T & H_{u_2u_2}^{j+1}-R_2 & \dots & H_{u_2u_n}^{j+1}\\
\vdots & \vdots & \ddots & \vdots\\
(H_{u_1u_n}^{j+1})^T & (H_{u_2u_n}^{j+1})^T & \dots & H_{u_nu_n}^{j+1}-R_n
\end{bmatrix}.
$$

Manipulating (2.39), one can get the following form:

$$
\hat{\theta}^j(k)\hat{L}^{j+1}=\hat{\rho}_k,
\tag{2.40}
$$

where

$$
\hat{\rho}_k=x_k^TQx_k+\sum_{i=1}^{n}u_{ik}^TR_iu_{ik}-\gamma^2 d_k^Td_k
$$

$$
\hat{L}^{j+1}=\left[(\mathrm{vec}(\hat{L}_{rz}^{j+1}))^T,\dots,(\mathrm{vec}(\hat{L}_{n+1,n+1}^{j+1}))^T\right]^T
$$

$$
\hat{\theta}^j(k)=\left[\hat{\theta}_{rz}^j,\dots,\hat{\theta}_{n+1,n+1}^j\right]
$$

with $r=0,1,2,\dots,n+1$, $z=r,r+1,r+2,\dots,n+1$. Besides,

$$
\begin{aligned}
&\hat{\theta}_{00}^j=x_k^T\otimes x_k^T-x_{k+1}^T\otimes x_{k+1}^T\\
&\hat{L}_{00}^{j+1}=H_{xx}^{j+1}\\
&\hat{\theta}_{ss}^j=-(K_s^jx_{k+1})^T\otimes(K_s^jx_{k+1})^T+u_s^T\otimes u_s^T\\
&\hat{L}_{ss}^{j+1}=H_{u_su_s}^{j+1}\\
&\hat{\theta}_{s+1,s+1}^j=-(K^jx_{k+1})^T\otimes(K^jx_{k+1})^T+d_k^T\otimes d_k^T\\
&\hat{L}_{s+1,s+1}^{j+1}=H_{dd}^{j+1}\\
&\hat{\theta}_{0s}^j=2x_{k+1}^T\otimes(K_s^jx_{k+1})^T+2x_k^T\otimes u_s^T\\
&\hat{L}_{0s}^{j+1}=H_{xu_s}^{j+1}\\
&\hat{\theta}_{0s+1}^j=2x_{k+1}^T\otimes(K^jx_{k+1})^T+2x_k^T\otimes d_k^T\\
&\hat{L}_{0s+1}^{j+1}=H_{xd}^{j+1}\\
&\hat{\theta}_{st}^j=-2(K_s^jx_{k+1})^T\otimes(K_t^jx_{k+1})^T+2u_s^T\otimes u_t^T
\end{aligned}
$$

$$
\begin{aligned}
\hat{L}_{st}^{j+1} &= H_{u_s u_t}^{j+1} \\
\hat{\theta}_{s,s+1}^{j} &= -2(K_s^j x_{k+1})^T \otimes (K^j x_{k+1})^T + 2u_s^T \otimes d_k^T \\
\hat{L}_{s,s+1}^{j+1} &= H_{u_s d}^{j+1}
\end{aligned}
$$

with $s \neq t$ and $s, t = 1, 2, \ldots, n$.

Based on the above work, $K_1^{j+1}, K_2^{j+1}, \ldots, K_n^{j+1}$ and K^{j+1} can be expressed as the form of $\hat{L}^{j+1}$

$$
K_i^{j+1} = (\hat{L}_{ii}^{j+1})^{-1} \left((\hat{L}_{0i}^{j+1})^T - \sum_{r=1, r\neq i}^{n+1} (\hat{L}_{ir}^{j+1})^T K_r^j \right) \tag{2.41}
$$

$$
K^{j+1} = (\hat{L}_{n+1,n+1}^{j+1})^{-1} \left((\hat{L}_{0,n+1}^{j+1})^T - \sum_{i=1}^{n} (\hat{L}_{n+1,i}^{j+1})^T K_i^j \right). \tag{2.42}
$$

Algorithm 2.2 Off-policy game Q-learning for the zero-sum game

1: Data collection: Collect system data x_k and store them in (2.40) by using (2.36);
2: Initialize the admissible control policies of multiple players $K_1^0, K_2^0, \ldots, K_n^0$ and disturbance policy gain K^0. Set the iteration index $j = 0$ and $i = 1$ represents player i ($i = 1, 2, \ldots, n+1$);
3: Performing the off-policy game Q-learning: use the recursive least-square method to solve the $\hat{L}^{j+1}$ in (2.40), and then K_i^{j+1} and K^{j+1} can be updated by (2.41) and (2.42);
4: If $i < n+1$, then $i = i+1$ and go back to Step 3. Otherwise, go to Step 5;
5: Stop when $\left\| K_i^{j+1} - K_i^j \right\| \leq \varepsilon$ ($i = 1, 2, \ldots, n+1$), the optimal control policy is obtained. Otherwise, $j = j+1$, $i = 1$, and go back to Step 3.

Theorem 2.2 *H^{j+1}, K_i^{j+1} and K^{j+1} are the solution to (2.40), (2.41) and (2.42) if and only if they are the solution to (2.25), (2.28) and (2.29).*

Proof We can see from the formula derivation that if H^{j+1}, K_i^{j+1} and K^{j+1} are solutions to (2.25), (2.28) and (2.29), then H^{j+1}, K_i^{j+1} and K^{j+1} will also satisfy (2.40), (2.41) and (2.42). Now we need to prove that the solutions to (2.40), (2.41) and (2.42) are the same as the solutions to (2.25), (2.28) and (2.29).

It can be seen that (2.40) is equivalent to the (2.38), so their solutions are also the same. Subtracting (2.38) from (2.37), one gets (2.25). Thus, one knows that the solution to (2.40), (2.41) and (2.42) is equal to (2.25), (2.28) and (2.29). This completes the proof. □

Remark 2.6 One thing to be noticed is that Algorithm 2.1 and Algorithm 2.2 have the same solution, which was proven in Theorem 2.1. Meanwhile, it concludes from Theorem 2.2 that, if $\hat{L}^{j+1}$ can be solved correctly, then the control policies K_i will converge to the optimal value since K_i learned by Algorithm 2.1 converge to the optimal values, that is, when $j \to \infty$, $u_i^j \to u_i^*$.

Remark 2.7 The game Q-learning in Algorithm 2.2 is indeed an off-policy Q-learning approach since the target control policies are updated but not applied to the real systems. In Algorithm 2.2, arbitrary behavior control policies u_i and behavior disturbance policy d_k that can make the system stable are used to generate data, thereby overcoming the shortcoming of insufficient system exploration, which is also the most basic characteristic of the off-policy learning (Li et al. 2017, 2019a; Vamvoudakis et al. 2017; Kiumarsi et al. 2017; Munos et al. 2016). While the policies $u_i^j = -K_i^j x_k$ and $d_k^j = -K^j x_k$ are the target policies, and they are updated using measured data.

2.1.3.2 No Bias Analysis of Solution Learned by Off-Policy Game Q-Learning Algorithm

We have proved that Algorithm 2.1 will have an impact on the learning results when probing noise is added. Next, we will prove the superiority of Algorithm 2.2 over Algorithm 2.1, that is, the probing noise will not affect the system, and the Nash equilibrium solution learned is without deviation.

Theorem 2.3 *Add probing noises to the behavior control policies in Algorithm 2.2. Let H^{j+1} be the solution to (2.37) with $e_i = 0$ and $\hat{H}^{j+1}$ be the solution to (2.37) with $e_i \neq 0$, then $\hat{H}^{j+1} = H^{j+1}$.*

Proof Probing noises are added to the behavior control policies $u_i + e_i$ ($e_i = 0$), the off-policy game Q-learning Eq. (2.37) is

$$\hat{x}_k^T (M^j)^T \hat{H}^{j+1} M^j \hat{x}_k = \hat{x}_k^T (M^j)^T \Lambda M^j \hat{x}_k + \hat{x}_k^T \left(A - \sum_{i=1}^{n} B_i K_i^j - E K^j \right)^T \times (M^j)^T \hat{H}^{j+1} M^j \left(A - \sum_{i=1}^{n} B_i K_i^j - E K^j \right) \hat{x}_k. \tag{2.43}$$

Notice that if adding probing noises $e_i \neq 0$ into system (2.36), then it becomes

$$\hat{x}_{k+1} = A_c \hat{x}_k + \sum_{i=1}^{n} B_i (u_{ik} + e_{ik} + K_i^j \hat{x}_k) + E(d_k + K^j \hat{x}_k). \tag{2.44}$$

In this case, (2.37) becomes

$$\hat{x}_k^T (M^j)^T \hat{H}^{j+1} M^j \hat{x}_k - \left(\hat{x}_{k+1} - \sum_{i=1}^{n} B_i(u_{ik} + e_{ik} + K_i^j \hat{x}_k) - E(d_k + K^j \hat{x}_k)\right)^T$$
$$\times (M^j)^T \hat{H}_i^{j+1} M^j \left(\hat{x}_{k+1} - \sum_{i=1}^{n} B_i(u_{ik} + e_{ik} + K_i^j \hat{x}_k) - E(d_k + K^j x_k)\right)$$
$$= \hat{x}_k^T (M^j)^T \Lambda M^j \hat{x}_k. \tag{2.45}$$

Substituting (2.44) into (2.45), (2.45) becomes (2.43). So the solution to (2.43) is the same as (2.37). Therefore, it is impossible for the off-policy game Q-learning algorithm to produce bias when adding probing noises. The unbiasedness of the off-policy game Q-learning method is proved. □

Remark 2.8 Different from Kiumarsi et al. (2017), where the off-policy RL method is used to solve standard H_∞ control problem for systems with single controller, Sect. 2.1.3 not only extends the result in Kiumarsi et al. (2017) to the case of H_∞ control for multi-player systems, but also no bias of solution using the off-policy game Q-learning and the biased solution using the on-policy game Q-learning have been rigorously proven.

Remark 2.9 Compared with Liu et al. (2014); Li et al. (2019a, 2017); Vamvoudakis and Lewis (2011); Li et al. (2019b) that ignored the negative impact of disturbance on the performance of multi-agent systems, this is the first time that the off-policy game Q-learning Algorithm is proposed and applied to H_∞ control of multi-player systems.

Remark 2.10 The proposed off-policy game Q-learning Algorithm 2.2 does not require any knowledge of the system dynamics, such that the control policies, with the guarantee of L_2 gain, can be found for unknown multi-player systems unlike (Vamvoudakis and Lewis 2012, 2010) which requires the accurate system model to be known.

2.1.4 Simulation Results

In this subsection, the effectiveness of the proposed off-policy game Q-learning Algorithm 2.2 is verified with the H_∞ control simulation results of three-player and five-player systems. Moreover, comparative simulation experiments with Liu et al. (2014); Vamvoudakis and Lewis (2011) where disturbance is ignored are carried out to show the advantages of the proposed off-policy game Q-learning algorithm for systems subject to external disturbance. It is assumed that the dynamics A, B_i and E are completely unknown during the learning process of the algorithm.

The state trajectories of the system are assumed to be subject to the following disturbance:

$$d_k = e^{-0.0001k} \sin(2.0k). \tag{2.46}$$

2.1.4.1 Comparison Results of On-Policy Learning with Off-Policy Learning

A three-player system is used as the object of simulation experiments when implementing on-policy game Q-learning Algorithm 2.1 and off-policy game Q-learning Algorithm 2.2.

Example 2.1 Consider the following linear DT system with three-player and disturbance input:

$$x_{k+1} = Ax_k + B_1u_1 + B_2u_2 + B_3u_3 + Ed_k, \tag{2.47}$$

where

$$A = \begin{bmatrix} 0.906488 & 0.0816012 & -0.0005 \\ 0.074349 & 0.90121 & -0.000708383 \\ 0 & 0 & 0.132655 \end{bmatrix}, \ B_1 = \begin{bmatrix} -0.00150808 \\ -0.0096 \\ 0.867345 \end{bmatrix}$$

$$B_2 = \begin{bmatrix} 0.00951892 \\ 0.00038373 \\ 0 \end{bmatrix}, \ B_3 = \begin{bmatrix} -0.00563451 \\ -0.08962 \\ 0.356478 \end{bmatrix}, \ E = \begin{bmatrix} 0.0123956 \\ 0.068 \\ -0.05673 \end{bmatrix}.$$

Choose $Q = \text{diag}(5, 5, 5)$ and $R_1 = R_2 = R_3 = 1$. The disturbance attenuation factor is selected to be $\gamma = 1$. Rewrite (2.22) as

$$H^* = \Lambda + G^T H^* G, \tag{2.48}$$

where

$$G = \begin{bmatrix} A & B_1 & B_2 & B_3 & E \\ -K_1A & -K_1B_1 & -K_1B_2 & -K_1B_3 & -K_1E \\ -K_2A & -K_2B_1 & -K_2B_2 & -K_2B_3 & -K_2E \\ -K_3A & -K_3B_1 & -K_3B_2 & -K_3B_3 & -K_3E \end{bmatrix}.$$

The theoretical solution to (2.48) can be obtained by using MATLAB. Thus, the optimal Q-function matrix H^* and the optimal controller gains (K_1^*, K_2^*, K_3^*) and the worst-case disturbance policy gain K^* can be obtained as follows:

$$
H^* = \begin{bmatrix}
47.2688 & 28.7098 & -0.0292 & -0.2510 & 0.4329 & -2.7108 & 2.4375 \\
28.7098 & 40.9565 & -0.0241 & -0.2922 & 0.2868 & -3.4598 & 2.8845 \\
-0.0292 & -0.0241 & 5.0883 & 0.5774 & -0.0003 & 0.2395 & -0.0398 \\
-0.2510 & -0.2922 & 0.5774 & 4.7762 & -0.0025 & 1.5791 & -0.2704 \\
0.4329 & 0.2868 & -0.0003 & -0.0025 & 1.0044 & -0.0270 & 0.0244 \\
-2.7108 & -3.4598 & 0.2395 & 1.5791 & -0.0270 & 1.9705 & -0.3786 \\
2.4375 & 2.8845 & -0.0398 & -0.2704 & 0.0244 & -0.3786 & -0.7515
\end{bmatrix}.
$$
$$
\begin{aligned}
K_1^* &= \begin{bmatrix}-0.5665 & -0.7300 & -0.1098\end{bmatrix}, \\
K_2^* &= \begin{bmatrix}-0.4271 & -0.2752 & -0.0009\end{bmatrix}, \\
K_3^* &= \begin{bmatrix}2.2643 & 2.8476 & -0.0329\end{bmatrix}, \\
K^* &= \begin{bmatrix}2.2925 & 2.6572 & 0.0032\end{bmatrix}.
\end{aligned} \tag{2.49}
$$

Three types of probing noises are used to demonstrate whether the on-policy game Q-learning Algorithm 2.1 and the off-policy game Q-learning Algorithm 2.2 would produce bias of solution to Q-function-based iterative Bellman Eqs. (2.25) and (2.40), when adding probing noises in control inputs to generate data. The following three probing noises are considered:

Case 1:

$$
e_i = \sum_j^{100} 0.5 * \sin(\text{noise}_{\text{feq}}(1, j) * k). \tag{2.50}
$$

Case 2:

$$
e_i = \sum_j^{100} 6 * \sin(\text{noise}_{\text{feq}}(1, j) * k). \tag{2.51}
$$

Case 3:

$$
e_i = \sum_j^{100} 10 * \sin(\text{noise}_{\text{feq}}(1, j) * k). \tag{2.52}
$$

where

$$
\text{noise}_{\text{feq}}(1, j) = 500 * \text{rand}(1, 100) - 200 * \text{ones}(1, 100). \tag{2.53}
$$

Table 2.1 shows the controller gains and the worst-case disturbance policy learned by Algorithms 2.1 and 2.2 under these three kinds of probing noises. It can be seen that Algorithm 2.1 is affected by probing noises, and its final controller gain is deviated from the theoretical value. However, Algorithm 2.2 is not affected by probing noises and can converge to the theoretical optimal value.

Under Case 1, the convergence process of matrix H^{j+1}, the control policy gains and disturbance policy gain (K_1^{j+1}, K_2^{j+1}, K_3^{j+1} and K^{j+1}) can be seen in Figs. 2.2 and 2.3. Figure 2.4 shows the system state when implementing Algorithm 2.1.

Figures 2.5 and 2.6 show the convergence state of matrix H^{j+1}, the control policy gain and disturbance policy gain (K_1^{j+1}, K_2^{j+1}, K_3^{j+1} and K^{j+1}) under Case 2, and

Table 2.1 H_∞ controller gains under three probing noises

Case	On-policy game Q-learning	Off-policy game Q-learning
1	$K_1 = \begin{bmatrix} -0.5665 & -0.7300 & -0.1098 \end{bmatrix}$ $K_2 = \begin{bmatrix} -0.4271 & -0.2753 & -0.0009 \end{bmatrix}$ $K_3 = \begin{bmatrix} 2.2645 & 2.8474 & -0.0329 \end{bmatrix}$ $K = \begin{bmatrix} 2.2926 & 2.6571 & 0.0032 \end{bmatrix}$	$K_1 = \begin{bmatrix} -0.5665 & -0.7300 & -0.1098 \end{bmatrix}$ $K_2 = \begin{bmatrix} -0.4271 & -0.2753 & -0.0009 \end{bmatrix}$ $K_3 = \begin{bmatrix} 2.2645 & 2.8474 & -0.0329 \end{bmatrix}$ $K = \begin{bmatrix} 2.2926 & 2.6571 & 0.0032 \end{bmatrix}$
2	$K_1 = \begin{bmatrix} -0.5685 & -0.7328 & -0.1098 \end{bmatrix}$ $K_2 = \begin{bmatrix} -0.4270 & -0.2759 & -0.0009 \end{bmatrix}$ $K_3 = \begin{bmatrix} 2.2712 & 2.8572 & -0.0329 \end{bmatrix}$ $K = \begin{bmatrix} 2.2982 & 2.6663 & 0.0032 \end{bmatrix}$	$K_1 = \begin{bmatrix} -0.5665 & -0.7300 & -0.1098 \end{bmatrix}$ $K_2 = \begin{bmatrix} -0.4271 & -0.2752 & -0.0009 \end{bmatrix}$ $K_3 = \begin{bmatrix} 2.2643 & 2.8476 & -0.0329 \end{bmatrix}$ $K = \begin{bmatrix} 2.2925 & 2.6572 & 0.0032 \end{bmatrix}$
3	$K_1 = \begin{bmatrix} -0.5103 & -0.7568 & -0.1107 \end{bmatrix}$ $K_2 = \begin{bmatrix} -0.4253 & -0.3143 & 0.0113 \end{bmatrix}$ $K_3 = \begin{bmatrix} 2.0908 & 2.9773 & -0.293 \end{bmatrix}$ $K = \begin{bmatrix} 2.3137 & 2.8585 & 0.0105 \end{bmatrix}$	$K_1 = \begin{bmatrix} -0.5665 & -0.7300 & -0.1098 \end{bmatrix}$ $K_2 = \begin{bmatrix} -0.4271 & -0.2752 & -0.0009 \end{bmatrix}$ $K_3 = \begin{bmatrix} 2.2643 & 2.8476 & -0.0329 \end{bmatrix}$ $K = \begin{bmatrix} 2.2925 & 2.6572 & 0.0032 \end{bmatrix}$

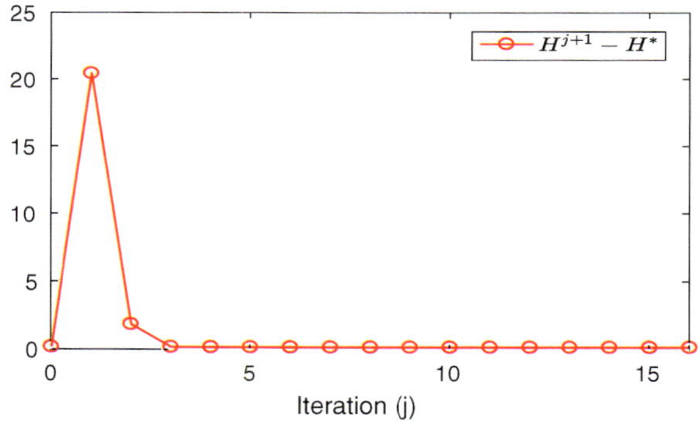

Fig. 2.2 Case 1: Convergence of H when implementing the on-policy game Q-learning

Figure 2.7 shows the evolution process of the system trajectory under the learned control policies using Algorithm 2.1.

Algorithm 2.1 was implemented under Case 3, Figs. 2.8 and 2.9 show the convergence processes of the matrix H^{j+1}, the control policy gains and disturbance policy gain K_1^{j+1}, K_2^{j+1}, K_3^{j+1}, K^{j+1}. Figure 2.10 shows the state responses of the system under the controller learned by Algorithm 2.1, and Fig. 2.11 shows the system performance J under the control policies learned by the Algorithm 2.1.

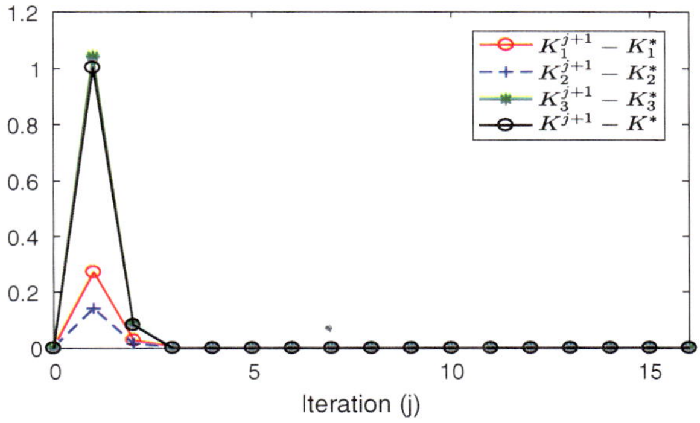

Fig. 2.3 Case 1: Convergence of $K_i (i = 1, 2, 3)$ and K when implementing the on-policy game Q-learning

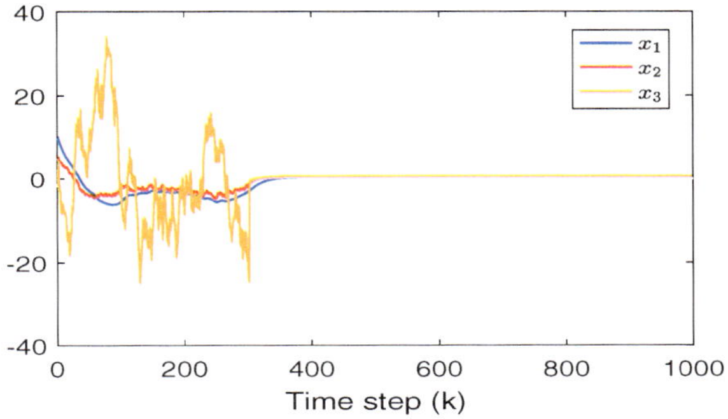

Fig. 2.4 Case 1: The states x of system when implementing the on-policy game Q-learning

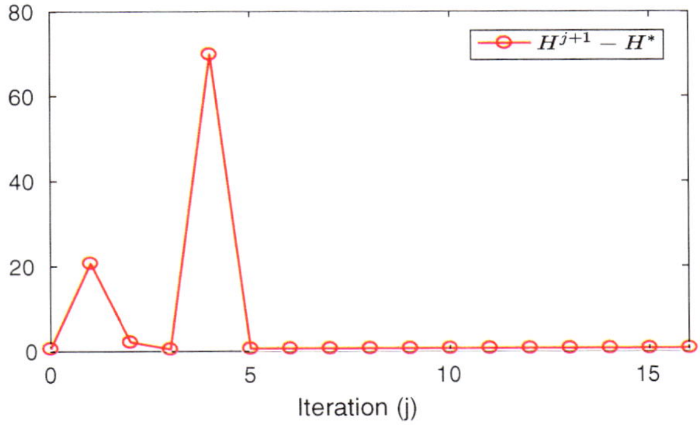

Fig. 2.5 Case 2: Convergence of H when implementing the on-policy game Q-learning

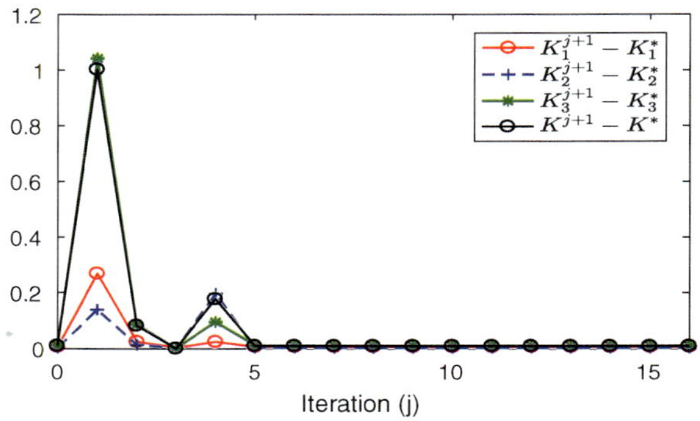

Fig. 2.6 Case 2: Convergence of $K_i (i = 1, 2, 3)$ and K when implementing the on-policy game Q-learning

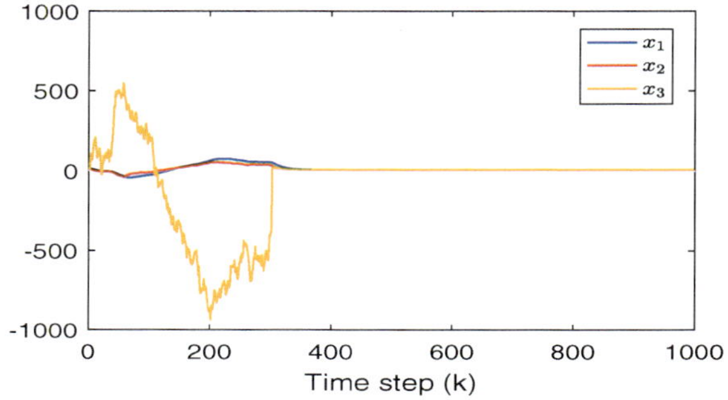

Fig. 2.7 Case 2: The states x of system when implementing the on-policy game Q-learning

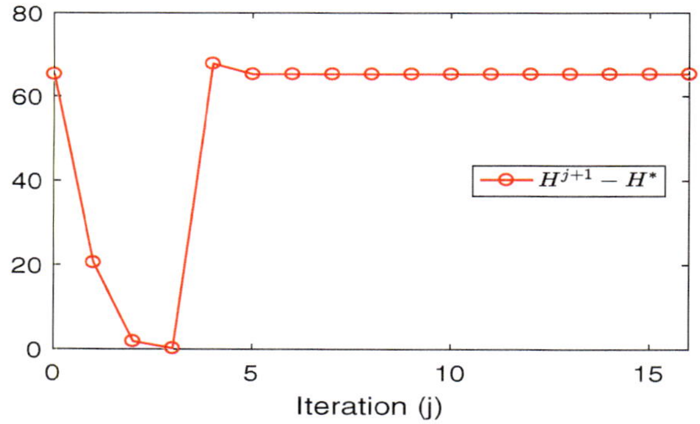

Fig. 2.8 Case 3: Convergence of H when implementing the on-policy game Q-learning

Now, we start to plot simulation results using the learned controller gains form Algorithm 2.2. The convergence results of matrix H^{j+1}, the control policy gains and disturbance policy gain (K_1^{j+1}, K_2^{j+1}, K_3^{j+1} and K^{j+1}) under Case 1 of probing noises are shown in Figs. 2.12 and 2.13. Figure 2.14 shows the state trajectory of the system, and Fig. 2.15 shows the performance J of the system when using Algorithm 2.2.

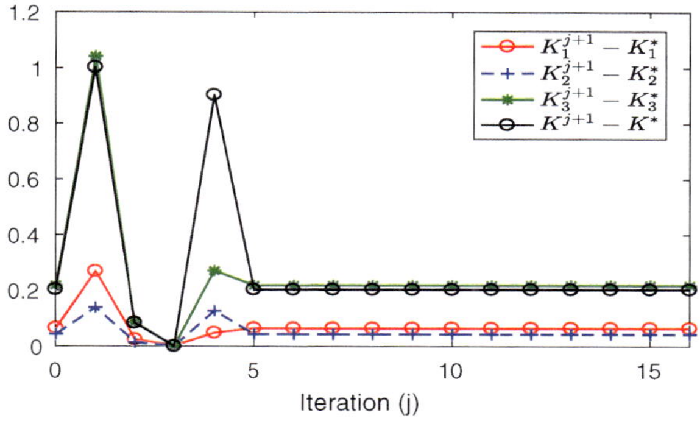

Fig. 2.9 Case 3: Convergence of $K_i (i = 1, 2, 3)$ and K when implementing the on-policy game Q-learning

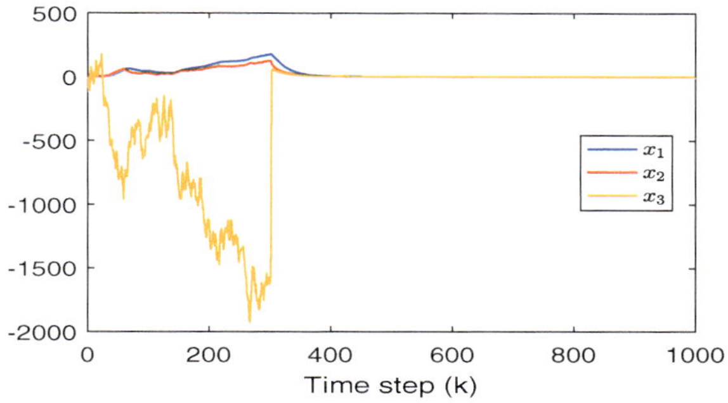

Fig. 2.10 Case 3: The states x of system when implementing the on-policy game Q-learning

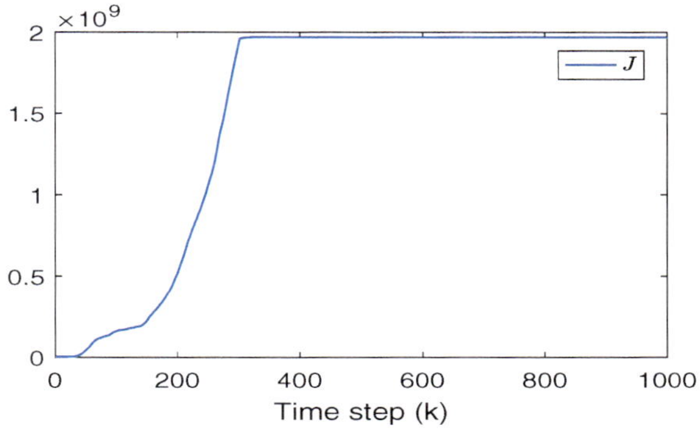

Fig. 2.11 Case 3: The cost J of system when implementing the on-policy game Q-learning

The signal to noise ratio is calculated as

$$\frac{\sum_{k=0}^{500}(x_k^T Q x_k + \sum_{i=1}^{3} u_i^T R_i u_i)}{\sum_{k=0}^{500} \|d_k\|^2} = 0.2992 < 1.$$

It can be seen that the disturbance attenuation condition (2.2) is satisfied.

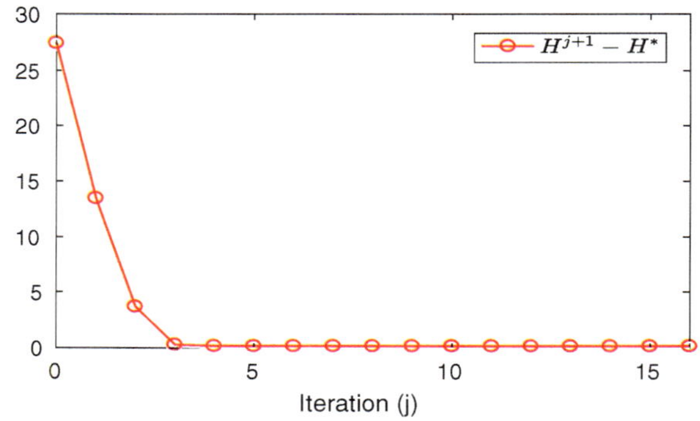

Fig. 2.12 Convergence of H when implementing the off-policy game Q-learning

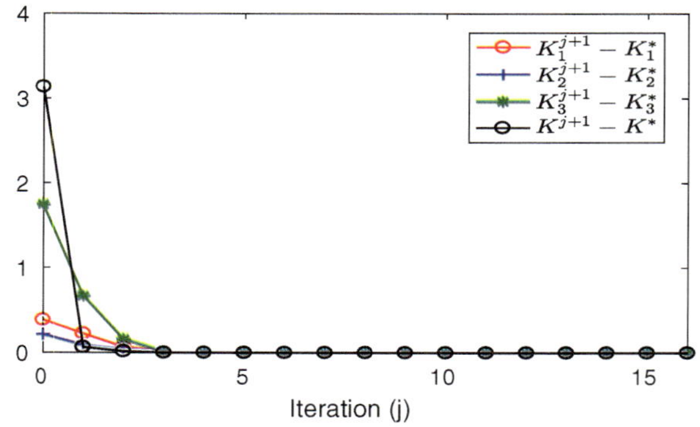

Fig. 2.13 Convergence of $K_i (i = 1, 2, 3)$ and K when implementing the off-policy game Q-learning

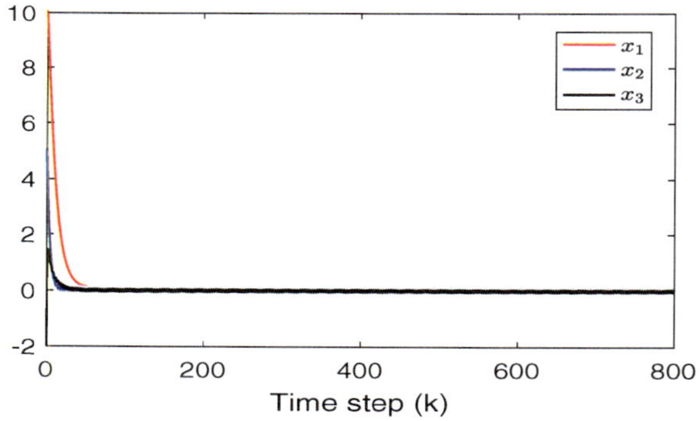

Fig. 2.14 The states x of system when implementing the off-policy game Q-learning

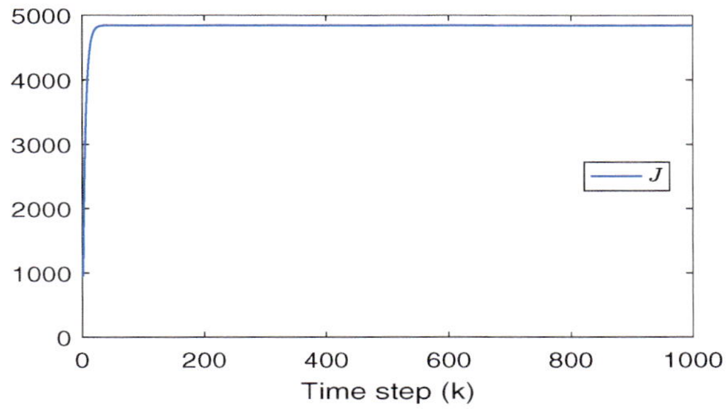

Fig. 2.15 The cost J of system when implementing the off-policy game Q-learning

As one can see in Figs. 2.4, 2.7 and 2.10, the state of system (2.47) has been obviously affected by adding probing noises when implementing the on-policy game Q-learning. However, as shown in Fig. 2.14, the system trajectory x can quickly converge to a stable state, and the system performance shown in Fig. 2.15 is smaller than that when utilizing Algorithm 2.1.

2.1.4.2 Verification of Anti-interference

We use a five-player system to further verify the validity of Algorithm 2.2 here. In addition, anti-interference of the proposed Algorithm 2.2 for H_∞ control of multi-player systems and the advantage over the methods without taking into external disturbance account is demonstrated.

Example 2.2 Consider the following linear DT system with five players and disturbance input:

$$x_{k+1} = Ax_k + B_1u_1 + B_2u_2 + B_3u_3 + B_4u_4 + B_5u_5 + Ed_k, \tag{2.54}$$

where

$$A = \begin{bmatrix} 0.906488 & 0.0816012 & -0.0005 \\ 0.074349 & 0.90121 & -0.000708383 \\ 0 & 0 & 0.132655 \end{bmatrix}$$

$$B_1 = \begin{bmatrix} -0.00150808 \\ -0.0096 \\ 0.867345 \end{bmatrix}, B_2 = \begin{bmatrix} 0.00951892 \\ 0.00038373 \\ 0 \end{bmatrix}, B_3 = \begin{bmatrix} -0.00563451 \\ -0.08962 \\ 0.356478 \end{bmatrix},$$

$$B_4 = \begin{bmatrix} 0.1568 \\ 0.006018 \\ -0.18235673 \end{bmatrix}, B_5 = \begin{bmatrix} -0.125 \\ 0.4 \\ -0.4898 \end{bmatrix}, E = \begin{bmatrix} 0.0123956 \\ 0.068 \\ -0.05673 \end{bmatrix}.$$

Choose $Q = \text{diag}(5, 5, 5)$ and $R_1 = R_2 = R_3 = R_4 = R_5 = 1$. The disturbance attenuation factor is selected to be $\gamma = 1$. Rewrite (2.22) as

$$H^* = \Lambda + \hat{G}^T H^* \hat{G} \tag{2.55}$$

where

$$\hat{G} = \begin{bmatrix} A & B_1 & \dots & B_5 & E \\ -K_1 A & -K_1 B_1 & \dots & -K_1 B_5 & -K_1 E \\ -K_2 A & -K_2 B_1 & \dots & -K_2 B_5 & -K_2 E \\ -K_3 A & -K_3 B_1 & \dots & -K_3 B_5 & -K_3 E \\ -K_4 A & -K_4 B_1 & \dots & -K_4 B_5 & -K_4 E \\ -K_5 A & -K_5 B_1 & \dots & -K_5 B_5 & -K_5 E \end{bmatrix}.$$

Similarly, the theoretical solution to (2.55) can be obtained using MATLAB software.

$$H^* = \begin{bmatrix} 17.2288 & 4.9395 & -0.0033 & -0.0201 & 0.1267 & -0.4418 & 2.0775 & 0.0509 & 0.4520 \\ 4.9395 & 13.7043 & 0.0036 & -0.0142 & 0.0482 & -0.8204 & 0.7744 & 3.0606 & 0.6807 \\ -0.0033 & 0.0036 & 5.0882 & 0.5769 & -0.00003 & 0.2368 & -0.1219 & -0.3236 & -0.0375 \\ -0.0201 & -0.0142 & 0.5769 & 4.7722 & -0.0002 & 1.5517 & -0.7964 & -2.1330 & -0.2479 \\ 0.1267 & 0.0482 & -0.00003 & -0.0002 & 1.0013 & -0.0043 & 0.0216 & -0.0008 & 0.0045 \\ -0.4418 & -0.8204 & 0.2368 & 1.5517 & -0.0043 & 1.7146 & -0.3948 & -1.1682 & -0.1654 \\ 2.0775 & 0.7744 & -0.1219 & -0.7964 & 0.0216 & -0.3948 & 1.5203 & 0.4272 & 0.1239 \\ 0.0509 & 3.0606 & -0.3236 & -2.1330 & -0.0008 & -1.1682 & 0.4272 & 3.5806 & 0.3669 \\ 0.4520 & 0.6807 & -0.0375 & -0.2479 & 0.0045 & -0.1654 & 0.1239 & 0.3669 & -0.9302 \end{bmatrix}$$

$$\begin{aligned} K_1^* &= \begin{bmatrix} -0.2870 & -0.6379 & -0.0098 \end{bmatrix}, \\ K_2^* &= \begin{bmatrix} -0.0950 & -0.0392 & -0.0004 \end{bmatrix}, \\ K_3^* &= \begin{bmatrix} 0.2282 & 0.1920 & -0.0316 \end{bmatrix}, \\ K_4^* &= \begin{bmatrix} -1.4928 & -0.4988 & 0.0144 \end{bmatrix}, \\ K_5^* &= \begin{bmatrix} 0.0330 & -1.1483 & 0.0188 \end{bmatrix}, \\ K^* &= \begin{bmatrix} 0.3356 & 0.3481 & 0.0012 \end{bmatrix}. \end{aligned} \tag{2.56}$$

Firstly, five players are used by the systems to verify the effectiveness of the proposed off-policy game Q-learning Algorithm 2.2. It can be seen from Figs. 2.16 and 2.17 that the learned matrix H^{j+1}, controller gains and disturbance policy gain

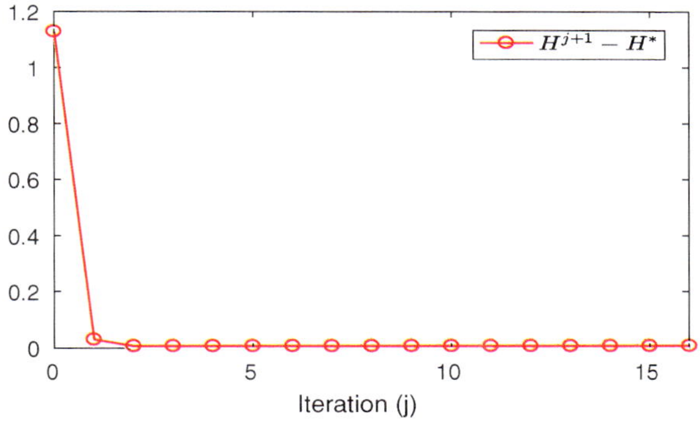

Fig. 2.16 Convergence of H when implementing the off-policy game Q-learning

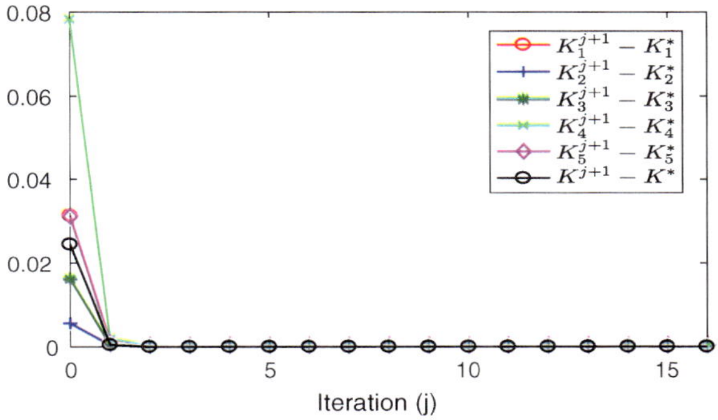

Fig. 2.17 Convergence of $K_i (i = 1, 2, \ldots, 5)$ and K when implementing the off-policy game Q-learning

Table 2.2 Controller gains when considering the disturbance or not

Probing noise	On-policy game Q-learning	Off-policy game Q-learning
Case 1/2/3	$K_1 = [-0.2870\ \ -0.6379\ \ -0.0098]$	$K_1 = [-0.2743\ \ -0.6278\ \ -0.0996]$
	$K_2 = [-0.0950\ \ -0.0392\ \ -0.0004]$	$K_2 = [-0.0926\ \ -0.0371\ \ -0.0004]$
	$K_3 = [0.2282\ 0.1920\ \ -0.0316]$	$K_3 = [0.2144\ 0.1790\ \ -0.0317]$
	$K_4 = [-1.4928\ \ -0.4988\ 0.0144]$	$K_4 = [-1.4563\ \ -0.4669\ 0.0147]$
	$K_5 = [0.0330\ \ -1.1483\ 0.0188]$	$K_5 = [0.0768\ \ -1.1065\ 0.0191]$
	$K = [0.3356\ 0.3481\ 0.0012]$	

(K_i^{j+1}, K^{j+1}) are not affected by the probing noise, and they can quickly converge to the theoretical optimal values without deviation as shown in Table 2.2. Figure 2.18 and 2.19 show the state trajectories and performance J of the system under the learned control policies using off-policy game Q-learning Algorithm 2.2.

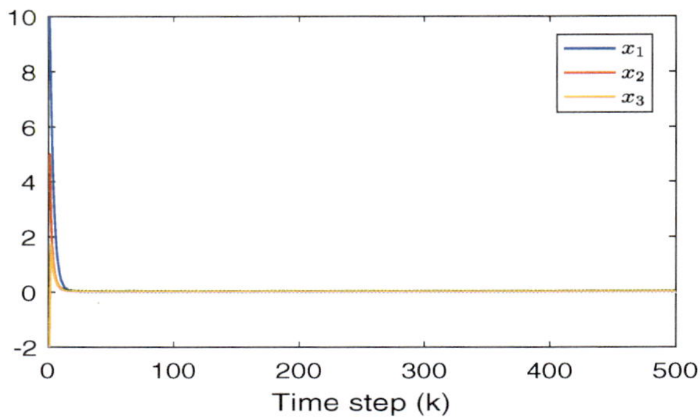

Fig. 2.18 The states x of system when implementing the off-policy game Q-learning

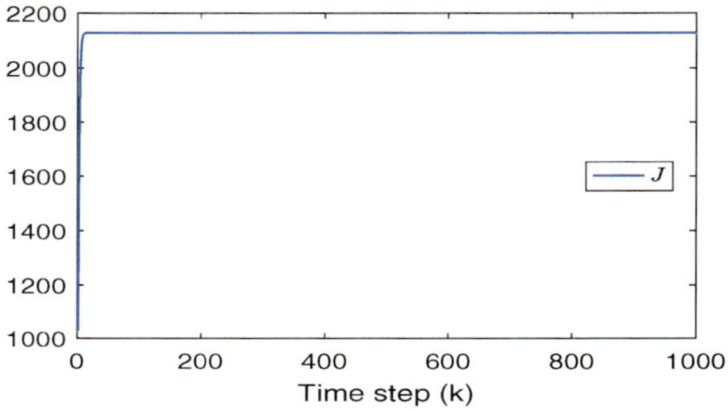

Fig. 2.19 The cost J of system when implementing the off-policy game Q-learning

The signal to noise ratio is calculated as

$$\frac{\sum_{k=0}^{500}(x_k^T Qx_k + \sum_{i=1}^{5} u_i^T R_i u_i)}{\sum_{k=0}^{500} \|d_k\|^2} = 0.4289 < 1.$$

It can be seen that the disturbance attenuation condition (2.2) is satisfied.

Then we assume $E = 0$ which means the external disturbance is not taken into account similar to the methods in Li et al. (2017); Liu et al. (2014); Vamvoudakis and Lewis (2011); Vamvoudakis et al. (2012, 2017), and implementing Algorithm 2.2 yields the optimal controller gains shown in Table 2.2. Simulation comparisons are going to be made under the following three kinds of external disturbances:

Case 1:

$$d_k = 0.5e^{-0.001k}\sin(2.0k). \tag{2.57}$$

Case 2:

$$d_k = e^{-0.001k}\sin(2.0k). \tag{2.58}$$

Case 3:

$$d_k = 5e^{-0.001k}\sin(2.0k). \tag{2.59}$$

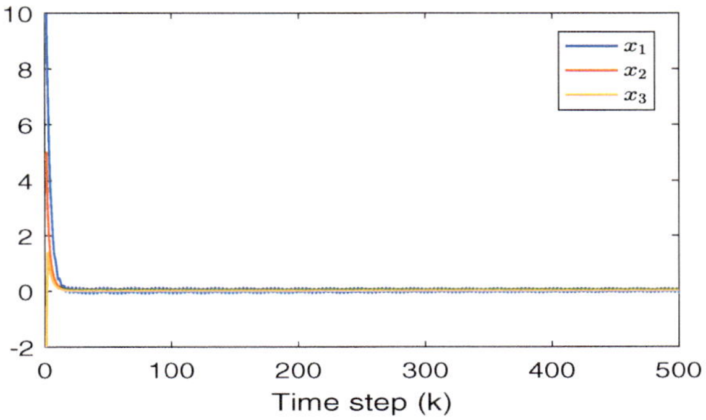

Fig. 2.20 Case 1: The states x of system when implementing the off-policy Q-learning($E = 0$)

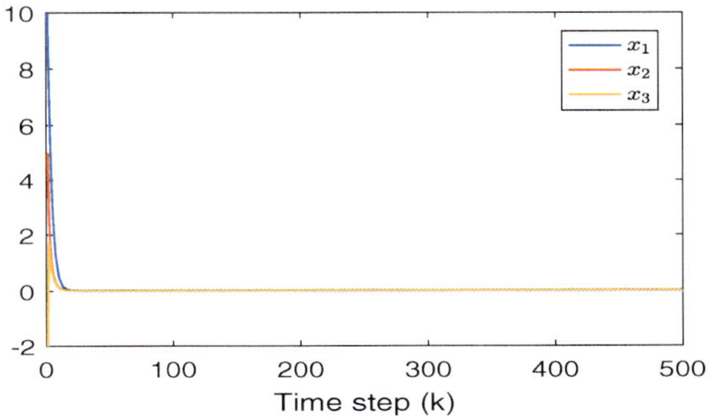

Fig. 2.21 Case 1: The states x of system when implementing the off-policy game Q-learning($E \neq 0$)

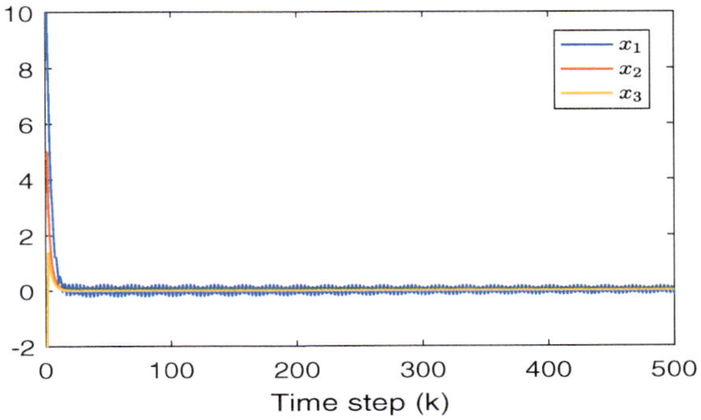

Fig. 2.22 Case 2: The states x of system when implementing the off-policy Q-learning($E = 0$)

Under the three kinds of disturbance, Figs. 2.20, 2.22 and 2.24 plot the state trajectories of the system under the controllers without considering the external disturbances. By comparing the results in Figs. 2.20, 2.22 and 2.24 with those in Figs. 2.21, 2.23 and 2.25, it can be seen that under the same external disturbance, the system state obtained by considering the interference in the learning process

Fig. 2.23 Case 2: The states x of system when implementing the off-policy game Q-learning($E \neq 0$)

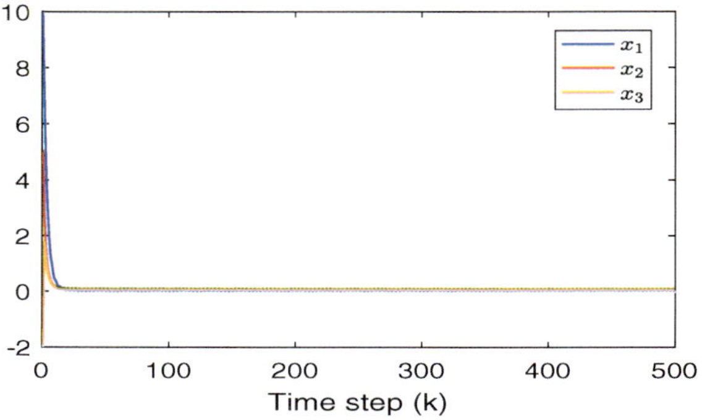

Fig. 2.24 Case 3: The states x of system when implementing the off-policy Q-learning($E = 0$)

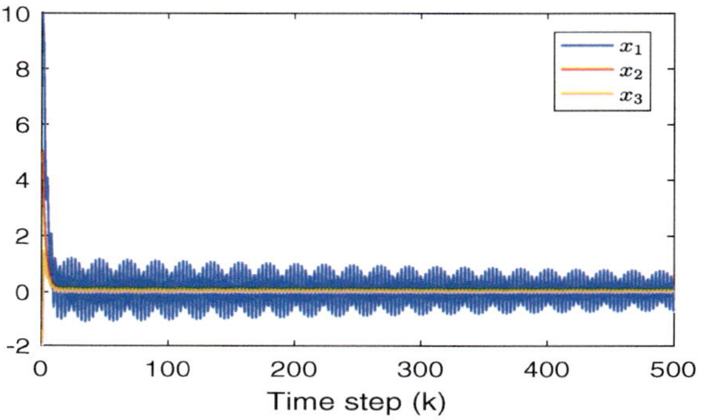

Fig. 2.25 Case 3: The states x of system when implementing the off-policy game Q-learning($E \neq 0$)

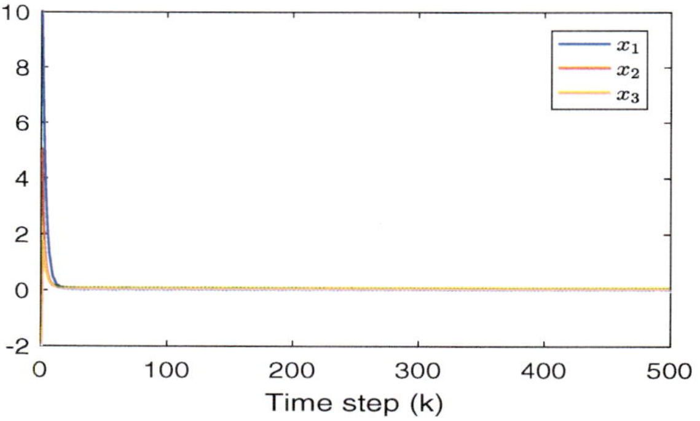

always tends to be stable, while the system state obtained without considering the interference will be greatly affected by the external disturbances.

2.2 H_∞ Output Feedback Control of Multi-player Systems

In Sect. 2.1, we have investigated how to use reinforcement learning methods to achieve H_∞ control of linear DT multi-player systems via state feedback. Thus, one issue is naturally raised, that is how to deal with anti-interference of multi-source disturbances of multi-player systems by intelligently interacting with each other of all players and all kinds of disturbances, since multi-source disturbances, such as time-varying temperature, raw material, price, inevitably exist in practice, such as power systems, large-scale industrial processes and circuit systems (Wei et al. 2014; Li et al. 2020; Ren et al. 2020). For multi-player systems with multi-source disturbances, Liu and Wei (2014) proposed a model-free adaptive algorithm based on ADP and neural networks (NNs) to solve the multi-player differential game problem for a class of nonlinear CT systems with uncertainties. Based on the off-policy learning and NNs, Song et al. (2017) proposed a synchronous solution approach of multi-player zero-sum games for CT systems without the need to know the system dynamics. Using a single network structure that is different from traditional NNs and iterative ADP algorithms, Jiang et al. (2018) investigated the multi-player zero-sum game problem in CT systems. Notice that the results (Kiumarsi et al. 2017; Modares et al. 2015; Peng et al. 2019; Li et al. 2018; Lv and Ren 2019; Liu and Wei 2014; Song et al. 2017; Jiang et al. 2018) about single-player systems, multi-player systems and multi-player systems with multi-source disturbances employed the state feedback control policy in solving the H_∞ control problem and the zero-sum game problem. However, the question is what if states of systems cannot be available? It is well known that states of systems are usually unmeasurable or are available at a great cost. Thus, for linear DT multi-player systems with multi-source disturbances and unmeasurable state, how to design output feedback controllers to achieve H_∞ control by RL will be our focus in this section.

Achieving a specific disturbance attenuation level under the mixed competitive and cooperative (MCC) relationship and output feedback controller design with a complete model-free approach for linear multi-player systems subject to multi-source disturbances are the main difficulties. To overcome these obstacles, a novel off-policy MCC game Q-learning algorithm is presented to learn the Nash equilibrium solution of the multi-player game, by integrating RL and game theory under the ADP framework. Moreover, a rigorous convergence proof is given for the Nash equilibrium solution to the multi-player MCC game learned by the proposed algorithm, along with a rigorous proof that the algorithm learning results remain unbiased when probing noises are added to the system to ensure the PE condition.

The rest of the section is organized as follows. Section 2.2.1 focuses on the formulation of the H_∞ control problem for linear DT systems with multiple players and multi-source disturbances. Section 2.2.2 gives the theoretical solution to the multi-player MCC game problem transformed from the H_∞ control problem and presents a model-free on-policy Q-learning algorithm. Section 2.2.3 develops a novel off-policy MCC game Q-learning algorithm, such that the output feedback controller can be found for resisting multi-source disturbances using measured data. Moreover, the proofs of convergence of the proposed algorithm and unbiasedness of the Nash equilibrium solution are provided. Section 2.2.4 verifies the effectiveness of the proposed algorithm through simulation results.

2.2.1 Problem Statement

In this section, the H_∞ control problem for linear DT multi-player systems is proposed first, and then it is converted into a multi-player MCC game problem. Thus, all players are divided into two groups, in which, one group works for minimizing the specific performance index, while the other group consisted of multi-source disturbances tries to make the performance worst on the opposite.

Consider the following linear DT system with multi-player and multiple exogenous disturbances:

$$\begin{aligned} x_{k+1} &= Ax_k + \sum_{i=1}^{s} B_i u_{ik} + \sum_{l=1}^{m} E_l d_{lk} \\ y_k &= Cx_k, \end{aligned} \tag{2.60}$$

where $x_k = x(k) \in \mathbb{R}^n$ represents the system state with initial state x_0, $u_{ik} = u_i(k) \in \mathbb{R}^{q_i}$ $(i = 1, ..., s)$ denotes the control input, $d_{lk} = d_l(k) \in \mathbb{R}^{p_l}$ $(l = 1, \ldots, m)$ is the external disturbance and $y_k \in \mathbb{R}^r$ is the system output. $A \in \mathbb{R}^{n\times n}$, $B_i \in \mathbb{R}^{n\times q_i}$, $E_l \in \mathbb{R}^{n\times p_l}$, $C \in \mathbb{R}^{r\times n}$ and k is the sampling time instant. The following definition about the H_∞ control holds.

Definition 2.4 (Kiumarsi et al. 2017) For the H_∞ control problem, the target is to find the control policies u_{ik} such that the following two points hold:

- If the following disturbance attenuation condition holds for all $d_{lk} \in L_2[0, \infty)$, then the L_2-gain of the system (2.60) is less than or equal to the attenuation factor γ.

$$\sum_{k=0}^{\infty} \left(y_k^T Q y_k + \sum_{i=1}^{s} u_{ik}^T R_i u_{ik} \right) \leq \gamma^2 \sum_{k=0}^{\infty} \left(\sum_{l=1}^{m} \|d_{lk}\|^2 \right), \tag{2.61}$$

  where $Q = Q^T \geq 0$, $R_i = R_i^T > 0$ and $\gamma \geq 0$.
- The system output y_k is asymptotically stable when all disturbances $d_{lk} = 0$ $(l = 1, \ldots, m)$.

The objective of this chapter is to develop an off-policy Q-learning algorithm for finding the output feedback H_∞ controllers $u_{ik} = -K_i y_k$ using system output data rather than system state. Two general assumptions like (Kiumarsi et al. 2017; Rizvi and Lin 2018) are established below for claiming that the feasibility and reasonability of finding the output feedback H_∞ controllers for system (2.60) in this research.

Actually, the controllability and observability of systems can be identified empirically or according to the background of practical applications (Kiumarsi et al. 2017; Rizvi and Lin 2018), even though the system matrices A, B_i, C are unknown.

Assumption 2.1 The pair (A, B_i) is stabilizable, $i = 1, 2, \ldots, s$. The pair (A, C) is observable.

Assumption 2.2 C is full column rank.

Solving the H_∞ control problem is equivalent to finding the solution of the following minimax problem (Kiumarsi et al. 2017):

$$J(x_0, U, D) = \min_U \max_D \sum_{k=0}^{\infty} \left(y_k^T Q y_k + \sum_{i=1}^{s} u_{ik}^T R_i u_{ik} - \gamma^2 \sum_{l=1}^{m} \|d_{lk}\|^2 \right), \quad (2.62)$$

where $U = \{u_{1k}, \ldots, u_{sk}\}$ and $D = \{d_{1k}, \ldots, d_{mk}\}$. All u_{ik} are viewed as the players in the group U where all players cooperatively work for minimizing performance index (2.62), and all d_{lk} could be considered as the disturbance players aiming at maximizing performance index (2.62). Therefore, this H_∞ control problem with multiple players and multiple disturbances can be considered as a multi-player MCC game problem according to game theory. For the multi-player game, there exists a saddle point (U^*, D^*) that makes the following equation hold:

$$J(x_0, U^*, D) \le J(x_0, U^*, D^*) \le J(x_0, U, D^*), \quad (2.63)$$

where $U^* = \left\{u_{1k}^*, \ldots, u_{sk}^*\right\}$ and $D^* = \left\{d_{1k}^*, \ldots, d_{mk}^*\right\}$. It means that for the MCC game problem to have a unique solution, the following Nash equilibrium condition must be satisfied:

$$\begin{aligned} &J(x_0, u_{1k}^*, \ldots, u_{sk}^*, d_{1k}^*, \ldots, d_{mk}^*) \\ &= \min_U \max_D J(x_0, U, D) = \max_D \min_U J(x_0, U, D). \end{aligned} \quad (2.64)$$

Therefore, to find the saddle point, we define the following optimal value function and the optimal Q-function by the admissible control policies u_{ik} and the disturbances d_{lk}, depending on the ADP method:

$$V^*(x_t) = \min_U \max_D \sum_{t=k}^{\infty} \left(y_t^T Q y_t + \sum_{i=1}^{s} u_{it}^T R_i u_{it} - \gamma^2 \sum_{l=1}^{m} \|d_{lt}\|^2 \right) \quad (2.65)$$

$$Q^*(x_k, U, D) = y_k^T Q y_k + \sum_{i=1}^{s} u_{ik}^T R_i u_{ik} - \gamma^2 \sum_{l=1}^{m} \|d_{lk}\|^2 + V^*(x_{k+1}). \quad (2.66)$$

Then, we have

$$V^*(x_k) = \min_U \max_D Q^*(x_k, U, D) = Q^*(x_k, U^*, D^*). \tag{2.67}$$

The definition of admissible control policies, Definition 2.5, is presented below.

Definition 2.5 (Lv and Ren 2019) The policies $\{u_{ik}\ (i = 1, \ldots, s), d_{lk}\ (l = 1, \ldots, m)\} \in \Psi(\Omega)$ are admissible with respect to (2.62) on the set Ω, if the policies $\{u_{ik}, d_{lk}\}$ are continuous on Ω, $u_i(0) = 0$ and $d_l(0) = 0$, $\{u_{ik}, d_{lk}\}$ can make the system (2.60) stable on Ω, and (2.62) is finite $\forall x_0 \in \Omega$.

Lemma 2.2 is used to indicate the quadratic form of the value function and the Q-function.

Lemma 2.2 *(Modares and Lewis 2014a) If there are admissible control policies $u_{ik} = -K_i y_k$ and disturbance policies $d_{lk} = -K_{d_l} y_k$, then the value function and the Q-function have the following quadratic form:*

$$V(x_k) = x_k^T P x_k \tag{2.68}$$
$$Q(x_k, U, D) = z_k^T H z_k, \tag{2.69}$$

where matrices $P > 0$ and $H > 0$ shown in (2.73), and

$$z_k = \begin{bmatrix} x_k^T & u_{1k}^T & \cdots & u_{sk}^T & d_{1k}^T & \cdots & d_{mk}^T \end{bmatrix}^T \tag{2.70}$$
$$P = M^T H M \tag{2.71}$$
$$M = \begin{bmatrix} I & -(K_1 C)^T & \cdots & -(K_s C)^T & -(K_{d_1} C)^T & \cdots & -(K_{d_m} C)^T \end{bmatrix}^T. \tag{2.72}$$

$$H = \begin{bmatrix} H_{xx} & H_{xu_1} & H_{xu_2} & \cdots & H_{xu_s} & H_{xd_1} & H_{xd_2} & \cdots & H_{xdm} \\ H_{xu_1}^T & H_{u_1u_1} & H_{u_1u_2} & \cdots & H_{u_1u_s} & H_{u_1d_1} & H_{u_1d_2} & \cdots & H_{u_1dm} \\ \vdots & \vdots & \vdots & \ddots & \vdots & \vdots & \vdots & \ddots & \vdots \\ H_{xu_s}^T & H_{u_1u_s}^T & H_{u_2u_s}^T & \cdots & H_{u_su_s} & H_{u_sd_1} & H_{u_sd_2} & \cdots & H_{u_sdm} \\ H_{xd_1}^T & H_{u_1d_1}^T & H_{u_2d_1}^T & \cdots & H_{u_sd_1}^T & H_{d_1d_1} & H_{d_1d_2} & \cdots & H_{d_1dm} \\ \vdots & \vdots & \vdots & \ddots & \vdots & \vdots & \vdots & \ddots & \vdots \\ H_{xdm}^T & H_{u_1dm}^T & H_{u_2dm}^T & \cdots & H_{u_sdm}^T & H_{d_1dm}^T & H_{d_2dm}^T & \cdots & H_{dmdm} \end{bmatrix}$$
$$= \begin{bmatrix} A^TPA + C^TQC & A^TPB_1 & \cdots & A^TPB_s & A^TPE_1 & \cdots & A^TPE_m \\ (A^TPB_1)^T & B_1^TPB_1 + R_1 & \cdots & B_1^TPB_s & B_1^TPE_1 & \cdots & B_1^TPE_m \\ \vdots & \vdots & \ddots & \vdots & \vdots & \ddots & \vdots \\ (A^TPB_s)^T & (B_1^TPB_s)^T & \cdots & B_s^TPB_s + R_s & B_s^TPE_1 & \cdots & B_s^TPE_m \\ (A^TPE_1)^T & (B_1^TPE_1)^T & \cdots & (B_sPE_1)^T & -\gamma^2 I + E_1^TPE_1 & \cdots & E_1^TPE_m \\ \vdots & \vdots & \ddots & \vdots & \vdots & \ddots & \vdots \\ (A^TPE_m)^T & (B_1^TPE_m)^T & \cdots & (B_s^TPE_m)^T & (E_1^TPE_m)^T & \cdots & -\gamma^2 I + E_m^TPE_m \end{bmatrix}. \tag{2.73}$$

Remark 2.11 By contrast to the standard H_∞ control problem (Kiumarsi et al. 2017), we extend it to the H_∞ control of linear DT systems with multiple players and multi-source disturbances. The proof of Lemma 2.2 can be similarly proved as (Modares and Lewis 2014a).

2.2.2 *Solving the Multi-player Zero-Sum Game*

This subsection presents the theoretically optimal control policies and the worst disturbances depending on system state and system output, respectively, assuming that all system dynamics information is known. Then we propose a model-free on-policy MCC game Q-learning algorithm to learn the output feedback controller gains.

2.2.2.1 Theoretical Solution

What will be presented in this section is how to solve the Nash equilibrium solution (2.64), that is, find the theoretically optimal control policies u_{ik}^* and the worst disturbances d_{lk}^* assumed to depend on the system state x_k. By Lemma 2.2, we have the following Q-function-based game Bellman equation:

$$z_k^T H z_k = y_k^T Q y_k + \sum_{i=1}^{s} u_{ik}^T R_i u_{ik} - \gamma^2 \sum_{l=1}^{m} \|d_{lk}\|^2 + z_{k+1}^T H z_{k+1}. \tag{2.74}$$

Based on the optimality principle, implementing $\frac{\partial Q^*(x_k,U,D)}{\partial u_i} = 0$ and $\frac{\partial Q^*(x_k,U,D)}{\partial d_l} = 0$ yields the optimal control policies and the worst disturbances as follows:

$$\begin{aligned} u_{ik}^* &= -K_i^* y_k \\ d_{lk}^* &= -K_{d_l}^* y_k, \end{aligned} \tag{2.75}$$

where

$$\begin{aligned} K_i^* C &= H_{u_i u_i}^{*,-1} \left[H_{xu_i}^{*,T} - \left(\sum\nolimits_{t=1,t\neq i}^{s} H_{u_i u_t}^* K_t^* C + \sum\nolimits_{r=1}^{m} H_{u_i d_r}^* K_{d_r}^* C \right) \right] \\ K_{d_l}^* C &= H_{d_l d_l}^{*,-1} \left[H_{xd_l}^{*,T} - \left(\sum\nolimits_{t=1}^{s} H_{d_l u_t}^* K_t^* C + \sum\nolimits_{v=1,v\neq l}^{m} H_{d_l d_v}^* K_{d_v}^* C \right) \right]. \end{aligned} \tag{2.76}$$

The matrix H^* satisfies the following algebraic Riccati equation (ARE) based on the Q-function:

$$z_k^T H^* z_k = y_k^T Q y_k + \sum_{i=1}^{s} u_{ik}^{*,T} R_i u_{ik}^* - \gamma^2 \sum_{l=1}^{m} \left\| d_{lk}^* \right\|^2 + z_{k+1}^T H^* z_{k+1}. \tag{2.77}$$

Referring to Kiumarsi et al. (2017), it can be noticed that u_{ik}^* and d_{lk}^* in (2.76) are the Nash equilibrium solutions that satisfy the multi-player MCC game problem expressed in (2.65), and (2.76) enables the disturbance attenuation condition (2.61) shown in Definition 2.4 to hold.

Remark 2.12 Notice that the optimal control policies and the worst disturbances in (2.76) depend on the system dynamic parameter C, and all controller gains K_i^* and disturbance policy gains $K_{d_l}^*$ are coupled to each other, which also increases the difficulty of solving them. Moreover, by reviewing the existing output feedback H_∞ control using RL (Valadbeigi et al. 2020; Moghadam and Lewis 2019), one can find that the matrix C is required to be known. Thus, we try to find a way to overcome this problem and the obstacle set by coupling controller gains and disturbance policy gains by some mathematical manipulations, game theory and the Q-learning technique.

2.2.2.2 Output Feedback Control Design

Define a new matrix $\bar{H}$ that satisfies the following equation:

$$H = \begin{bmatrix} C & 0 & 0 & \dots & 0 \\ 0 & I & 0 & \dots & 0 \\ 0 & 0 & I & \dots & 0 \\ \vdots & \vdots & \vdots & \ddots & \vdots \\ 0 & 0 & 0 & \dots & I \end{bmatrix}^T \bar{H} \begin{bmatrix} C & 0 & 0 & \dots & 0 \\ 0 & I & 0 & \dots & 0 \\ 0 & 0 & I & \dots & 0 \\ \vdots & \vdots & \vdots & \ddots & \vdots \\ 0 & 0 & 0 & \dots & I \end{bmatrix}. \tag{2.78}$$

By Lemma 2.2 and (2.78), we have

$$\begin{aligned} H_{xx} = C^T \bar{H}_{yy} C,\ H_{xu_i} &= C^T \bar{H}_{yu_i},\ H_{xd_l} = C^T \bar{H}_{yd_l} \\ H_{u_iu_i} = \bar{H}_{u_iu_i},\ H_{u_id_l} &= \bar{H}_{u_id_l},\ H_{d_ld_l} = \bar{H}_{d_ld_l}, \end{aligned} \tag{2.79}$$

and the matrix $\bar{H}$ shown in (2.80) with $\bar{C} = (C^T C)^{-1} C^T$.

$$\bar{H} = \begin{bmatrix} \bar{H}_{yy} & \bar{H}_{yu_1} & \bar{H}_{yu_2} & \dots & \bar{H}_{yu_s} & \bar{H}_{yd_1} & \bar{H}_{yd_2} & \dots & \bar{H}_{yd_m} \\ \bar{H}_{yu_1}^T & \bar{H}_{u_1u_1} & \bar{H}_{u_1u_2} & \dots & \bar{H}_{u_1u_s} & \bar{H}_{u_1d_1} & \bar{H}_{u_1d_2} & \dots & \bar{H}_{u_1d_m} \\ \vdots & \vdots & \vdots & \ddots & \vdots & \vdots & \vdots & \ddots & \vdots \\ \bar{H}_{yu_s}^T & \bar{H}_{u_1u_s}^T & \bar{H}_{u_2u_s}^T & \dots & \bar{H}_{u_su_s} & \bar{H}_{u_sd_1} & \bar{H}_{u_sd_2} & \dots & \bar{H}_{u_sd_m} \\ \bar{H}_{yd_1}^T & \bar{H}_{u_1d_1}^T & \bar{H}_{u_2d_1}^T & \dots & \bar{H}_{u_sd_1}^T & \bar{H}_{d_1d_1} & \bar{H}_{d_1d_2} & \dots & \bar{H}_{d_1dm} \\ \vdots & \vdots & \vdots & \ddots & \vdots & \vdots & \vdots & \ddots & \vdots \\ \bar{H}_{yd_m}^T & \bar{H}_{u_1d_m}^T & \bar{H}_{u_2d_m}^T & \dots & \bar{H}_{u_sd_m}^T & \bar{H}_{d_1d_m}^T & \bar{H}_{d_2d_m}^T & \dots & \bar{H}_{d_md_m} \end{bmatrix}$$

$$
= \begin{bmatrix}
\bar{C}^T(A^TPA + C^TQC)\bar{C} & \bar{C}^TA^TPB_1 & \dots & \bar{C}^TA^TPB_s & \dots & \bar{C}^TA^TPE_m \\
(\bar{C}^TA^TPB_1)^T & B_1^TPB_1 + R_1 & \dots & B_1^TPB_s & \dots & B_1^TPE_m \\
\vdots & \vdots & \ddots & \vdots & \ddots & \vdots \\
(\bar{C}^TA^TPB_s)^T & (B_1^TPB_s)^T & \dots & B_s^TPB_s + R_s & \dots & B_s^TPE_m \\
(\bar{C}^TA^TPE_1)^T & (B_1^TPE_1)^T & \dots & (B_sPE_1)^T & \dots & E_1^TPE_m \\
\vdots & \vdots & \ddots & \vdots & \ddots & \vdots \\
(\bar{C}^TA^TPE_m)^T & (B_1^TPE_m)^T & \dots & (B_sPE_m)^T & \dots & -\gamma^2 I + E_m^TPE_m
\end{bmatrix}. \tag{2.80}
$$

Thus, the Q-function-based Bellman Eq. (2.74) can be rewritten as

$$
\bar{z}_k^T \bar{H} \bar{z}_k = \bar{z}_k^T \Lambda \bar{z}_k + \bar{z}_{k+1}^T \bar{H} \bar{z}_{k+1}, \tag{2.81}
$$

where

$$
\begin{aligned}
\bar{z}_k &= \begin{bmatrix} y_k^T & u_{1k}^T & \dots & u_{sk}^T & d_{1k}^T & \dots & d_{mk}^T \end{bmatrix}^T \\
\Lambda &= \mathrm{diag}(Q, R_1, \dots, R_s, -\gamma^2 I, \dots, -\gamma^2 I).
\end{aligned}
$$

Based on the Q-function and matrix $\bar{H}$, we have the following optimal output feedback control policies and the worst disturbances:

$$
\begin{aligned}
u_{ik}^* &= -K_i^* y_k = \bar{H}_{u_iu_i}^{-1}\left[\bar{H}_{yu_i}^T - \left(\sum\nolimits_{t=1, t\neq i}^{s} \bar{H}_{u_iu_t} K_t^* + \sum\nolimits_{r=1}^{m} \bar{H}_{u_id_r} K_{d_r}^*\right)\right] y_k \\
d_{lk}^* &= -K_{d_l}^* y_k = \bar{H}_{d_ld_l}^{-1}\left[\bar{H}_{yd_l}^T - \left(\sum\nolimits_{t=1}^{s} \bar{H}_{d_lu_t} K_t^* + \sum\nolimits_{v=1, v\neq l}^{m} \bar{H}_{d_ld_v} K_{d_v}^*\right)\right] y_k.
\end{aligned} \tag{2.82}
$$

Theorem 2.4 is used to prove the uniqueness of the control policies and the disturbances in (2.82).

Theorem 2.4 *Given Assumption 2.2, then there is a unique matrix $\bar{H}$ that makes (2.81) hold, and the unique optimal control policies and the worst disturbances will be solved by (2.82). Moreover, (2.82) is the same as (2.75).*

Proof We first show that there is a unique matrix $\bar{H}$ satisfying (2.81) and the optimal control policies and the worst disturbances obtained by (2.82) based on the matrix $\bar{H}$ are also unique. If there are two different matrices, matrix $\bar{H}$ and matrix $\bar{N}$, such that (2.81) holds, then we can obtain matrix H and the following matrix N according to (2.78):

$$
N = F^T \bar{N} F. \tag{2.83}
$$

Then, from (2.81), (2.78) and (2.83), it can conclude that both the matrix H and the matrix N satisfy (2.74). However, it has been shown in Lewis et al. (2012) that there is a unique matrix H for (2.74), thus the contradiction is generated. This means that the assumption that there are two matrices $\bar{H}$ and $\bar{N}$ that satisfy (2.81) cannot be true. Therefore, there exists a unique matrix $\bar{H}$ to (2.81) and (2.82) is also unique.

Next, we shall show that (2.82) is equal to (2.75). Utilizing (2.79), (2.82) can be the form of (2.75). This completes the proof. □

2.2.2.3 On-Policy Game Q-Learning Algorithm

Reviewing on-policy learning techniques, we present an on-policy MCC game Q-learning algorithm to solve the H_∞ control problem for linear DT systems with multi-players and multi-source disturbances.

Algorithm 2.3 On-policy MCC game Q-learning

1: Initialization: Set the initial stabilizing control policies u_i^0 $(i = 1, \ldots, s)$ and disturbances d_l^0 $(l = 1, \ldots, m)$. Let $j = 0$ be the iteration index and set $i = 1$ and $l = 1$;

2: Policy evaluation: solving the matrix $\bar{H}^{j+1}$:

$$\bar{z}_k^T \bar{H}^{j+1} \bar{z}_k = y_k^T Q y_k + \sum_{i=1}^{n} u_{ik}^{j,T} R_i u_{ik}^j - \gamma^2 \sum_{l=1}^{m} \left\| d_k^j \right\|^2 + \bar{z}_{k+1}^T \bar{H}^{j+1} \bar{z}_{k+1}; \tag{2.84}$$

3: Policy improvement:

3.1: Updating the control policies:

$$\begin{aligned} u_{ik}^{j+1} &= -K_i^{j+1} y_k \\ &= \bar{H}_{u_i u_i}^{j+1,-1} \left[\bar{H}_{y u_i}^{j+1,T} - \left(\sum\nolimits_{t=1, t \neq i}^{s} \bar{H}_{u_i u_t}^{j+1} K_t^j + \sum\nolimits_{r=1}^{m} \bar{H}_{u_i d_r}^{j+1} K_{d_r}^j \right) \right] y_k; \end{aligned} \tag{2.85}$$

3.2: If $i < s$, then $i = i + 1$ and go back to Step 3.1. Otherwise go to Step 3.3;

3.3: Updating the disturbances:

$$\begin{aligned} d_{lk}^{j+1} &= -K_{d_l}^{j+1} y_k \\ &= \bar{H}_{d_l d_l}^{j+1,-1} \left[\bar{H}_{y d_l}^{j+1,T} - \left(\sum\nolimits_{t=1}^{s} \bar{H}_{d_l u_t}^{j+1} K_t^j + \sum\nolimits_{v=1, v \neq l}^{m} \bar{H}_{d_l d_v}^{j+1} K_{d_v}^j \right) \right] y_k; \end{aligned} \tag{2.86}$$

3.4: If $l < m$, then $l = l + 1$ and go back to Step 3.3. Otherwise go to Step 4;

4: If $\left\| \bar{H}^{j+1} - \bar{H}^j \right\| \leq \varepsilon$ for a small constant ε $(\varepsilon > 0)$, then stop the iteration and the gains $(K_i^{j+1}, K_{d_l}^{j+1})$ have been obtained. Otherwise $j = j + 1, i = 1, l = 1$ and go back to Step 2.

Remark 2.13 Algorithm 2.3 is a typical on-policy learning approach because the constantly updated control policies and disturbances are used to generate data for further policy updating. Moreover, the disturbances need to be updated in a prescribed manner, which is impractical in practical industrial applications. Besides, it has been shown in Kiumarsi et al. (2017) that the learning results of the on-policy learning algorithm are susceptible due to adding probing noises, i.e., the learning results of the

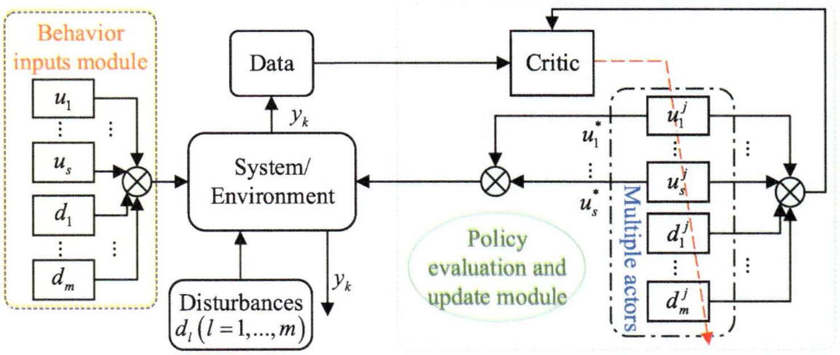

Fig. 2.26 The structure of the off-policy MCC game Q-learning for H_∞ control

on-policy learning algorithm may be biased. Therefore, subsequently, we will discuss how to deal with the H_∞ control problem for linear DT systems with multiple players and multi-source disturbances via the data-driven off-policy Q-learning method.

2.2.3 *Off-Policy Game Q-Learning Technique*

In this subsection, a novel data-driven off-policy MCC game Q-learning algorithm is proposed to solve the multi-player MCC game problem with multi-source disturbances, and the convergence of the learning results is proved. Moreover, we show that the Nash equilibrium solution learned by the proposed algorithm is unbiased, even if the probing noises are added to the system. Figure 2.26 shows the architecture of the proposed algorithm in solving multi-player MCC game problems for multi-player systems with multi-source disturbances.

2.2.3.1 Off-Policy MCC Game Q-Learning Algorithm

To derive the off-policy MCC game Q-learning algorithm, we introduce the auxiliary variables $u_{ik}^j = -K_i^j y_k$ and $d_{lk}^j = -K_{d_l}^j y_k$ as the target policies, hence the initial system (2.60) is written as follows:

$$x_{k+1} = \left(Ax_k + \sum_{i=1}^{s} B_i u_{ik}^j + \sum_{l=1}^{m} E_l d_{lk}^j\right) + \sum_{i=1}^{s} B_i (u_{ik} - u_{ik}^j) + \sum_{l=1}^{m} E_l (d_{lk} - d_{lk}^j), \tag{2.87}$$

where u_{ik} and d_{lk} are the behavioral policies used to generate the data needed for algorithm learning. By following the system trajectory (2.87), we have

$$
\begin{aligned}
&x_k^T(M^j)^T H^{j+1} M^j x_k - \left(x_{k+1} - \sum_{i=1}^{s} B_i(u_{ik} - u_{ik}^j) - \sum_{l=1}^{m} E_l(d_l - d_l^j)\right)^T (M^j)^T \\
&\quad \times H^{j+1} M^j \left(x_{k+1} - \sum_{i=1}^{s} B_i(u_{ik} - u_{ik}^j) - \sum_{l=1}^{m} E(d_l - d_l^j)\right) \\
&\quad = z_k^T F^T \bar{H}^{j+1} F z_k - \left(x_{k+1} - \sum_{i=1}^{s} B_i(u_i - u_i^j) - \sum_{l=1}^{m} E_l(d_l - d_l^j)\right)^T (M^j)^T \\
&\quad \times F^T \bar{H}^{j+1} F M^j \left(x_{k+1} - \sum_{i=1}^{s} B_i(u_i - u_i^j) - \sum_{l=1}^{m} E_l(d_l - d_l^j)\right) \\
&\quad = x_k^T (M^j)^T F^T \Lambda F M^j x_k \\
&\quad = \bar{z}_k^T \Lambda \bar{z}_k.
\end{aligned} \tag{2.88}
$$

Then, one has

$$
\begin{aligned}
&\bar{z}_k^T \bar{H}^{j+1} \bar{z}_k - \bar{z}_{k+1}^T \bar{H}^{j+1} \bar{z}_{k+1} + 2\left(Ax_k + \sum_{i=1}^{s} B_i u_i + \sum_{l=1}^{m} E_l d_l\right)^T (M^j)^T F^T \bar{H}^{j+1} \\
&\quad \times F M^j \sum_{i=1}^{s} B_i(u_i - u_i^j) + 2\left(Ax_k + \sum_{i=1}^{s} B_i u_i + \sum_{l=1}^{m} E_l d_l\right)^T (M^j)^T F^T \bar{H}^{j+1} \\
&\quad \times F M^j \sum_{l=1}^{m} E_l(d_l - d_l^j) - \left(\sum_{i=1}^{s} B_i(u_i - u_i^j)\right)^T P^{j+1} \sum_{i=1}^{s} B_i(u_i - u_i^j) \\
&\quad - 2\left(\sum_{i=1}^{s} B_i(u_i - u_i^j)\right)^T P^{j+1} \sum_{l=1}^{m} E_l(d_l - d_l^j) \\
&\quad - \left(\sum_{l=1}^{m} E_l(d_l - d_l^j)\right)^T P^{j+1} \sum_{l=1}^{m} E_l(d_l - d_l^j) \\
&\quad = \bar{z}_k^T \Lambda \bar{z}_k.
\end{aligned} \tag{2.89}
$$

By Lemma 2.2, one has

$$
\begin{aligned}
&\bar{z}_k^T \bar{H}^{j+1} \bar{z}_k - \bar{z}_{k+1}^T \bar{H}^{j+1} \bar{z}_{k+1} + 2x_k^T [H_{xu_1}^{j+1}, \ldots, H_{xu_s}^{j+1}] \sum_{i=1}^{s} (u_i - u_i^j) \\
&\quad + 2x_k^T [H_{xd_1}^{j+1}, \ldots, H_{xd_m}^{j+1}] \sum_{l=1}^{m} (d_l - d_l^j) + 2\sum_{l=1}^{m} d_l^T E_l^T P^{j+1} \sum_{i=1}^{s} B_i(u_i - u_i^j) \\
&\quad + \sum_{i=1}^{s} (u_i + u_i^j)^T B_i^T P^{j+1} \sum_{i=1}^{s} B_i(u_i - u_i^j) + \sum_{l=1}^{m} (d_l + d_l^j) E_l^T P^{j+1} \sum_{l=1}^{m} E_l(d_l - d_l^j)
\end{aligned}
$$

$$
\begin{aligned}
&+2\sum_{i=1}^{s}(u_i^j)^T B_i^T P^{j+1}\sum_{l=1}^{m} E_l(d_l-d_l^j)\\
&=\bar{z}_k^T\Lambda\bar{z}_k. \qquad (2.90)
\end{aligned}
$$

According to (2.73), (2.80) and (2.79), (2.90) becomes

$$
\begin{aligned}
&\bar{z}_k^T\bar{H}^{j+1}\bar{z}_k-\bar{z}_{k+1}^T\bar{H}^{j+1}\bar{z}_{k+1}+2y_k^T[\bar{H}_{yu_1}^{j+1},\ldots,\bar{H}_{yu_s}^{j+1}]\sum_{i=1}^{s}(u_i-u_i^j)\\
&+2y_k^T[\bar{H}_{yd_1}^{j+1},\ldots,\bar{H}_{yd_m}^{j+1}]\sum_{l=1}^{m}(d_l-d_l^j)+2\sum_{l=1}^{m}d_l^T(\bar{H}_{u_id_l}^{j+1})^T\sum_{i=1}^{s}(u_i-u_i^j)\\
&+\sum_{i=1}^{s}(u_i+u_i^j)^T G_1\sum_{i=1}^{s}(u_i-u_i^j)+\sum_{l=1}^{m}(d_l+d_l^j)^T G_2\sum_{l=1}^{m}(d_l-d_l^j)\\
&+2\sum_{i=1}^{s}(u_i^j)^T\bar{H}_{u_id_l}^{j+1}\sum_{l=1}^{m}(d_l-d_l^j)\\
&=\bar{z}_k^T\Lambda\bar{z}_k, \qquad (2.91)
\end{aligned}
$$

where

$$
G_1=\begin{bmatrix}
\bar{H}_{u_1u_1}^{j+1}-R_1 & \bar{H}_{u_1u_2}^{j+1} & \ldots & \bar{H}_{u_1u_s}^{j+1}\\
(\bar{H}_{u_1u_2}^{j+1})^T & \bar{H}_{u_2u_2}^{j+1}-R_2 & \ldots & \bar{H}_{u_2u_s}^{j+1}\\
\vdots & \vdots & \ddots & \vdots\\
(\bar{H}_{u_1u_s}^{j+1})^T & (\bar{H}_{u_2u_s}^{j+1})^T & \ldots & \bar{H}_{u_su_s}^{j+1}-R_s
\end{bmatrix}
$$

$$
G_2=\begin{bmatrix}
\bar{H}_{d_1d_1}^{j+1}-\gamma^2 I & \bar{H}_{d_1d_2}^{j+1} & \ldots & \bar{H}_{d_1d_m}^{j+1}\\
(\bar{H}_{d_1d_2}^{j+1})^T & \bar{H}_{d_2d_2}^{j+1}-\gamma^2 I & \ldots & \bar{H}_{d_2d_m}^{j+1}\\
\vdots & \vdots & \ddots & \vdots\\
(\bar{H}_{d_1d_m}^{j+1})^T & (\bar{H}_{d_2d_m}^{j+1})^T & \ldots & \bar{H}_{d_md_m}^{j+1}-\gamma^2 I
\end{bmatrix}.
$$

Based on the form of Kronecker product, (2.91) is rewritten as follows:

$$
\hat{\theta}^j(k)\hat{L}^{j+1}=\hat{\rho}_k^j, \qquad (2.92)
$$

where

$$
\begin{aligned}
\hat{\rho}_k&=y_k^TQy_k+\sum_{i=1}^{s}u_{ik}^TR_iu_{ik}-\gamma^2\sum_{l=1}^{m}d_{lk}^Td_{lk}\\
\hat{L}^{j+1}&=\Big[(\mathrm{vec}(\hat{L}_{00}^{j+1}))^T\ (\mathrm{vec}(\hat{L}_{0i}^{j+1}))^T\ (\mathrm{vec}(\hat{L}_{0l}^{j+1}))^T\ (\mathrm{vec}(\hat{L}_{ii}^{j+1}))^T
\end{aligned}
$$

$$(\mathrm{vec}(\hat{L}_{ll}^{j+1}))^T\ (\mathrm{vec}(\hat{L}_{ir}^{j+1}))^T\ (\mathrm{vec}(\hat{L}_{il}^{j+1}))^T\ (\mathrm{vec}(\hat{L}_{lt}^{j+1}))^T\Big]^T$$
$$\hat{\theta}^j(k) = \left[\hat{\theta}_{00}^j,\ \hat{\theta}_{0i}^j,\ \hat{\theta}_{0l}^j,\ \hat{\theta}_{ii}^j,\ \hat{\theta}_{ll}^j,\ \hat{\theta}_{ir}^j,\ \hat{\theta}_{il}^j,\ \hat{\theta}_{lt}^j\right].$$

and

$$\begin{aligned}
\hat{\theta}_{00}^j &= y_k^T \otimes y_k^T - y_{k+1}^T \otimes x_{k+1}^T\\
\hat{\theta}_{0i}^j &= 2y_{k+1}^T \otimes (K_i^j y_{k+1})^T + 2y_k^T \otimes u_i^T\\
\hat{\theta}_{0l}^j &= 2y_{k+1}^T \otimes (K_{d_l}^j y_{k+1})^T + 2y_k^T \otimes d_l^T\\
\hat{\theta}_{ii}^j &= -(K_i^j y_{k+1})^T \otimes (K_i^j y_{k+1})^T + u_i^T \otimes u_i^T\\
\hat{\theta}_{ll}^j &= -(K_{d_l}^j y_{k+1})^T \otimes (K_{d_l}^j y_{k+1})^T + d_l^T \otimes d_l^T\\
\hat{\theta}_{ir}^j &= -2(K_i^j y_{k+1})^T \otimes (K_r^j y_{k+1})^T + 2u_i^T \otimes u_r^T\\
\hat{\theta}_{il}^j &= -2(K_i^j y_{k+1})^T \otimes (K_{d_l}^j y_{k+1})^T + 2u_i^T \otimes d_l^T\\
\hat{\theta}_{lt}^j &= -2(K_{d_l}^j y_{k+1})^T \otimes (K_{d_t}^j y_{k+1})^T + 2d_l^T \otimes d_t^T\\
\hat{L}_{00}^{j+1} &= \bar{H}_{yy}^{j+1},\ \hat{L}_{0i}^{j+1} = \bar{H}_{yu_i}^{j+1},\ \hat{L}_{0l}^{j+1} = \bar{H}_{yd_l}^{j+1},\ \hat{L}_{ii}^{j+1} = \bar{H}_{u_iu_i}^{j+1}\\
\hat{L}_{ll}^{j+1} &= \bar{H}_{d_ld_l}^{j+1},\ \hat{L}_{ir}^{j+1} = \bar{H}_{u_iu_r}^{j+1},\ \hat{L}_{il}^{j+1} = \bar{H}_{u_id_l}^{j+1},\ \hat{L}_{lt}^{j+1} = \bar{H}_{d_ld_t}^{j+1}
\end{aligned}$$

with $i = 1, \ldots, s$, $l = 1, \ldots, m$, $r = i + 1, \ldots, s - 1$ and $t = l + 1, \ldots, m - 1$.

Therefore, the controller and disturbance gains $(K_i,\ K_{d_l})$ based on the output feedback control can be calculated as

$$\begin{aligned}
K_i^{j+1} &= (\hat{L}_{ii}^{j+1})^{-1}\left((\hat{L}_{0i}^{j+1})^T - \left[\sum\nolimits_{r=1,r\neq i}^{s} \hat{L}_{ir}^{j+1} K_r^j + \sum\nolimits_{l=1}^{m} \hat{L}_{il}^{j+1} K_{d_l}^j\right]\right)\\
K_{d_l}^{j+1} &= (\hat{L}_{ll}^{j+1})^{-1}\left((\hat{L}_{0l}^{j+1})^T - \left[\sum\nolimits_{i=1}^{s} (\hat{L}_{il}^{j+1})^T K_i^j + \sum\nolimits_{t=1,t\neq l}^{m} \hat{L}_{lt}^{j+1} K_{d_t}^j\right]\right).
\end{aligned} \tag{2.93}$$

Now we propose a data-driven off-policy MCC game Q-learning algorithm to solve (2.64), that is, learning the Nash Equilibrium solution to the MCC game of multi-player systems with multi-source disturbances.

Theorem 2.5 is given below to show the convergence of Algorithm 2.4.

Theorem 2.5 *The matrices $\bar{H}^{j+1}$, K_i^{j+1} and $K_{d_l}^{j+1}$ learned by Algorithm 2.4 converge to the theoretical optimum values $\bar{H}^*$, K_i^* and $K_{d_l}^*$ as $j \to \infty$.*

Proof It can be easily derived from the derivations that if matrices H^{j+1}, K_i^{j+1} and $K_{d_l}^{j+1}$ are the solutions to (2.84)–(2.86), then they also make (2.92) and (2.93) hold. Next, we will show that the solutions to (2.92) and (2.93) also make sure that (2.84)–(2.86) hold.

Notice that (2.92) is the same thing as (2.89), so they have the same solution. Moreover, if $u_i = -K_i^j y_k$ and $d_l = -K_{d_l}^j y_k$, then (2.84) can be obtained by sub-

Algorithm 2.4 Off-policy MCC game Q-learning

1: Data collection: Collecting system data y_k generated by using behavior control policies u_i $(i = 1, \ldots, s)$ and behavior disturbances d_l $(l = 1, \ldots, m)$ and storing them in the data set $(\hat{\theta}^j(k), \hat{\rho}_k^j)$ of (2.92);

2: Initialization: Select K_i^0 $(i = 1, \ldots, s)$ and $K_{d_l}^0$ $(l = 1, \ldots, m)$ as the initial controller gains and disturbance gains. Let $j = 0$ be the iteration index and set $i = 1$ and $l = 1$;

3: Implementing the MCC game Q-learning:

3.1: Policy Evaluation: Solve the $\hat{L}^{j+1}$ in (2.92) by employing the least-squares method;

3.2: Policy update: Calculate the K_i^{j+1} according to (2.93);

3.3: If $i < s$, then $i = i + 1$ and go back to Step 3.2. Otherwise go to Step 3.4;

3.4: Policy update: Calculate the $K_{d_l}^{j+1}$ according to (2.93);

3.5: If $l < m$, then $l = l + 1$ and go back to Step 3.4. Otherwise go to Step 4;

4: If $\left\| K_i^{j+1} - K_i^j \right\| \leq \varepsilon$ and $\left\| K_{d_l}^{j+1} - K_{d_l}^j \right\| \leq \varepsilon$ for each of controllers and disturbances, then the iteration is stopped. Otherwise, $j = j + 1$, $i = 1$, $l = 1$ and go back to Step 3.

tracting (2.89) from (2.88). Thus, the solutions to (2.92) and (2.93) and the solutions to (2.84)–(2.86) are equivalent. Besides, it has been shown in Li et al. (2018) that the matrices $\bar{H}^{j+1}$, K_i^{j+1} and $K_{d_l}^{j+1}$ learned by Algorithm 2.3 can converge to the optimal values. Therefore, the convergence of Algorithm 2.4 is confirmed, i.e., $\lim_{j\to\infty} \bar{H}^{j+1} = \bar{H}^*$, $\lim_{j\to\infty} K_i^{j+1} = K_i^*$ and $\lim_{j\to\infty} K_{d_l}^{j+1} = K_{d_l}^*$. □

Remark 2.14 Algorithm 2.4 separates the data generation and the control policies learning by introducing behavior control policies and behavior disturbances, which is one of the biggest differences with the on-policy learning algorithm and also overcomes the shortcomings of the on-policy learning approach as listed in Li et al. (2019a). The initial controller gains K_i^0 and disturbance gains $K_{d_l}^0$, as well as the behavior control policies and behavior disturbances, can be chosen empirically or utilize the robust control method like (Modares and Lewis 2014a) and (Li et al. 2018).

Remark 2.15 It is worth pointing out that Algorithm 2.4 presents a completely model-free approach to fix the output feedback H_∞ control problem by finding the Nash equilibrium solution using off-policy Q-learning combined with game theory, which is the first time attempt for multi-player systems resisting multi-source disturbances under the foundation laid by Song et al. (2017); Jiang et al. (2018); Liu and Wei (2014) with data-driven RL algorithms for CT systems, and Li and Xiao (2020) with RL algorithm worked for state feedback controller design of multi-player systems with single-source disturbance. Moreover, we extend the H_∞ output feedback controller design (Rizvi and Lin 2018; Valadbeigi et al. 2020; Moghadam and Lewis 2019) for single-controller systems into the case of multi-player and multi-source disturbance systems by developing a novel off-policy MCC game Q-learning algorithm. The challenges brought out by coupling of control policies and disturbance policies and the unknown matrix C that is assumed to be known in Valadbeigi et al. (2020) and Moghadam and Lewis (2019) to design the output feedback H_∞ controller have been successfully worked out by introducing a matrix $\bar{H}$, corresponding mathematical manipulation, Q-learning and game theory.

Remark 2.16 Indeed, the proposed off-policy MCC game Q-learning for output feedback H_∞ control of linear systems can be extended to the case of nonlinear systems by following the idea shown in Sect. 2.2.3. However, solving this problem for nonlinear systems will be even more challenging than that for linear systems, because the natural complexity of nonlinearity makes the Q-function be nonlinear rather than quadratic form (Liu and Wei 2014; Li et al. 2019a). In addition to those difficulties, the coupling of control policies u_{ik}^* and worst disturbances d_{lk}^* definitely set the obstacle for model-free Q-learning of output feedback H_∞ control of nonlinear systems.

2.2.3.2 No Bias Analysis of Solutions

For ensuring the PE condition, probing noises that generally are white noises are put into the behavior control policies and behavior disturbances. Next, we shall show by Theorem 2.6 that the learning results of the proposed Algorithm 2.4 are still unbiased even though adding probing noises.

Theorem 2.6 *Suppose the probing noises e_{ik} $(i = 1, \ldots, s)$ and e_{lk} $(l = 1, \ldots, m)$ are added to the behavior control policies u_{ik} and behavior disturbances d_{lk} of Algorithm 2.4. Set the matrix $\bar{H}^{j+1}$ to be the solution to (2.88) in the case where $e_{ik} = 0$ and set the $\hat{H}^{j+1}$ to be the solution to (2.88) with $e_{ik} \neq 0$, respectively. Then, $\bar{H}^{j+1} = \hat{H}^{j+1}$.*

Proof Equation (2.88) can be rewritten as the following form:

$$
\begin{aligned}
y_k^T (\hat{M}^j)^T \hat{H}^{j+1} \hat{M}^j y_k = {} & y_k^T (\hat{M}^j)^T \Lambda \hat{M}^j y_k + \left(A x_k + \sum_{i=1}^{s} B_i u_{ik}^j + \sum_{l=1}^{m} E_l d_{lk}^j \right)^T \\
& \times (M^j)^T F^T \hat{H}^{j+1} F M^j \left(A x_k + \sum_{i=1}^{s} B_i u_{ik}^j + \sum_{l=1}^{m} E_l d_{lk}^j \right),
\end{aligned} \tag{2.94}
$$

where

$$
\hat{M}^j = \begin{bmatrix} I & -(K_1^j)^T & \ldots & -(K_s^j)^T & -(K_{d_1}^j)^T & \ldots & -(K_{d_m}^j)^T \end{bmatrix}^T .
$$

Adding the probing noises to system (2.87) yields

$$
\begin{aligned}
x_{k+1} = & \left(Ax_k + \sum_{i=1}^{s} B_i u_{ik}^j + \sum_{l=1}^{m} E_l d_{lk}^j\right) \\
& + \sum_{i=1}^{s} B_i (u_{ik} + e_{ik} - u_{ik}^j) + \sum_{l=1}^{m} E_l (d_{lk} + e_{lk} - d_{lk}^j).
\end{aligned} \tag{2.95}
$$

In the existence of probing noises, (2.88) is changed to the following form:

$$
\begin{aligned}
& z_k^T F^T \hat{H}^{j+1} F z_k - \left(x_{k+1} - \sum_{i=1}^{s} B_i (u_i + e_i - u_i^j) - \sum_{l=1}^{m} E_l (d_l + e_l - d_l^j)\right)^T (M^j)^T \\
& \times F^T \hat{H}^{j+1} F M^j \left(x_{k+1} - \sum_{i=1}^{s} B_i (u_i + e_i - u_i^j) - \sum_{l=1}^{m} E_l (d_l + e_l - d_l^j)\right) \\
& = x_k^T (M^j)^T F^T \Lambda F M^j x_k \\
& = \bar{z}_k^T \Lambda \bar{z}_k.
\end{aligned} \tag{2.96}
$$

It can be noted that if we substitute (2.95) into (2.96), (2.96) becomes (2.94). Moreover, the solution to (2.94) is equivalent to that of (2.88), that is, $\bar{H}^{j+1} = \hat{H}^{j+1}$. Hence, the learning results of Algorithm 2.4 will not be biased, even if the probing noises are added to the system. The proof is completed. □

Remark 2.17 The traditional method of proving the unbiasedness of the off-policy learning algorithms is to add probing noises only to the behavior control policy (Kiumarsi et al. 2017). In this section, we prove the unbiasedness of the proposed algorithm by adding probing noises to both the behavior control policies and the behavior disturbances for sufficiently exciting systems, which is a significant difference compared with the existing proof method.

Remark 2.18 The proposed off-policy MCC game Q-learning can work for DT linear multi-player systems resisting multi-source disturbances. Referring to robust adaptivity and H_∞ control of CT systems with model-based (Chakrabortty and Arcak 2009; Gadewadikar et al. 2006) and model-free RL algorithms (Li et al. 2014; Modares et al. 2015; Moghadam and Lewis 2019), a simple derivation of CT multi-player systems is made here for showing how it works if using our idea.

Consider the following CT multi-player systems subject to multi-source disturbances:

$$
\begin{aligned}
\dot{x} &= Ax + \sum_{i=1}^{s} B_i u_i + \sum_{l=1}^{m} E_l d_l \\
y &= Cx.
\end{aligned} \tag{2.97}
$$

Corresponding to (2.65) and (2.66), the optimal value function and the optimal Q-function can be defined as follows:

$$V^*(x_t) = \min_U \max_D \int_t^{\infty} \left(y^T Q y + \sum_{i=1}^{s} u_i^T R_i u_i - \gamma^2 \sum_{l=1}^{m} \|d_l\|^2 \right) \mathrm{d}\tau \quad (2.98)$$

$$Q^*(x_t, U, D) = V^*(x_{t+T}) + \int_t^{t+T} \left(y^T Q y + \sum_{i=1}^{s} u_i^T R_i u_i - \gamma^2 \sum_{l=1}^{m} \|d_l\|^2 \right) \mathrm{d}\tau. \quad (2.99)$$

Similar to Modares et al. (2015); Liu and Wei (2014); Song et al. (2017); Jiang et al. (2018), it is not hard to prove $V(x_t) = x_t^T P x_t$ and $Q(x_t, U, D) = z_t^T H z_t$. Following the similar steps, (2.79)–(2.82) yields the Q-function- based Bellman equation as follows:

$$\begin{aligned} \bar{z}_t^T \bar{H}^{j+1} \bar{z}_t = & \int_t^{t+T} \left(y^T Q y + \sum_{i=1}^{s} u_i^T R_i u_i - \gamma^2 \sum_{l=1}^{m} \|d_l\|^2 \right) \mathrm{d}\tau \\ & + \bar{z}_{t+T}^T \bar{H}^{j+1} \bar{z}_{t+t} \end{aligned} \quad (2.100)$$

and

$$\begin{aligned} u_i^* &= -K_i^* y = -\bar{H}_{u_i u_i}^{-1} \bar{H}_{y u_i}^T y \\ d_l^* &= -K_{d_l}^* y = -\bar{H}_{d_l d_l}^{-1} \bar{H}_{y d_l}^T y. \end{aligned} \quad (2.101)$$

By comparing (2.100) and (2.101) with (2.81) and (2.82), derived in this section for solving output feedback H_∞ control of DT systems, one can easily find that solving (2.81) and (2.82) seems to be more difficult than (2.100) and (2.101), since no coupling relationship of control policy gains and disturbance policy gains is in (2.100) and (2.101), which is caused by the natural difference of DT systems from CT systems.

2.2.4 Simulation Results

In this subsection, we consider a practical application, a winding machine system (Noura et al. 2000) with some modification, to verify the effectiveness of the proposed data-driven off-policy MCC game Q-learning algorithm dealing with the H_∞ control problem for linear DT systems with multi-players and multi-source disturbances. As shown in Fig. 2.27, the winding machine system has three motors corresponding to three control inputs u_1, u_2 and u_3, which represent control strip tension T_1, angular velocity Ω_2 and strip tension T_3, respectively. Without loss of generality, we assume that there are also two disturbance sources d_1 and d_2. The probing noises are added to satisfy the PE condition.

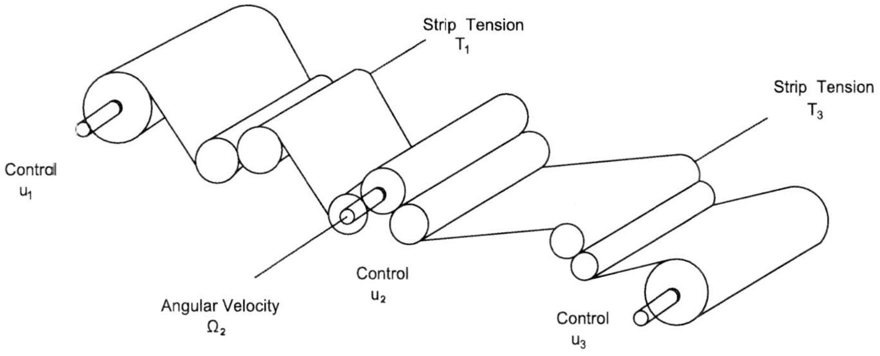

Fig. 2.27 The structure of winding machine systems

Example 2.3 The winding machine system can be described as follows:

$$
\begin{aligned}
x_{k+1} &= Ax_k + \sum_{i=1}^{3} B_i u_i + \sum_{l=1}^{2} E_l d_l \\
y_k &= Cx_k,
\end{aligned}
\tag{2.102}
$$

where

$$
A = \begin{bmatrix} 0.4126 & 0 & -0.0196 \\ 0.0333 & 0.5207 & -0.0413 \\ -0.0101 & 0 & 0.2571 \end{bmatrix}, \; B_1 = \begin{bmatrix} -1.7734 \\ 0.0928 \\ -0.0424 \end{bmatrix}, \; B_2 = \begin{bmatrix} 0.0696 \\ 0.4658 \\ -0.0930 \end{bmatrix}
$$

$$
B_3 = \begin{bmatrix} 0.0734 \\ 0.1051 \\ 2.0752 \end{bmatrix}, \; E_1 = \begin{bmatrix} 0.123956 \\ -0.37895 \\ 0.12698 \end{bmatrix}, \; E_2 = \begin{bmatrix} -0.08564 \\ 0.1268 \\ 0.036984 \end{bmatrix}, \; C = \begin{bmatrix} 1 & 0 & -1 \\ -2 & 1 & 0 \\ 1 & 1 & 1 \end{bmatrix}.
$$

Choose $Q = \mathrm{diag}(10, 10, 10)$, $R_1 = R_2 = R_3 = 10$, the disturbance attenuation $\gamma = 5$ and the initial state $y_0 = \begin{bmatrix} 0.2 & -0.65 & 1.55 \end{bmatrix}^T$.

Firstly, we assume that the system matrices A, B_i, E_l and C are all accurately known, then the following theoretical solution $\bar{H}$, optimal controller gains and worst disturbance gains according to the method in Sect. 2.2.2 can be obtained:

$$
\bar{H}^* = \begin{bmatrix}
11.9034 & 0.7048 & 1.3912 & -7.0643 & 2.2639 & -1.3920 & -1.1326 & 0.1113 \\
0.7048 & 12.4743 & 0.9395 & 15.6356 & 2.3394 & 2.7991 & -3.3714 & 1.6084 \\
1.3912 & 0.9395 & 12.0526 & -5.6924 & 2.5654 & 8.6504 & -1.2766 & 0.5746 \\
-7.0643 & 15.6356 & -5.6924 & 203.7291 & 1.3469 & -5.6241 & -20.6349 & 11.6972 \\
2.2639 & 2.3394 & 2.5654 & 1.3469 & 14.1173 & 6.4873 & -3.2727 & 1.3211 \\
-1.3920 & 2.7991 & 8.6504 & -5.6241 & 6.4873 & 101.4629 & -2.4835 & 4.0903 \\
-1.1326 & -3.3714 & -1.2766 & -20.6349 & -3.2727 & -2.4835 & -20.3743 & -2.1329 \\
0.1113 & 1.6084 & 0.5746 & 11.6972 & 1.3211 & 4.0903 & -2.1329 & -23.8448
\end{bmatrix}
\tag{2.103}
$$

$$
\begin{aligned}
K_1^* &= \begin{bmatrix}0.0295 & -0.0829 & 0.0209\end{bmatrix}\\
K_2^* &= \begin{bmatrix}-0.1904 & -0.1612 & -0.1622\end{bmatrix}\\
K_3^* &= \begin{bmatrix}0.0253 & -0.0240 & -0.0756\end{bmatrix}\\
K_{d_1}^* &= \begin{bmatrix}-0.0599 & -0.0546 & -0.0503\end{bmatrix}\\
K_{d_2}^* &= \begin{bmatrix}0.0183 & 0.0186 & 0.0169\end{bmatrix}.
\end{aligned}
\tag{2.104}
$$

Let the initial gains be

$$
\begin{aligned}
K_1^0 = \begin{bmatrix}-10.3603 & -8.8495 & -5.3501\end{bmatrix},\ K_2^0 = \begin{bmatrix}7.1171 & 8.5319 & 4.8977\end{bmatrix}\\
K_3^0 = \begin{bmatrix}0.1674 & -6.2478 & -4.4080\end{bmatrix},\ K_{d_1}^0 = \begin{bmatrix}-7.8473 & 2.1523 & 3.1386\end{bmatrix}\\
K_{d_2}^0 = \begin{bmatrix}11.4279 & 5.9016 & 2.3473\end{bmatrix}.
\end{aligned}
$$

The gains of behavior control policies and behavior disturbance policies are chosen as

$$
\begin{aligned}
K_1 = \begin{bmatrix}0.0296 & -0.0828 & 0.0210\end{bmatrix},\ K_2 = \begin{bmatrix}-0.1835 & -0.1555 & -0.1566\end{bmatrix}\\
K_3 = \begin{bmatrix}0.0248 & -0.0244 & -0.0761\end{bmatrix},\ K_{d_1} = \begin{bmatrix}-0.0577 & -0.0528 & -0.0485\end{bmatrix}\\
K_{d_2} = \begin{bmatrix}0.0176 & 0.0180 & 0.0163\end{bmatrix}.
\end{aligned}
$$

Implementing the proposed data-driven Algorithm 2.4, we can obtain the controller gains (K_1^{j+1}, K_2^{j+1}, K_3^{j+1}) and the disturbance gains ($K_{d_1}^{j+1}$, $K_{d_2}^{j+1}$). Figures 2.28 and 2.29 show the convergence results of the proposed algorithm. From Fig. 2.29, it can be found that after 6 iterations, the controller gains and disturbance gains converge to the following values:

$$
\begin{aligned}
K_1^6 &= \begin{bmatrix}0.0295 & -0.0829 & 0.0209\end{bmatrix},\\
K_2^6 &= \begin{bmatrix}-0.1904 & -0.1612 & -0.1622\end{bmatrix},\\
K_3^6 &= \begin{bmatrix}0.0253 & -0.0240 & -0.0756\end{bmatrix},\\
K_{d_1}^6 &= \begin{bmatrix}-0.0599 & -0.0546 & -0.0503\end{bmatrix},\\
K_{d_2}^6 &= \begin{bmatrix}0.0183 & 0.0186 & 0.0169\end{bmatrix}.
\end{aligned}
\tag{2.105}
$$

Next, we test the anti-interference ability of the system using the learned controllers from Algorithm 2.4 under the following disturbances:

$$
\begin{aligned}
d_{1k} &= \sin(0.1k)e^{-0.01k}\\
d_{2k} &= \sin(2.1k)e^{-0.1k}.
\end{aligned}
\tag{2.106}
$$

Figure 2.30 depicts the output trajectories of the system subject to the disturbances in (2.106). The performance of the system is presented in Fig. 2.31. The actual disturbance attenuation can be calculated as

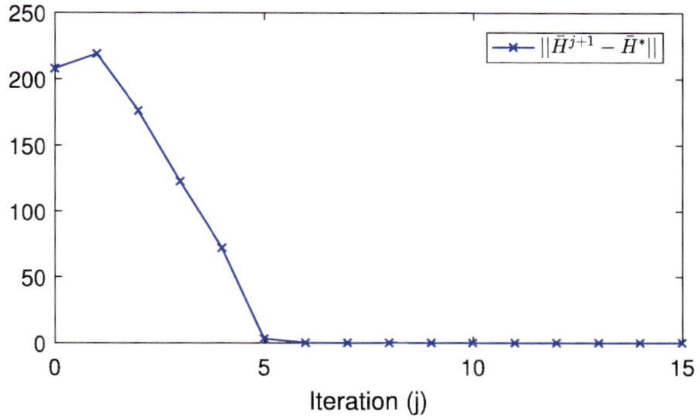

Fig. 2.28 Convergence of matrix $\bar{H}^{j+1}$ using Algorithm 2.4

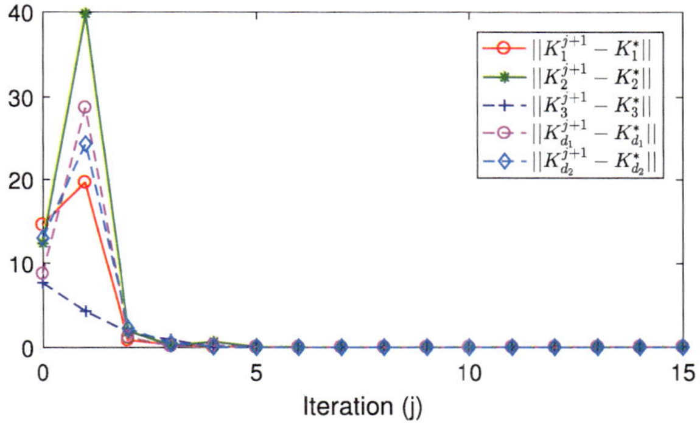

Fig. 2.29 Convergence of matrices K_1^{j+1}, K_2^{j+1}, K_3^{j+1}, $K_{d_1}^{j+1}$ and $K_{d_2}^{j+1}$ using Algorithm 2.4

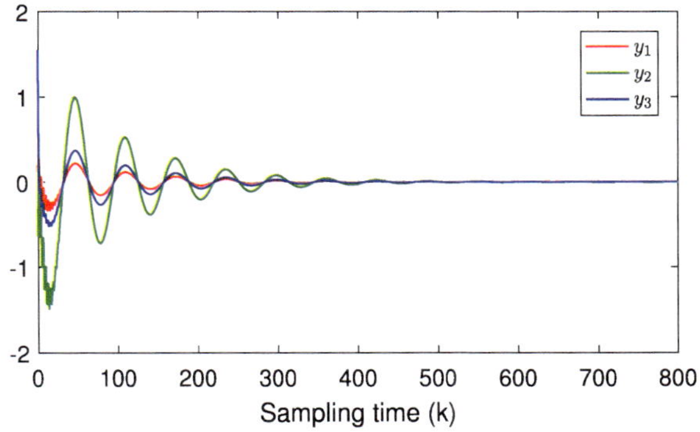

Fig. 2.30 Trajectories of the system state using Algorithm 2.4

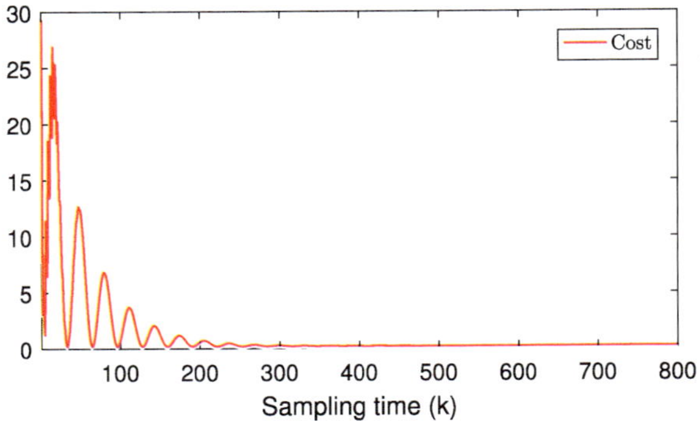

Fig. 2.31 Cost performance of the system using Algorithm 2.4

$$\sqrt{\frac{\sum_{k=0}^{1000}(y_k^T Q y_k + \sum_{i=1}^{3} u_i^T R_i u_i)}{\sum_{k=0}^{1000}(\sum_{l=1}^{2} \|d_{lk}\|^2)}} = 2.6317 < 5. \tag{2.107}$$

From Figs. 2.28 and 2.29, it can be concluded that the matrices $\bar{H}^{j+1}$, K_1^{j+1}, K_2^{j+1}, K_3^{j+1}, $K_{d_1}^{j+1}$ and $K_{d_2}^{j+1}$ learned by Algorithm 2.4 can converge to the theoretically optimal solutions. Moreover, according to Figs. 2.30, 2.31 and (2.107), we can see that the output feedback controllers learned by Algorithm 2.4 can be applied to the system to achieve the H_∞ control and satisfy the disturbance attenuation condition.

2.3 Conclusion

A novel off-policy game Q-learning algorithm is presented to solve the H_∞ control problem for multi-player linear DT systems. The proposed algorithm does not need to know the dynamics model of the system in advance, such that the state feedback controllers can be learned via a completely data-driven way. Then, we further developed a novel off-policy MCC game Q-learning algorithm based on the ADP method and the game theory to solve the H_∞ control problem for linear DT systems with multiple players and multi-source disturbances. The proposed algorithm learns the Nash equilibrium solutions of the system under the zero-sum game framework, i.e., the output feedback controllers and the worst disturbances, using data only. Moreover, we give rigorous proofs on the unbiasedness of solutions to Nash equilibrium and the convergence of the proposed algorithm. The simulation results show the effectiveness of the proposed approach.

References

Al-Tamimi A, Abu-Khalaf M, Lewis FL (2007) Adaptive critic designs for discrete-time zero-sum games with application to H_∞ control. IEEE Trans Syst Man Cybern Part B (Cybern) 37(1):240–247

Al-Tamimi A, Lewis FL, Abu-Khalaf M (2007) Model-free Q-learning designs for linear discrete-time zero-sum games with application to H-infinity control. Automatica 43(3):473–481

Al-Tamimi A, Lewis FL, Abu-Khalaf M (2008) Discrete-time nonlinear hjb solution using approximate dynamic programming: Convergence proof. IEEE Trans Syst Man Cybern Part B (Cybern) 38(4):943–949

Amini F, Samani MZ (2014) A wavelet-based adaptive pole assignment method for structural control. Comput-Aided Civil Infrastruct Eng 29(6):464–477

Basar T (2008) H_∞ optimal control and related minimax design problems. Birkhäuser Boston

Byers R, Nash SG (1989) Approaches to robust pole assignment. Int J Control 49(1):97–117

Cao YY, Lam J, Sun YX (1998) Static output feedback stabilization: an ilmi approach. Automatica 34(12):1641–1645

Chakrabortty A, Arcak M (2009) Robust stabilization and performance recovery of nonlinear systems with unmodeled dynamics. IEEE Trans Autom Control 54(6):1351–1356

El Ghaoui L, Oustry F, AitRami M (1997) A cone complementarity linearization algorithm for static output-feedback and related problems. IEEE Trans Autom Control 42(8):1171–1176

Fu Y, Chai T (2016) Online solution of two-player zero-sum games for continuous-time nonlinear systems with completely unknown dynamics. IEEE Trans Neural Networks Learn Syst 27(12):2577–2587

Gadewadikar J, Lewis FL, Abu-Khalaf M (2006) Necessary and sufficient conditions for H-infinity static output-feedback control. J Guidance Control Dyn 29(4):915–920

Geromel JC, De Souza C, Skelton R (1998) Static output feedback controllers: Stability and convexity. IEEE Trans Autom Control 43(1):120–125

Isidori A (1994) H_∞ control via measurement feedback for affine nonlinear systems, vol 4. Wiley Online Library

Jiang H, Zhang H, Luo Y, Cui X (2017) H_∞ control with constrained input for completely unknown nonlinear systems using data-driven reinforcement learning method. Neurocomputing 237:226–234

Jiang H, Zhang H, Han J, Zhang K (2018) Iterative adaptive dynamic programming methods with neural network implementation for multi-player zero-sum games. Neurocomputing 307:54–60

Jiang Y, Jiang ZP (2012) Computational adaptive optimal control for continuous-time linear systems with completely unknown dynamics. Automatica 48(10):2699–2704

Jiao Q, Modares H, Xu S, Lewis FL, Vamvoudakis KG (2016) Multi-agent zero-sum differential graphical games for disturbance rejection in distributed control. Automatica 69:24–34

Kautsky J, Nichols NK, Van Dooren P (1985) Robust pole assignment in linear state feedback. Int J Control 41(5):1129–1155

Kim JH, Lewis FL (2010) Model-free H_∞ control design for unknown linear discrete-time systems via Q-learning with LMI, vol 46. Elsevier

Kiumarsi B, Lewis FL, Modares H, Karimpour A, Naghibi-Sistani MB (2014) Reinforcement Q-learning for optimal tracking control of linear discrete-time systems with unknown dynamics. Automatica 50(4):1167–1175

Kiumarsi B, Kang W, Lewis FL (2016) H_∞ control of nonaffine aerial systems using off-policy reinforcement learning. Unmanned Syst 4(01):51–60

Kiumarsi B, Lewis FL, Jiang Z (2017) H_∞ control of linear discrete-time systems: Off-policy reinforcement learning. Automatica 78:144–152

Lee JY, Park JB, Choi YH (2012) Integral Q-learning and explorized policy iteration for adaptive optimal control of continuous-time linear systems. Automatica 48(11):2850–2859

Lewis FL, Vrabie D, Syrmos VL (2012) Optimal control. John Wiley & Sons, New York, NY, USA

Li H, Liu D, Wang D (2014) Integral reinforcement learning for linear continuous-time zero-sum games with completely unknown dynamics. IEEE Trans Autom Sci Eng 11(3):706–714
Li J, Xiao Z (2020) H_∞ control for discrete-time multi-player systems via off-policy Q-learning. IEEE Access 8:28831–28846
Li J, Modares H, Chai T, Lewis FL, Xie L (2017) Off-policy reinforcement learning for synchronization in Multi-agent graphical games. IEEE Trans Neural Networks Learn Syst 28(10):2434–2445
Li J, Chai T, Lewis FL, Fan J, Ding Z, Ding J (2018) Off-policy Q-learning: set-point design for optimizing dual-rate rougher flotation operational processes. IEEE Trans Ind Electron 65(5):4092–4102
Li J, Chai T, Lewis FL, Ding Z, Jiang Y (2019) Off-policy interleaved Q-learning: Optimal control for affine nonlinear discrete-time systems. IEEE Trans Neural Networks Learn Syst 30(5):1308–1320
Li J, Xiao Z, Li P (2019) Discrete-time multi-player games based on off-policy q-learning. IEEE Access 7:134647–134659
Li J, Ding J, Chai T, Lewis FL (2020) Nonzero-sum game reinforcement learning for performance optimization in large-scale industrial processes. IEEE Trans Cybern 50(9):4132–4145
Liu D, Wei Q (2014) Multiperson zero-sum differential games for a class of uncertain nonlinear systems. Int J Adapt Control Signal Process 28(3–5):205–231
Liu D, Li H, Wang D (2014) Online synchronous approximate optimal learning algorithm for multi-player non-zero-sum games with unknown dynamics. IEEE Trans Syst Man Cybern: Syst 44(8):1015–1027
Luo B, Wu HN, Huang T (2014) Off-policy reinforcement learning for H_∞ control design. IEEE Trans Cybern 45(1):65–76
Luo B, Huang T, Wu HN, Yang X (2015) Data-driven H_∞ control for nonlinear distributed parameter systems. IEEE Trans Neural Networks Learn Syst 26(11):2949–2961
Lv Y, Ren X (2019) Approximate nash solutions for multiplayer mixed-zero-sum game with reinforcement learning. IEEE Trans Syst Man Cybern: Syst 49(12):2739–2750
Modares H, Lewis FL (2014) Linear quadratic tracking control of partially-unknown continuous-time systems using reinforcement learning. IEEE Trans Autom Control 59(11):3051–3056
Modares H, Lewis FL (2014) Optimal tracking control of nonlinear partially-unknown constrained-input systems using integral reinforcement learning. Automatica 50(7):1780–1792
Modares H, Lewis FL, Sistani MBN (2014) Online solution of nonquadratic two-player zero-sum games arising in the H_∞ control of constrained input systems. Int J Adapt Control Signal Process 28(3–5):232–254
Modares H, Lewis FL, Jiang ZP (2015) H_∞ tracking control of completely unknown continuous-time systems via off-policy reinforcement learning. IEEE Trans Neural Networks Learn Syst 26(10):2550–2562
Moghadam R, Lewis FL (2019) Output-feedback H_∞ quadratic tracking control of linear systems using reinforcement learning. Int J Adapt Control Signal Process 33(2):300–314
Munos R, Stepleton T, Harutyunyan A, Bellemare M (2016) Safe and efficient off-policy reinforcement learning. Adv Neural Inf Process Syst 29
Noura H, Sauter D, Hamelin F, Theilliol D (2000) Fault-tolerant control in dynamic systems: Application to a winding machine. IEEE Control Syst Mag 20(1):33–49
Peng Y, Chen Q, Sun W (2019) Reinforcement Q-learning algorithm for H_∞ tracking control of unknown discrete-time linear systems. IEEE Trans Syst Man Cybern: Syst 50(11):4109–4122
Ren L, Zhang G, Mu C (2020) Data-based H_∞ control for the constrained-input nonlinear systems and its applications in chaotic circuit systems. IEEE Trans Circuits Syst I: Regul Pap 67(8):2791–2802
Rizvi SAA, Lin Z (2018) Output feedback Q-learning for discrete-time linear zero-sum games with application to the H-infinity control. Automatica 95:213–221
Song R, Wei Q, Song B (2017) Neural-network-based synchronous iteration learning method for multi-player zero-sum games. Neurocomputing 242:73–82

Valadbeigi AP, Sedigh AK, Lewis FL (2020) H_∞ static output-feedback control design for discrete-time systems using reinforcement learning. IEEE Trans Neural Networks Learn Syst 31(2):396–406

Vamvoudakis KG (2017) Q-learning for continuous-time graphical games on large networks with completely unknown linear system dynamics. Int J Robust Nonlinear Control 27(16):2900–2920

Vamvoudakis KG (2017) Q-learning for continuous-time linear systems: A model-free infinite horizon optimal control approach. Syst Control Lett 100:14–20

Vamvoudakis KG, Lewis FL (2010) Online actor-critic algorithm to solve the continuous-time infinite horizon optimal control problem. Automatica 46(5):878–888

Vamvoudakis KG, Lewis FL (2011) Multi-player non-zero-sum games: Online adaptive learning solution of coupled Hamilton-Jacobi equations. Automatica 47(8):1556–1569

Vamvoudakis KG, Lewis FL (2012) Online solution of nonlinear two-player zero-sum games using synchronous policy iteration. Int J Robust Nonlinear Control 22(13):1460–1483

Vamvoudakis KG, Lewis FL, Hudas GR (2012) Multi-agent differential graphical games: Online adaptive learning solution for synchronization with optimality. Automatica 48(8):1598–1611

Vamvoudakis KG, Modares H, Kiumarsi B, Lewis FL (2017) Game theory-based control system algorithms with real-time reinforcement learning: How to solve multiplayer games online. IEEE Control Syst Mag 37(1):33–52

Van Der Schaft AJ (1992) L_2-gain analysis of nonlinear systems and nonlinear state feedback H_∞ control, vol 37

Vrabie D (2010) Online adaptive optimal control for continuous-time systems

Wang C, Zuo Z, Qi Z, Ding Z (2018) Predictor-based extended-state-observer design for consensus of mass with delays and disturbances. IEEE Trans Cybern 49(4):1259–1269

Wei Q, Liu D, Shi G (2014) A novel dual iterative Q-learning method for optimal battery management in smart residential environments. IEEE Trans Ind Electron 62(4):2509–2518

Wei Q, Lewis FL, Sun Q, Yan P, Song R (2017) Discrete-time deterministic Q-learning: A novel convergence analysis. IEEE Trans Cybern 47(5):1224–1237

Werbos P (1992) Approximate dynamic programming for realtime control and neural modelling. Neural Fuzzy Adapt Approach, Handb Intell Control, pp 493–525

Chapter 3
Robust Tracking Control and Output Regulation

In this chapter, we mainly focus on the robust tracking control problem for linear discrete-time multi-player systems using reinforcement learning combined with output regulation. A novel off-policy cooperative game Q-learning algorithm is proposed for achieving optimal tracking control of linear discrete-time multi-player systems suffering from exogenous dynamic disturbance. The key strategy, for the first time, is to integrate reinforcement learning and cooperative games with output regulation under the discrete-time sampling framework for achieving data-driven optimal tracking control and disturbance rejection. Without the information of state and input matrices of multi-player systems, as well as the dynamics of exogenous disturbance and command generator, the coordination equilibrium solution and the steady-state control laws are learned using data by a novel off-policy Q-learning approach, such that multi-player systems have the capability of tolerating disturbance and follow the reference signal via the optimal approach. Moreover, the rigorous theoretical proofs of unbiasedness of coordination equilibrium solution and convergence of the proposed algorithm are presented. Simulation results are given to show the efficacy of the developed approach.

Optimal tracking control has been intensively investigated by academic and industry scientists as it has potential benefits and profits for many real-world applications (Kwatny et al. 1975; ÇImen and Banks 2004; Tan et al. 2007; Preitl et al. 2007). The fundamental task of optimal tracking control is to design a feedback controller in order to reject the dynamic disturbance while forcing systems to track the reference signals via an optimal approach (Adib Yaghmaie et al. 2018). Over the last decade, a wide variety of data-driven optimization control methods have been reported to improve the specific performance concerned by practical applications (Tan et al. 2007; Preitl et al. 2007; Ye and Hu 2016; Gil et al. 2018), in which ADPRL, as one of the powerful data-driven adaptive optimal control techniques, has a capability of enforcing performance of systems by updating control policy when interacting with unknown and uncertain environments (Luo et al. 2019; Li et al. 2020).

J. Li et al., *Reinforcement Learning*, Advances in Industrial Control,
https://doi.org/10.1007/978-3-031-28394-9_3

Regarding the data-driven optimal tracking game problem in which multi-player or multi-agent systems are subject to exogenous disturbances, the considerable results of ADPRL-based optimal tracking control indeed laid a solid foundation for an intensive research of robust tracking control for complex large-scale systems and showed the relevance between RL-based optimal control and game theory (Jiao et al. 2016; Luy 2018; Zhang et al. 2015; Gao et al. 2019; Odekunle et al. 2020). Notice that the results in the work of Zhang et al. (2015), Gao et al. (2019) and Odekunle et al. (2020) focused on only CT multi-player and multi-agent systems rather than DT systems with digital controllers. However, the DT systems converted by employing digital sampling controllers become more popular since digital sampling controllers are preferred from the perspective of industrial applications with the rapid progress of digital technologies. The most relevant result (Xiong et al. 2020) reported a kind of model-free adaptive robust control technique for DT multi-agent systems, while performance optimality cannot be guaranteed. In addition, the information on the reference signal dynamics still needs to be known a priori in Zhang et al. (2015), Gao et al. (2019) and Odekunle et al. (2020) when learning the optimal tracking control policies for systems with multiple players or agents.

Therefore, from the motivations mentioned before, this chapter integrates ADPRL and output regulation with game theory to find the solutions to coordination Nash equilibrium and output regulator equations using measured data, resulting in the optimal tracking control policies for DT multi-player systems. These control policies force all players to follow the reference signal and resist the exogenous disturbance via an optimal approach. The faced intractable obstacles come from the requirement of data-driven learning instead of complete model-based controller design, and another challenging issue is the distinct nature of DT systems and CT systems. In addition, the complex multi-player game makes it more difficult for us to find the optimal tracking control policies. This is the first time to integrate the output regulation technique, game theory and ADPRL method for handling the robust optimal tracking control problem of DT multi-player systems using data, and the proposed algorithm is more general and applicable in the sense of simultaneously considering unknown dynamics of exogenous disturbances and reference signals, and unknown state matrix and input matrices. Introducing multiple behavior control policies for the dynamics of all players, therefore, technically builds the relationship between dynamic process and steady-state process, which intuitively derives an off-policy cooperative game (OPCG) Q-learning algorithm. This manipulation successfully deals with learning the coordination equilibrium solution and solving the regulator equations, resulting in the robust optimal tracking control policies of all players.

This chapter is organized as follows. In Sect. 3.1, we formulate the robust optimal tracking control problem, together with some assumptions, definitions and a relevant lemma. Section 3.2 is devoted to solve the robust optimal tracking control problem respectively from the steady-state and dynamic aspects. Section 3.3 first proposes the OPCG Q-learning algorithm, and then rigorous proofs of the convergence of the algorithm and the unbiasedness of the learning results are derived. Illustrative examples are given to show the effectiveness of the proposed approach in Sect. 3.4. Finally, conclusions are contained in Sect. 3.5.

3.1 Optimal Robust Control Problem Statement

In this section, the optimal tracking problem of the linear DT multi-player systems with exogenous disturbance is proposed first. Then, based on the ADP method and game theory, we divide the robust tracking problem into two optimization control problems by using the output regulation method and the optimal control theory.

Consider a class of linear DT systems with multiple control actions described as follows:

$$\begin{aligned} x_{k+1} &= Ax_k + \sum_{i=1}^{s} B_i u_{ik} + d_k \\ y_k &= Cx_k, \end{aligned} \tag{3.1}$$

where $x_k = x(k) \in \mathbb{R}^p$ is the system state, $u_{ik} = u_i(k) \in \mathbb{R}^{m_i}$ is the control input of each player i ($i = 1, ..., s$) and $y_k \in \mathbb{R}^q$ is the system output. $A \in \mathbb{R}^{p\times p}$, $B_i \in \mathbb{R}^{p\times m_i}$, $C \in \mathbb{R}^{q\times p}$ and k ($k = 0, 1, \ldots$) is the sampling time instant. $d_k \in \mathbb{R}^d$ denotes the disturbance from an external system represented as

$$\begin{aligned} w_{k+1} &= Ew_k \\ d_k &= Sw_k, \end{aligned} \tag{3.2}$$

where $w_k \in \mathbb{R}^w$ is the state of the external system (3.2). $E \in \mathbb{R}^{w\times w}$ and $S \in \mathbb{R}^{d\times w}$.

The target of this paper is to design robust optimal tracking control policies taken by all players using measurable data, which implicates that the designed control policies can (a) force the output of system (3.1) to follow a dynamical reference signal even suffering from disturbance (3.2); and (b) optimize a specific performance of system (3.1). Here a linear reference signal generator of consideration is given as follows:

$$r_{k+1} = Fr_k, \tag{3.3}$$

where $r_k \in \mathbb{R}^r$ is the state of the reference signal system (3.3) and $F \in \mathbb{R}^{r\times r}$. Let $e_k = y_k - r_k$ denote the tracking error.

Compacting the state variable w_k of disturbance system (3.2) and the reference signal and integrating three systems (3.1)–(3.3) yield the following augmented system:

$$\begin{aligned} x_{k+1} &= Ax_k + \sum_{i=1}^{s} B_i u_{ik} + \bar{S}\bar{v}_k \\ \bar{v}_{k+1} &= W\bar{v}_k \\ e_k &= Cx_k - \bar{F}\bar{v}_k, \end{aligned} \tag{3.4}$$

where $W = \begin{bmatrix} E & 0 \\ 0 & F \end{bmatrix}$, $\bar{S} = [S\ \mathbf{0}]$, $\bar{F} = [\mathbf{0}\ I]$ and I is an identity matrix with appropriate dimensions. $\bar{v}_k = \begin{bmatrix} w_k^T & r_k^T \end{bmatrix}^T$ is a compacted vector. The following general assumptions are made throughout this chapter.

Assumption 3.1 The pair (A, B_i) is stabilizable, $i = 1, 2, \ldots, s$. The pair (A, C) is observable.

Assumption 3.2 $\text{rank}\left(\begin{bmatrix} A - \lambda I & B \\ C & 0 \end{bmatrix}\right) = p + q$, $\forall \lambda \in \sigma(W)$, where $B = [B_1\ \ B_2\ \ \ldots\ \ B_s]$.

To achieve the target of this chapter as we said before, the output regulation technique and optimal control theory are going to be employed and integrated. Recalling the output regulation technique in the work of Francis (1977), one can conclude that Assumption 3.2 can make sure the solvability of the following regulator equations for arbitrary matrices $\bar{S}$ and $\bar{F}$:

$$\begin{aligned} XW &= AX + \sum_{i=1}^{s} B_i U_i + \bar{S} \\ 0 &= CX - \bar{F}, \end{aligned} \tag{3.5}$$

where $X \in \mathbb{R}^{p \times (w+r)}$ and $U_i \in \mathbb{R}^{m_i \times (w+r)}$. Referring to the work of Huang (2004) and Krener (1992), one solution to the above regulator equations can be derived by minimizing an optimization problem below.

Problem 3.1

$$\begin{cases} \min\limits_{(X,U)} \text{Tr}(X^T \bar{Q} X + \sum_{i=1}^{s} U_i^T \bar{R}_i U_i) \\ s.t.\ (3.5) \end{cases}, \tag{3.6}$$

where $U = \{U_1, U_2, \ldots, U_s\}$, $\bar{Q} = \bar{Q}^T \geq 0$ and $\bar{R}_i = \bar{R}_i^T > 0$.

Let $z_k = x_k - X\bar{v}_k$, $v_{ik} = u_{ik} - U_i \bar{v}_k$, system (3.4) can be directly transformed into the following error system:

$$\begin{aligned} z_{k+1} &= Az_k + \sum_{i=1}^{s} B_i v_{ik} \\ e_k &= Cz_k. \end{aligned} \tag{3.7}$$

Now we define a common cost function for all players as

$$J = \frac{1}{2} \sum_{k=0}^{\infty} \left(z_k^T Q z_k + \sum_{i=1}^{s} v_{ik}^T R_i v_{ik} \right), \tag{3.8}$$

where $Q = Q^T \geq 0$ and $R_i = R_i^T > 0$. All players work hard together to achieve the optimality of common performance defined by (3.8) with respect to the error system (3.7). Thus, the cooperative game of all players can be formulated as follows.

Problem 3.2

$$\begin{cases} \min\limits_{V} \frac{1}{2} \sum_{k=0}^{\infty} \left(z_k^T Q z_k + \sum_{i=1}^{s} v_{ik}^T R_i v_{ik} \right) \\ s.t.\ (3.7) \end{cases}, \tag{3.9}$$

where $V = \{v_{1k}, v_{2k}, \ldots, v_{sk}\}$ and suppose $v_{ik} = -K_i z_k$.

Remark 3.1 In general, the variables v_{ik} are called the transient controllers, and the steady-state control law of each player could be $U_i \bar{v}_k$. From the derivations of Problems 3.1 and 3.2, the robust optimal tracking problem is decomposed into two optimization problems, i.e., Problems 3.1 and 3.2.

Now we are in the position to solve Problems 3.1 and 3.2. To this end, game theory is briefly reviewed since Problem 3.2 is indeed a cooperative game of multiple players. The following definitions are useful for solving the game problem.

Definition 3.1 (Başar and Olsder 1998) (Coordination Equilibrium) Suppose the s-tuple control policies $\left\{v_{1k}^*, v_{2k}^*, \ldots, v_{sk}^*\right\}$ makes system (3.7) achieve the optimal performance, i.e.,

$$J^* \triangleq J(z_0, v_{1k}^*, v_{2k}^*, \ldots, v_{sk}^*) = \min_{V} \frac{1}{2} \sum_{k=0}^{\infty} \left(z_k^T Q z_k + \sum_{i=1}^{s} v_{ik}^T R_i v_{ik} \right). \tag{3.10}$$

Then the s-tuple control policies $\left\{v_{1k}^*, v_{2k}^*, \ldots, v_{sk}^*\right\}$ is called the coordination equilibrium solution, and J^* is the coordination equilibrium outcome, which means

$$J(v_{1k}^*, v_{2k}^*, \ldots, v_{sk}^*) \leq \min\left[J(v_{1k}, v_{2k}^*, \ldots, v_{sk}^*), \ldots, J(v_{1k}^*, v_{2k}^*, \ldots, v_{sk})\right]. \tag{3.11}$$

Definition 3.2 (Vamvoudakis and Lewis 2011) (Admissible Control Policies) A set of control policies $v_i(x_k)$ $(i = 1, 2, \ldots, s)$ is called admissible on compact $x_k \in \Omega \in \mathbb{R}^p$, denoted by $v_i(x_k) \in \Psi(\Omega)$, if $v_i(x_k)$ $(i = 1, 2, \ldots, s)$ are continuous on Ω, $v_i(0) = 0$ $(i = 1, 2, \ldots, s)$ and (3.8) is finite $\forall x_0 \in \Omega$.

Referring to optimal control theory and game theory, the optimal value function and the optimal Q-function in terms of the cost function (3.9) can be defined below

$$V^*(z_k) = \min_{V} \frac{1}{2} \sum_{q=k}^{\infty} \left(z_q^T Q z_q + \sum_{i=1}^{s} v_{iq}^T R_i v_{iq} \right) \tag{3.12}$$

and

$$Q^*(z_k, V) = z_k^T Q z_k + \sum_{i=1}^{s} v_{ik}^T R_i v_{ik} + V^*(z_{k+1}). \tag{3.13}$$

The relationship between $V^*(z_k)$ and $Q^*(z_k, v_{ik})$ can be expressed as

$$V^*(z_k) = \min_V Q^*(z_k, V) = Q^*(z_k, v_{1k}^*, v_{2k}^*, \dots, v_{sk}^*). \tag{3.14}$$

Further, the conclusion that the optimal value and the Q-function are quadratic can be made similar to Kiumarsi et al. (2014).

Lemma 3.1 *For any admissible control policies $v_{ik} = -K_i z_k$, the optimal value function and the Q-function have the quadratic forms of*

$$V^*(z_k) = \frac{1}{2} z_k^T P z_k \tag{3.15}$$

and

$$Q^*(z_k, V) = \frac{1}{2} \xi_k^T H \xi_k, \tag{3.16}$$

where P and H are positive definite matrices, and

$$\xi_k = \begin{bmatrix} z_k^T & v_{1k}^T & v_{2k}^T & \dots & v_{sk}^T \end{bmatrix}^T \tag{3.17}$$

$$P = M^T H M \tag{3.18}$$

$$M = \begin{bmatrix} I & -K_1^T & -K_2^T & \dots & -K_s^T \end{bmatrix}^T$$

$$H = \begin{bmatrix} H_{zz} & H_{zv_1} & H_{zv_2} & \dots & H_{zv_s} \\ H_{zv_1}^T & H_{v_1v_1} & H_{v_1v_2} & \dots & H_{v_1v_s} \\ H_{zv_2}^T & H_{v_1v_2}^T & H_{v_2v_2} & \dots & H_{v_2v_s} \\ \vdots & \vdots & \vdots & \ddots & \vdots \\ H_{zv_s}^T & H_{v_1v_s}^T & H_{v_2v_s}^T & \dots & H_{v_sv_s} \end{bmatrix}$$
$$= \begin{bmatrix} A^T P A + Q & A^T P B_1 & A^T P B_2 & \dots & A^T P B_s \\ (A^T P B_1)^T & B_1^T P B_1 + R_1 & B_1^T P B_2 & \dots & B_1^T P B_s \\ (A^T P B_2)^T & (B_1^T P B_2)^T & B_2^T P B_2 + R_2 & \dots & B_2^T P B_s \\ \vdots & \vdots & \vdots & \ddots & \vdots \\ (A^T P B_s)^T & (B_1^T P B_s)^T & (B_2^T P B_s)^T & \dots & B_s^T P B_s + R_s \end{bmatrix}. \tag{3.19}$$

3.2 Theoretical Solutions

In this section, we will devote to solving the robust optimal tracking control problem formulated in Sect. 3.1. The theoretical solutions to the regulator equations and the cooperative game can be derived if all information of the augmented system (3.4) is assumed to be known.

3.2.1 Solving the Regulator Equations with Known Dynamics

Before solving Problem 3.1, we first find the way to solve regulator equations when assuming the dynamics of systems (3.1)–(3.3) to be known. Define a map $\Theta: \mathbb{R}^{p\times r} \rightarrow \mathbb{R}^{p\times r}$ by

$$\Theta(X) = XW - AX. \tag{3.20}$$

Choose two constant matrices, one is the zero matrix X_0, and the other is the matrix X_1 satisfying $CX_1 = \bar{F}$. Then we choose matrices X_l $(l = 2, 3, \ldots, m+1)$ and all the vectors $\mathrm{vec}(X_l)$ from a basis for the $\ker(I_{w+r} \otimes C)$, where $m = (p-r) \times (p+r)$. Thus, the matrix X in (3.5) can be expressed by a sequence of $\alpha_l \in \mathbb{R}$ as follows (Gao and Jiang, 2016):

$$X = X_1 + \sum_{l=2}^{m+1} \alpha_l X_l. \tag{3.21}$$

The following lemma is presented to provide a solution to the regulator equations in (3.5).

Lemma 3.2 *(Odekunle et al. 2020) The pair $(X, U_1, \ldots, U_s)$ is the solution to (3.5) if and only if it can make sure the following equation holds:*

$$\omega\chi = \psi, \tag{3.22}$$

where

$$\begin{aligned}\omega &= \begin{bmatrix} vec(\Theta(X_2)) & \ldots & vec(\Theta(X_{m+1})) & \boldsymbol{0} & -I_{p+r}\otimes B_1 & \ldots & -I_{p+r}\otimes B_s \\ vec(X_2) & \ldots & vec(X_{m+1}) & -I_{p\times(p+r)} & \boldsymbol{0} & \ldots & \boldsymbol{0} \end{bmatrix}\\ \chi &= \begin{bmatrix}\alpha_2 & \ldots & \alpha_{m+1} & vec(X)^T & vec(U_1)^T & \ldots & vec(U_s)^T\end{bmatrix}^T\\ \psi &= \begin{bmatrix} vec(-\Theta(X_1) + \bar{S}) \\ -vec(X_1)\end{bmatrix}.\end{aligned} \tag{3.23}$$

3.2.2 Solving Problem 3.2 with Known Dynamics

To solve Problem 3.2, utilizing Lemma 3.1 and (3.13) yields the following Q-function-based Bellman equation:

$$\xi_k^T H \xi_k = z_k^T Q z_k + \sum_{i=1}^{s} v_{ik}^T R_i v_{ik} + \xi_{k+1}^T H \xi_{k+1}. \tag{3.24}$$

According to the relationship between matrix P and matrix H shown in (3.18), eliminating the same variable z_k on both sides of (3.24) yields

$$M^T H M = M^T \Lambda M + (A - \sum_{i=1}^{s} B_i K_i)^T M^T H M (A - \sum_{i=1}^{s} B_i K_i), \tag{3.25}$$

where $\Lambda = \text{diag}(Q, R_1, R_2, \ldots, R_s)$. Based on the necessary condition for optimality, implementing $\frac{\partial Q^*(z_k, V)}{\partial v_{ik}} = 0$ yields

$$v_{ik}^* = -H_{v_i v_i}^{-1} \left(H_{z v_i}^T z_k + \sum\nolimits_{i=1, q \neq i}^{s} H_{v_i v_q} v_{qk} \right). \tag{3.26}$$

Combining (3.25) and (3.26) derives the DT cooperative game algebraic Riccati equation (ARE) below

$$(M^*)^T H^* M^* = (M^*)^T \Lambda M^* + (A - \sum_{i=1}^{s} B_i K_i^*)^T (M^*)^T H^* M^* (A - \sum_{i=1}^{s} B_i K_i^*), \tag{3.27}$$

where

$$K_i^* = (H_{v_i v_i}^*)^{-1} ((H_{z v_i}^*)^T - \sum\nolimits_{i=1, q \neq i}^{s} H_{v_i v_q}^* K_q^*)$$
$$M^* = \begin{bmatrix} I & -(K_1^*)^T & -(K_2^*)^T & \ldots & -(K_s^*)^T \end{bmatrix}^T. \tag{3.28}$$

Remark 3.2 The s-tuple of control policies $\{v_{1k}^*, v_{2k}^*, \ldots, v_{sk}^*\}$ represented by (3.26) is essentially the coordination equilibrium solution to Problem 3.2, which can be easily proven like Odekunle et al. (2020), Kiumarsi et al. (2014). Similar model-based policy iteration algorithms to Odekunle et al. (2020) are adopted below, i.e., Algorithm 3.1, for solving the coupled game ARE equation (3.27).

Algorithm 3.1 Model-based policy iteration

1: Initialize each player's controller gain K_i^0 $(i = 1, 2, \ldots, s)$. Set the iteration index $j = 0$ and $i = 1$;

2: Policy evaluation by calculating H^{j+1}:

$$(M^j)^T H^{j+1} M^j = (M^j)^T \Lambda M^j + (A - \sum_{i=1}^{s} B_i K_i^j)^T (M^j)^T H^{j+1} M^j (A - \sum_{i=1}^{s} B_i K_i^j); \tag{3.29}$$

3: Policy updating:

$$K_i^{j+1} = (H_{v_i v_i}^{j+1})^{-1} \left((H_{z v_i}^{j+1})^T - \sum\nolimits_{i=1, q \neq i}^{s} H_{v_i v_q}^{j+1} K_q^j \right); \tag{3.30}$$

4: If $i < s$, then $i = i + 1$ and go back to Step 3. Otherwise go to Step 5;

5: If $\left\| H^{j+1} - H^j \right\| \leq \varepsilon$ (ε is a small positive number), then stop the iteration and output the controller gains K_i^{j+1} as the approximate values of K_i^*. Otherwise, set $i = 1$, $j = j + 1$ and go back to Step 2.

Remark 3.3 There exists an obstacle when implementing Algorithm 3.1, that is, the matrices A, B_i are unknown. In the sequel, a data-driven OPCG Q-learning algorithm will be developed for fixing this obstacle.

3.3 Data-Driven Solutions

This section develops a novel data-driven OPCG Q-learning algorithm to solve the robust optimal tracking control problem and rigorously proves the convergence of the proposed algorithm. Moreover, it also presents proof of the unbiasedness of solutions learned by the proposed algorithm, even though the probing noises are added for satisfying the PE condition. The detailed schematic of solving the robust optimal tracking control can be found in Fig. 3.1.

3.3.1 Data-Driven OPCG Q-Learning

Define $z_k(l) = x_k - X_l \bar{v}_k$ $(l = 0, 1, \ldots, m+1)$. From (3.7), one has

$$z_{k+1}(l) = x_{k+1} - X_l \bar{v}_{k+1} = Ax_k + \sum_{i=1}^{s} B_i u_{ik} + (\bar{S} - X_l W)\bar{v}_k. \tag{3.31}$$

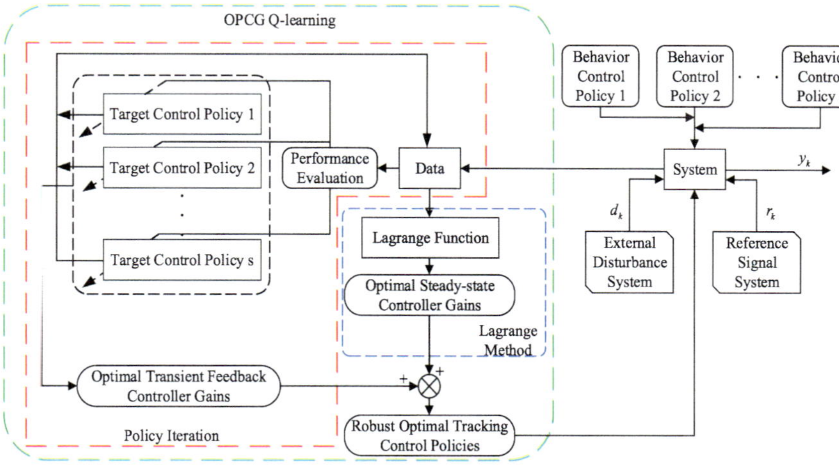

Fig. 3.1 Schematic of the robust optimal tracking control

Adding and abstracting the target control policies $v_{ik}^{j}(l) = -K_i^j z_k(l)$ to system (3.31) yields

$$\begin{aligned} z_{k+1}(l) &= A(z_k(l) + X_l\bar{v}_k) + \sum_{i=1}^{s} B_i u_{ik} + \sum_{i=1}^{s} B_i v_{ik}^{j}(l) - \sum_{i=1}^{s} B_i v_{ik}^{j}(l) + (\bar{S} - X_l W)\bar{v}_k \\ &= \left(A - \sum_{i=1}^{s} B_i K_i^j\right) z_k(l) + \sum_{i=1}^{s} B_i(u_{ik} + K_i^j z_k(l)) + (AX_l + \bar{S} - X_l W)\bar{v}_k \\ &= A_c^j z_k(l) + \sum_{i=1}^{s} B_i w_i^j(l) - \Pi(X_l)\bar{v}_k, \end{aligned} \tag{3.32}$$

where u_{ik} $(i = 1, 2, \ldots, s)$ are viewed as the behavior control policies used to generate data, and

$$A_c^j = A - \sum_{i=1}^{s} B_i K_i^j,\ w_i^j(l) = u_{ik} + K_i^j z_k(l),\ \Pi(X_l) = \Theta(X_l) - \bar{S}.$$

Then, the following manipulation is made

$$\begin{aligned} &\xi_{k+1}^T(l) H^{j+1} \xi_{k+1}(l) - \xi_k^T(l) H^{j+1} \xi_k(l) \\ &= z_{k+1}^T(l)(M^j)^T H^{j+1} M^j z_{k+1}(l) - z_k^T(l)(M^j)^T H^{j+1} M^j z_k(l) \\ &= \left(A_c^j z_k(l) + \sum_{i=1}^{s} B_i w_i^j(l) - \Pi(X_l)\bar{v}_k\right)^T (M^j)^T H^{j+1} M^j \\ &\times \left(A_c^j z_k(l) + \sum_{i=1}^{s} B_i w_i^j(l) - \Pi(X_l)\bar{v}_k\right) - z_k^T(l)(M^j)^T H^{j+1} M^j z_k(l). \end{aligned} \tag{3.33}$$

Due to (3.27), (3.33) can be rewritten based on Lemma 3.1 as

$$\begin{aligned} &\xi_{k+1}^T(l) H^{j+1} \xi_{k+1}(l) - \xi_k^T(l) H^{j+1} \xi_k(l) \\ &= -z_k^T(l)(M^j)^T \Lambda M^j z_k(l) + 2 z_k^T(l) A^T P^{j+1} \sum_{i=1}^{s} B_i w_i^j(l) \\ &- 2 z_k^T(l) \left(\sum_{i=1}^{s} B_i K_i^j\right)^T P^{j+1} \left(\sum_{i=1}^{s} B_i w_i^j(l)\right) - 2 \sum_{i=1}^{s} (w_i^j(l))^T B_i^T P^{j+1} \Pi(X_l)\bar{v}_k \\ &- 2 z_k^T(l) A^T P^{j+1} \Pi(X_l)\bar{v}_k + 2 z_k^T(l) \left(\sum_{i=1}^{s} B_i K_i^j\right)^T P^{j+1} \Pi(X_l)\bar{v}_k \\ &+ \bar{v}_k^T \Pi(X_l)^T P^{j+1} \Pi(X_l)\bar{v}_k + \sum_{i=1}^{s} (w_i^j(l))^T B_i^T P^{j+1} \sum_{i=1}^{s} B_i w_i^j(l), \end{aligned} \tag{3.34}$$

where

$$
\begin{aligned}
&2z_k^T(l)\left(\sum_{i=1}^{s} B_i K_i^j\right)^T P^{j+1}\Pi(X_l)\bar{v}_k - 2\sum_{i=1}^{s}(w_i^j(l))^T B_i^T P^{j+1}\Pi(X_l)\bar{v}_k \\
&\quad = -2\sum_{i=1}^{s} u_{ik}^T B_i^T P^{j+1}\Pi(X_l)\bar{v}_k \\
&\sum_{i=1}^{s}(w_i^j(l))^T B_i^T P^{j+1}\sum_{i=1}^{s} B_i w_i^j(l) - 2z_k^T(l)\left(\sum_{i=1}^{s} B_i K_i^j\right)^T P^{j+1}\left(\sum_{i=1}^{s} B_i w_i^j(l)\right) \\
&\quad = \sum_{i=1}^{s}(u_{ik} - K_i^j z_k(l))^T B_i^T P^{j+1}\sum_{i=1}^{s} B_i w_i^j(l).
\end{aligned} \tag{3.35}
$$

By (3.35), (3.34) becomes

$$
\begin{aligned}
&\xi_{k+1}^T(l)H^{j+1}\xi_{k+1}(l) - \xi_k^T(l)H^{j+1}\xi_k(l) \\
&\quad = -z_k^T(l)(M^j)^T \Lambda M^j z_k(l) + 2z_k^T(l)[H_{z_l v_1}^{j+1}, \ldots, H_{z_l v_s}^{j+1}]\sum_{i=1}^{s} w_i^j(l) \\
&\quad + \sum_{i=1}^{s}(u_{ik} - K_i^j z_k(l))^T G \sum_{i=1}^{s} w_i^j(l) - 2z_k^T(l)A^T P^{j+1}\Pi(X_l)\bar{v}_k \\
&\quad - 2\sum_{i=1}^{s} u_{ik}^T B_i^T P^{j+1}\Pi(X_l)\bar{v}_k + \bar{v}_k^T \Pi(X_l)^T P^{j+1}\Pi(X_l)\bar{v}_k,
\end{aligned} \tag{3.36}
$$

where

$$
G(l) = \begin{bmatrix}
H_{v_1v_1}^{j+1} - R_1 & H_{v_1v_2}^{j+1} & H_{v_1v_3}^{j+1} & \ldots & H_{v_1v_s}^{j+1} \\
(H_{v_1v_2}^{j+1})^T & H_{v_2v_2}^{j+1} - R_2 & H_{v_2v_3}^{j+1} & \ldots & H_{v_2v_s}^{j+1} \\
(H_{v_1v_3}^{j+1})^T & (H_{v_2v_3}^{j+1})^T & H_{v_3v_3}^{j+1} - R_3 & \ldots & H_{v_3v_s}^{j+1} \\
\vdots & \vdots & \vdots & \ddots & \vdots \\
(H_{v_1v_s}^{j+1})^T & (H_{v_2v_s}^{j+1})^T & (H_{v_3v_s}^{j+1})^T & \ldots & H_{v_sv_s}^{j+1} - R_s
\end{bmatrix}.
$$

Then, (3.36) can be further rewritten as

$$
\hat{\theta}_k^j(l)\hat{L}^{j+1}(l) = \hat{\rho}_k^j(l), \tag{3.37}
$$

where

$$
\begin{aligned}
\hat{\rho}_k^j(l) &= -z_k^T(l)Qz_k(l) - \sum_{i=1}^{s} u_{ik}^T R_i u_{ik} \\
\hat{\theta}_k^j(l) &= \left[\hat{\theta}_{00}^j(l)\ \hat{\theta}_{0i}^j(l)\ \hat{\theta}_{ii}^j(l)\ \hat{\theta}_{im}^j(l)\ \hat{\theta}_{s1}^j(l)\ \hat{\theta}_{s2,i}^j(l)\ \hat{\theta}_{s3}^j(l)\right] \\
\hat{L}^{j+1}(l) &= \Big[(\hat{L}_{00}^{j+1}(l))^T\ (\hat{L}_{0i}^{j+1}(l))^T\ (\hat{L}_{ii}^{j+1}(l))^T\ (\hat{L}_{io}^{j+1}(l))^T \\
&\quad (\hat{L}_{s1}^{j+1}(l))^T\ (\hat{L}_{s2,i}^{j+1}(l))^T\ (\hat{L}_{s3}^{j+1}(l))^T\Big]^T \\
\hat{L}_{00}^{j+1}(l) &= \mathrm{vecs}(H_{z_l z_l}^{j+1}),\ \hat{L}_{0i}^{j+1}(l) = \mathrm{vec}(H_{z_l v_i}^{j+1}),\ \hat{L}_{ii}^{j+1}(l) = \mathrm{vec}(H_{v_i v_i}^{j+1}) \\
\hat{L}_{io}^{j+1}(l) &= \mathrm{vec}(H_{v_i v_o}^{j+1}),\ \hat{L}_{s1}^{j+1}(l) = \mathrm{vec}(A^T P^{j+1}\Pi(X_l)) \\
\hat{L}_{s2,i}^{j+1}(l) &= \mathrm{vec}(B_i^T P^{j+1}\Pi(X_l)),\ \hat{L}_{s3}^{j+1}(l) = \mathrm{vecs}(\Pi(X_l)^T P^{j+1}\Pi(X_l))
\end{aligned}
$$

and

$$
\begin{aligned}
\hat{\theta}_{00}^j(l) &= z_{k+1}^T(l) \otimes z_{k+1}^T(l) - z_k^T(l) \otimes z_k^T(l) \\
\hat{\theta}_{0i}^j(l) &= 2z_{k+1}^T(l) \otimes (-K_i^j z_{k+1}(l))^T - 2z_k^T(l) \otimes (-K_i^j z_k(l))^T - 2z_k^T(l) \times (w_i^j(l))^T \\
\hat{\theta}_{ii}^j(l) &= (-K_i^j z_{k+1}(l))^T \otimes (-K_i^j z_{k+1}(l))^T - u_{ik}^T \otimes u_{ik}^T \\
\hat{\theta}_{io}^j(l) &= 2(-K_i^j z_{k+1}(l))^T \otimes (-K_o^j z_{k+1}(l))^T - 2u_{ik}^T \otimes u_{ok}^T \\
\hat{\theta}_{s1}^j(l) &= 2z_k^T(l) \otimes \bar{v}_k^T \\
\hat{\theta}_{s2,i}^j(l) &= 2u_{ik}^T \otimes \bar{v}_k^T \\
\hat{\theta}_{s3}^j(l) &= -\bar{v}_k^T \otimes \bar{v}_k^T\ (i = 1, 2, \ldots, s, o = i+1, i+2, \ldots, s-1).
\end{aligned}
$$

Thus, the transient feedback controller gains can be calculated as

$$
K_i^{j+1} = (\hat{L}_{ii}^{j+1}(l))^{-1}\left((\hat{L}_{0i}^{j+1}(l))^T - \sum_{t=1,t\neq i}^{s} \hat{L}_{it}^{j+1}(l)K_t^j\right)(l=0). \tag{3.38}
$$

Recalling Sect. 3.2.1 where the model-based solution to the regulator equations is presented, we shall study how to solve Problem 3.1 using the results from (3.37). Thus, the solution to the regulator equations is finally derived using only data.

Firstly, we can define

$$
\bar{\Pi}(X) = A^T P^{j+1}\Pi(X) \tag{3.39}
$$

$$
\bar{\Theta}(X) = A^T P^{j+1}\Theta(X). \tag{3.40}
$$

Then, one has

$$
\bar{\Pi}(X_l) = A^T P^{j+1}\Pi(X_l) = \hat{L}_{s1}^{j+1}(l) \tag{3.41}
$$

$$
\bar{\Pi}(X_0) = -A^T P^{j+1}\bar{S} = \hat{L}_{s1}^{j+1}(0) \tag{3.42}
$$

$$
\bar{\Theta}(X_l) = A^T P^{j+1}(\Pi(X_l) - \Pi(X_0)) = \hat{L}_{s1}^{j+1}(l) - \hat{L}_{s1}^{j+1}(0). \tag{3.43}
$$

By Lemma 3.2, we can rewrite (3.22) as

$$\bar{\omega}\chi = \bar{\psi}, \tag{3.44}$$

where

$$\begin{aligned}
\bar{\omega} &= \begin{bmatrix} \mathrm{vec}(\bar{\Theta}(X_2)) \dots \mathrm{vec}(\bar{\Theta}(X_{m+1})) & \mathbf{0} & -I_{p+r} \otimes A^T P^{j+1} B_1 \dots -I_{p+r} \otimes A^T P^{j+1} B_s \\ \mathrm{vec}(X_2) \quad \dots \quad \mathrm{vec}(X_{m+1}) & -I_{p\times(p+r)} & \mathbf{0} \qquad \dots \qquad \mathbf{0} \end{bmatrix} \\
&= \begin{bmatrix} \mathrm{vec}(\hat{L}_{s1}^{j+1}(2) - \hat{L}_{s1}^{j+1}(0)) \dots & \mathbf{0} & -I_{p+r} \otimes \hat{L}_{01}^{j+1}(0) \dots -I_{p+r} \otimes \hat{L}_{0s}^{j+1}(0) \\ \mathrm{vec}(X_2) & \dots -I_{p\times(p+r)} & \mathbf{0} \qquad \dots \qquad \mathbf{0} \end{bmatrix} \\
&= \begin{bmatrix} \bar{\omega}_{11} & \bar{\omega}_{12} \\ \bar{\omega}_{21} & \bar{\omega}_{22} \end{bmatrix} \\
\chi &= \begin{bmatrix} \alpha_2 \dots \alpha_{m+1} \ \mathrm{vec}(X)^T \ \mathrm{vec}(U_1)^T \dots \mathrm{vec}(U_s)^T \end{bmatrix}^T = \begin{bmatrix} \chi_1 \ \chi_2 \end{bmatrix}^T \\
\bar{\psi} &= \begin{bmatrix} \mathrm{vec}(-\bar{\Theta}(X_1) - \bar{\Pi}(X_0)) \\ -\mathrm{vec}(X_1) \end{bmatrix} = \begin{bmatrix} \mathrm{vec}(-\hat{L}_{s1}^{j+1}(1)) \\ -\mathrm{vec}(X_1) \end{bmatrix} = \begin{bmatrix} \bar{\psi}_1 \\ \bar{\psi}_2 \end{bmatrix}.
\end{aligned} \tag{3.45}$$

Thus, we have

$$\begin{aligned}
\bar{\omega}_{11}\chi_1 + \bar{\omega}_{12}\chi_2 &= \bar{\psi}_1 \\
\bar{\omega}_{21}\chi_1 + \bar{\omega}_{22}\chi_2 &= \bar{\psi}_2.
\end{aligned} \tag{3.46}$$

From (3.46), one has

$$\bar{\zeta}\begin{bmatrix} \mathrm{vec}(X)^T \ \mathrm{vec}(U_1)^T \dots \mathrm{vec}(U_s)^T \end{bmatrix}^T = \bar{\zeta}\chi_2^T = \bar{b}, \tag{3.47}$$

where $\bar{\zeta} = \bar{\omega}_{12} - \bar{\omega}_{11}\bar{\omega}_{21}^{-1}\bar{\omega}_{22}$, $\bar{b} = \bar{\psi}_1 - \bar{\omega}_{11}\bar{\omega}_{21}^{-1}\bar{\psi}_1$. Utilizing the Lagrange Multiplier representation, the solution to Problem 3.1 can be directly derived by defining a Lagrange function as follows:

$$\begin{aligned}
L = &\begin{bmatrix} \mathrm{vec}(X) \\ \mathrm{vec}(U_1) \\ \vdots \\ \mathrm{vec}(U_s) \end{bmatrix}^T \begin{bmatrix} I_{p+r} \otimes \bar{Q} & \mathbf{0} & \dots & \mathbf{0} \\ \mathbf{0} & I_{p+r} \otimes \bar{R}_1 & \dots & \mathbf{0} \\ \vdots & \vdots & \ddots & \vdots \\ \mathbf{0} & \mathbf{0} & \dots & I_{p+r} \otimes \bar{R}_s \end{bmatrix} \begin{bmatrix} \mathrm{vec}(X) \\ \mathrm{vec}(U_1) \\ \vdots \\ \mathrm{vec}(U_s) \end{bmatrix} \\
&+ \lambda^T \mathrm{vec}\left(\bar{\zeta} \begin{bmatrix} \mathrm{vec}(X) \\ \mathrm{vec}(U_1) \\ \vdots \\ \mathrm{vec}(U_s) \end{bmatrix} - \bar{b} \right).
\end{aligned} \tag{3.48}$$

Taking the partial derivation of L with respect to (X, U_{ik}) and λ yields

$$\mathcal{M}\boldsymbol{\varphi} = \mathcal{O}, \tag{3.49}$$

where

$$\mathcal{M} = \begin{bmatrix} 2\begin{bmatrix} I_{p+r}\otimes\bar{Q} & \mathbf{0} & \cdots & \mathbf{0} \\ \mathbf{0} & I_{p+r}\otimes\bar{R}_1 & \cdots & \mathbf{0} \\ \vdots & \vdots & \ddots & \vdots \\ \mathbf{0} & \mathbf{0} & \cdots & I_{p+r}\otimes\bar{R}_s \end{bmatrix} & \bar{\zeta} \\ \bar{\zeta}^T & \mathbf{0} \end{bmatrix}, \ \varphi = \begin{bmatrix} \begin{bmatrix} \mathrm{vec}(X) \\ \mathrm{vec}(U_1) \\ \vdots \\ \mathrm{vec}(U_s) \end{bmatrix} \\ \lambda \end{bmatrix}, \ \mathcal{O} = \begin{bmatrix} \begin{bmatrix} \mathbf{0} \\ \mathbf{0} \\ \vdots \\ \mathbf{0} \end{bmatrix} \\ \bar{b} \end{bmatrix}.$$

By solving (3.49) correctly, the steady-state controller gains ($U_1, \ldots, U_s$) and the matrix X can be obtained. Now, a data-driven OPCG Q-learning algorithm is presented, such that the optimal transient feedback controller gains, i.e., coordination equilibrium solution, and the steady-state controller gains can be learned using the measured data.

Theorem 3.1 is provided to show the convergence of Algorithm 3.2.

Theorem 3.1 *The matrices H^{j+1} and K_i^{j+1} learned by Algorithm 3.2 converge to H^* and K_i^* when $j \to \infty$, respectively.*

Proof It is not difficult to see from the derivation that if H^{j+1} and K_i^{j+1} are solutions to (3.29) and (3.30), then they can also make (3.37) and (3.38) hold. Next, it is going to prove that the solutions to (3.37) and (3.38) also satisfy (3.29) and (3.30).

Equation (3.37) is equivalent to (3.36), so their solutions are the same. (3.29) can be obtained by subtracting (3.36) from (3.34). Therefore, the solution to (3.37) is also the solution to (3.29) resulting in the value of (3.38) being the same as (3.30). Moreover, the convergence of the solutions to (3.29) and (3.30) in Algorithm 3.1 has been proved in Song et al. (2016), that is, the $H^{j+1} \to H^*$ and $K_i^{j+1} \to K_i^*$ as $j \to \infty$. Therefore, the convergence of Algorithm 3.2 is guaranteed. This completes the proof. □

So far, if the steady-state controller gains ($U_1, \ldots, U_s$), the matrix X and the transient feedback controller gains K_i^{j+1} ($i = 1, 2, \ldots, s$) are derived, then the robust optimal tracking control policies u_{ik} for all players can be found, i.e.,

$$\begin{aligned} u_{ik} &= v_{ik} + U_i\bar{v}_k = -K_i^* z_k + U_i\bar{v}_k \\ &= -K_i^*(x_k - X\bar{v}_k) + U_i\bar{v}_k \\ &= -K_i^* x_k + (K_i^* X + U_i)\bar{v}_k. \end{aligned} \tag{3.50}$$

Remark 3.4 Unlike the work of Zhang et al. (2015), Gao et al. (2019) and Odekunle et al. (2020) requiring the knowledge including the output matrix C, the dynamics of disturbances and reference signals, the proposed Algorithm 3.2 does not need to know the matrices of state and input, the dynamics of external disturbances and reference signals any longer except for the output matrix C when finding the robust optimal tracking control policies of all players. In addition, this is the first time to develop a data-driven RL method combined with output regulation and game theory for optimal tracking control problem with disturbance rejection for DT multi-player systems.

Algorithm 3.2 OPCG Q-learning

1: Data collection: Compute matrices X_l $(l = 0, 1, \ldots, m+1)$, then collect the system data $z_k(l)$ and store them in the sample sets $\hat{\theta}_k^j(l)$ and $\hat{\rho}_k^j(l)$ by using the behavior control policies u_{ik} $(i = 1, 2, \ldots, s)$ and matrices X_l;

2: Initiation: Choose the initial controller gains K_i^0 $(i = 1, 2, \ldots, s)$. Set $l = 0$, $i = 1$, $j = 0$ and j represents the iteration index;

3: Implementing Q-learning:

(a) Policy evaluation by using recursive least-square or batch least-squares method for (3.37), $\hat{L}^{j+1}(l)$ can be estimated using the collected data in Step 1;

(b) If $l < m+1$, then $l = l+1$ and go back to Step 3(a). Otherwise go to Step 3(c);

(c) Update the controller gains K_i^{j+1} in terms of (3.38);

4: If $i < s$, then $i = i+1$ and go back to Step 3(c). Otherwise go to Step 5;

5: If $\left\| K_i^{j+1} - K_i^j \right\| \leq \varepsilon$ $(i = 1, 2, \ldots, s)$ (ε is a small positive number) for all players, then stop the iteration and the coordination equilibrium $\left\{ v_{1k}^{j+1}, v_{2k}^{j+1}, \ldots, v_{sk}^{j+1} \right\}$ where $v_{ik}^{j+1} = -K_i^{j+1} z_k$ can be obtained. Otherwise, $i = 1$, $l = 0$, $j = j+1$ and go back to Step 3(a);

6: Obtain $(X, U_1, \ldots, U_s)$ by solving (3.49) based on $\hat{L}_{oi}^{j+1}(l)$ and $\hat{L}_{s1}^{j+1}(l)$ when the iteration stops in Step 5.

3.3.2 *No Bias Analysis of the Solution for the Proposed Algorithm*

It is well known that probing noises need to be added to the behavior policies due to satisfying the PE condition when implementing the off-policy RL. Now, we suppose that the control policies u_{ik} become the following form due to adding probing noises, then the unbiasedness of coordination equilibrium is rigorously proved in Theorem 3.2.

$$\hat{u}_{ik} = u_{ik} + e_{ik} \tag{3.51}$$

where e_{ik} is called probing noise.

Theorem 3.2 *Let H^{j+1} be the solution to (3.34) with $e_{ik} = 0$ and $\hat{H}^{j+1}$ be the solution to (3.34) with $e_{ik} \neq 0$. Then, $H^{j+1} = \hat{H}^{j+1}$.*

Proof Substituting (3.51) with $e_i \neq 0$ into (3.33) yields

$$\begin{aligned}
&\xi_{k+1}^T(l)\hat{H}^{j+1}\xi_{k+1}(l) - \xi_k^T(l)\hat{H}^{j+1}\xi_k(l) \\
&= \left(A_c^j z_k(l) + \sum_{i=1}^{s} B_i w_i^j(l) + \sum_{i=1}^{s} B_i e_{ik} - \Pi(X_l)\bar{v}_k \right)^T \left(M^j\right)^T \hat{H}^{j+1} M^j \\
&\times \left(A_c^j z_k(l) + \sum_{i=1}^{s} B_i w_i^j(l) + \sum_{i=1}^{s} B_i e_{ik} - \Pi(X_l)\bar{v}_k \right) - z_k^T(l) \left(M^j\right)^T \hat{H}^{j+1} M^j z_k(l)
\end{aligned}$$

$$
\begin{aligned}
&= \left(z_{k+1}(l) + \sum_{i=1}^{s} B_i e_{ik}\right)^T (M^j)^T \hat{H}^{j+1} M^j \left(z_{k+1}(l) + \sum_{i=1}^{s} B_i e_{ik}\right) \\
&\quad - z_k^T(l) \left(M^j\right)^T \hat{H}^{j+1} M^j z_k(l) \\
&= z_{k+1}^T(l)(M^j)^T \hat{H}^{j+1} M^j z_{k+1}(l) + 2 z_{k+1}^T(l)(M^j)^T \hat{H}^{j+1} M^j \left(\sum_{i=1}^{s} B_i e_{ik}\right) \\
&\quad + \left(\sum_{i=1}^{s} B_i e_{ik}\right)^T (M^j)^T \hat{H}^{j+1} M^j \left(\sum_{i=1}^{s} B_i e_{ik}\right) - z_k^T(l) \left(M^j\right)^T \hat{H}^{j+1} M^j z_k(l) \\
&= -z_k^T(l)(M^j)^T \Lambda M^j z_k(l) + 2 z_k^T(l)(A_c^j)^T \hat{P}^{j+1} \left(\sum_{i=1}^{s} B_i (w_i^j(l) + e_{ik})\right) \\
&\quad + \left(\sum_{i=1}^{s} B_i \left(w_i^j(l) + e_{ik}\right)\right)^T \hat{P}^{j+1} \left(\sum_{i=1}^{s} B_i \left(w_i^j(l) + e_{ik}\right)\right) \\
&\quad - 2 \left(\sum_{i=1}^{s} B_i (w_i^j(l) + e_{ik})\right)^T (M^j)^T \hat{H}^{j+1} M^j \Pi(X_l)\bar{v}_k \\
&\quad - 2 z_k^T(l)(A_c^j)^T \hat{P}^{j+1} \Pi(X_l)\bar{v}_k + (\Pi(X_l)\bar{v}_k)^T \hat{P}^{j+1} \Pi(X_l)\bar{v}_k.
\end{aligned}
\tag{3.52}
$$

By (3.32), one has

$$
\begin{aligned}
z_{k+1}^T(l)\hat{P}^{j+1} \sum_{i=1}^{s} B_i e_{ik} &= z_k^T(l)(A_c^j)^T \hat{P}^{j+1} \sum_{i=1}^{s} B_i e_{ik} + \sum_{i=1}^{s} (w_i^j(l))^T B_i^T \hat{P}^{j+1} \sum_{i=1}^{s} B_i e_{ik} \\
&\quad - (\Pi(X_l)\bar{v}_k)^T \hat{P}^{j+1} \sum_{i=1}^{s} B_i e_{ik}.
\end{aligned}
\tag{3.53}
$$

Substituting (3.53) into (3.52) yields

$$
\begin{aligned}
&\xi_{k+1}^T(l)\hat{H}^{j+1}\xi_{k+1}(l) - \xi_k^T(l)\hat{H}^{j+1}\xi_k(l) \\
&\quad = -z_k^T(l)(M^j)^T \Lambda M^j z_k(l) + 2 z_k^T(l)(A_c^j)^T \hat{P}^{j+1} \left(\sum_{i=1}^{s} B_i w_i^j(l)\right) \\
&\quad + \left(\sum_{i=1}^{s} B_i \left(w_i^j(l)\right)\right)^T \hat{P}^{j+1} \left(\sum_{i=1}^{s} B_i \left(w_i^j(l)\right)\right) + (\Pi(X_l)\bar{v}_k)^T \hat{P}^{j+1} \Pi(X_l)\bar{v}_k \\
&\quad - 2 \left(\sum_{i=1}^{s} B_i (w_i^j(l))\right)^T (M^j)^T \hat{H}^{j+1} M^j \Pi(X_l)\bar{v}_k - 2 z_k^T(l)(A_c^j)^T \hat{P}^{j+1} \Pi(X_l)\bar{v}_k.
\end{aligned}
\tag{3.54}
$$

Notice that (3.54) is equal to (3.34), then their solutions are the same, that is, $H^{j+1} = \hat{H}^{j+1}$. Therefore, when probing noise is added to the data collection process of the proposed Algorithm 3.2, the results learned by Algorithm 3.2 cannot produce bias. This completes the proof.

Remark 3.5 The proof of the unbiasedness of solutions caused by the added probing noises is given to clearly show the difference and similarity of the proofs shown in Jiang et al. (2019), Kiumarsi et al. (2017) and Li et al. (2019), since using output regulation brings out the different system dynamics from general multi-player systems or single player systems when adding the probing noise.

Remark 3.6 It should be pointed out that finding probing noises for satisfying the PE condition is not going to be easy if the systems' order is high. Designing u_{ik} for each player instead of compacting all control inputs into one vector provides an alternative to overcome this difficulty caused by high dimensions. If the order of u_{ik} is still too high for each player, adding different probing noise to each element of vector u_{ik} may be another way to ensure the PE condition.

3.4 Illustrative Examples

In this section, two examples are given to verify the effectiveness and efficacy of the proposed OPCG Q-learning algorithm. In each example, the robust optimal control policies are first calculated by solving the regulator equations and Problem 3.2 using model-based methods shown in Sects. 3.2.1 and 3.2.2. Then, the data-driven Algorithm 3.2, which only requires the known output matrix, is implemented to learn the robust optimal control policies for all players.

Example 3.1 Consider the systems (3.1)–(3.3) with two players, where

$$A = \begin{bmatrix} -1 & 2 \\ 2.2 & 1.7 \end{bmatrix},\ B_1 = \begin{bmatrix} 2 \\ 1.6 \end{bmatrix},\ B_2 = \begin{bmatrix} 0.951892 \\ 0.38373 \end{bmatrix},\ C = \begin{bmatrix} 1 & 2 \end{bmatrix}$$

$$S = \begin{bmatrix} \cos(0.1) & \sin(0.1) \\ -\sin(0.1) & \cos(0.1) \end{bmatrix},\ E = \begin{bmatrix} \cos(1) & \sin(1) \\ -\sin(1) & \cos(1) \end{bmatrix},\ F = -1. \tag{3.55}$$

Choose $Q = \bar{Q} = \text{diag}(5, 5)$, $R_1 = R_2 = \bar{R}_1 = \bar{R}_2 = 1$, $r_0 = 40$ and

$$X_1 = \begin{bmatrix} 0 & 0 & 0.2 \\ 0 & 0 & 0.4 \end{bmatrix},\ X_2 = \begin{bmatrix} 0.8 & -0.4 & 0 \\ -0.4 & 0.2 & 0 \end{bmatrix}$$
$$X_3 = \begin{bmatrix} -0.1789 & -0.3578 & 0.8 \\ 0.0894 & 0.1789 & -0.4 \end{bmatrix},\ X_4 = \begin{bmatrix} -0.3578 & -0.7155 & -0.4 \\ 0.1789 & 0.3578 & 0.2 \end{bmatrix}.$$

First, we assume that the dynamics of systems (3.1)–(3.3) are known, then the optimal transient feedback controller gains (K_1^*, K_2^*), the optimal steady-state controller gains (U_1^*, U_2^*) and the matrix X^* can be solved according to Algorithm 3.1 and Lemma 3.2.

$$K_1^* = \begin{bmatrix} -1.1974 & -0.9525 \end{bmatrix}, \; K_2^* = \begin{bmatrix} 3.2469 & -0.1154 \end{bmatrix}$$
$$U_1^* = \begin{bmatrix} -0.1378 & -0.3426 & -0.6851 \end{bmatrix}, \; U_2^* = \begin{bmatrix} -0.1020 & -0.0573 & 0.0510 \end{bmatrix}$$
$$X^* = \begin{bmatrix} 0.1449 & -0.2999 & -0.3217 \\ -0.0725 & 0.1499 & 0.6608 \end{bmatrix}.$$

Then, we shall implement the data-driven Algorithm 3.2. The Q-function matrix H^{j+1}, the transient feedback controller gains (K_1^{j+1}, K_2^{j+1}), the steady-state controller gains (U_1, U_2) and the matrix X can be learned. Figure 3.2 shows the evolution process of H^{j+1}, K_1^{j+1} and K_2^{j+1} learned by Algorithm 3.2. One can find that the matrices K_1^{j+1} and K_2^{j+1} respectively converge to the following values after 5 iterations from Fig. 3.2.

$$K_1^5 = \begin{bmatrix} -1.1974 & -0.9525 \end{bmatrix}, \; K_2^5 = \begin{bmatrix} 3.2469 & -0.1154 \end{bmatrix}. \tag{3.56}$$

The matrices U_1, U_2 and X are

$$\begin{aligned} U_1 = \begin{bmatrix} -0.1378 & -0.3426 & -0.6851 \end{bmatrix}, \; U_2 = \begin{bmatrix} -0.1020 & -0.0573 & 0.0510 \end{bmatrix} \\ X = \begin{bmatrix} 0.1449 & -0.2999 & -0.3217 \\ -0.0725 & 0.1499 & 0.6608 \end{bmatrix}. \end{aligned} \tag{3.57}$$

Figure 3.3 shows the trajectories of system output y_k and reference signal r_k. Figure 3.4 depicts the corresponding tracking error e_k. The cost performance is plotted in Fig. 3.5. As shown in Figs. 3.2, 3.3 and 3.4, the transient feedback controller gains learned by Algorithm 3.2 can quickly converge to the theoretically optimal values, and the satisfactory tracking result can be obtained even though there exists disturbance in the system. Moreover, it can be seen from Fig. 3.5 that the consumption of system performance quickly converges to zero.

For the purpose of comparison with the work where the external disturbance is ignored when designing the optimal tracking controller, for example, Luo et al. (2019), we pretend no disturbance exists and correspondingly revise the model constructed in (3.4). Then K_1, K_2, U_1, U_2 and X are learned by implementing the proposed method and listed in (3.58). Using them yields the tracking results of the system plotted in Figs. 3.6 and 3.7. It should be pointed out that the simulation results plotted in Figs. 3.6 and 3.7 are obtained under the same disturbance shown in (3.55) with Figs. 3.3 and 3.4 where disturbance rejection is adopted. Moreover, by comparisons, one can find that the tracking would be affected if the disturbance is not considered when designing controllers

$$\begin{aligned} K_1 = \begin{bmatrix} -1.1974 & -0.9525 \end{bmatrix}, \; K_2 = \begin{bmatrix} 3.2469 & -0.1154 \end{bmatrix} \\ U_1 = \begin{bmatrix} 0.0284 & 0.0567 & -0.6887 \end{bmatrix}, \; U_2 = \begin{bmatrix} -0.0630 & -0.1260 & 0.0609 \end{bmatrix} \\ X = \begin{bmatrix} 0.0189 & 0.0379 & -0.3194 \\ -0.0095 & -0.0189 & 0.6597 \end{bmatrix}. \end{aligned} \tag{3.58}$$

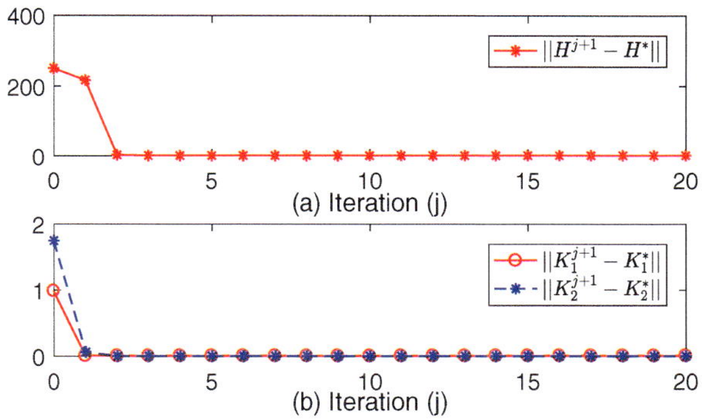

Fig. 3.2 Convergence of the H^{j+1} and (K_1^{j+1}, K_2^{j+1}) by Algorithm 3.2

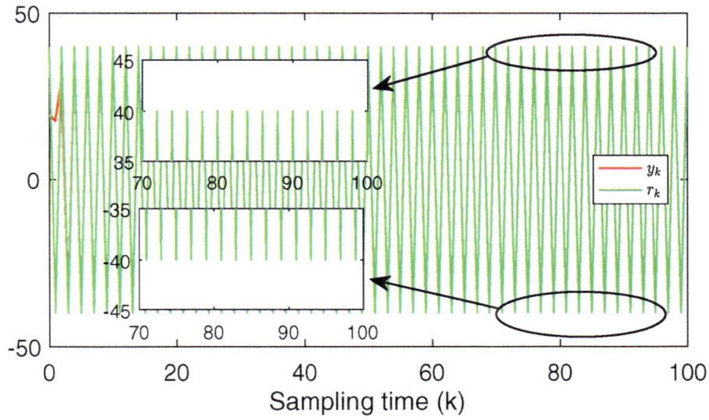

Fig. 3.3 Trajectories of the output y_k and the reference signal r_k using Algorithm 3.2

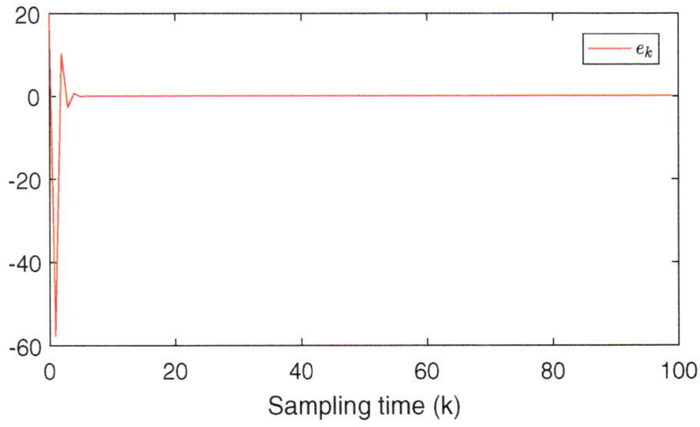

Fig. 3.4 Trajectories the tracking error e_k using Algorithm 3.2

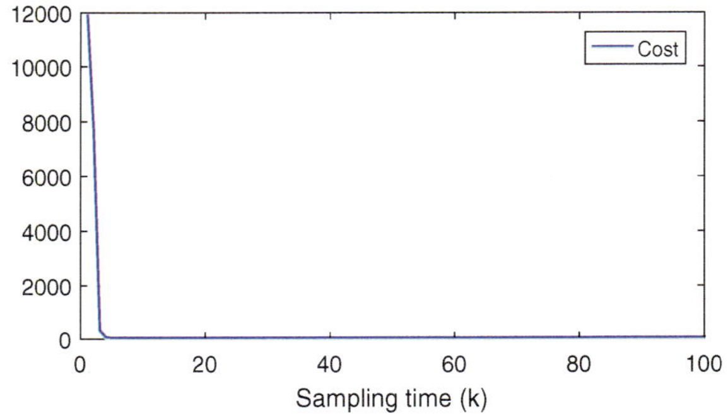

Fig. 3.5 Cost performance using Algorithm 3.2

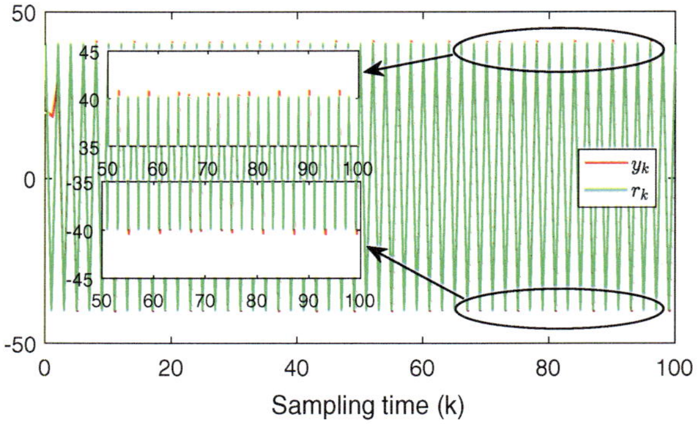

Fig. 3.6 Trajectories of the output y_k and the reference signal r_k (without consider d_k in learning process)

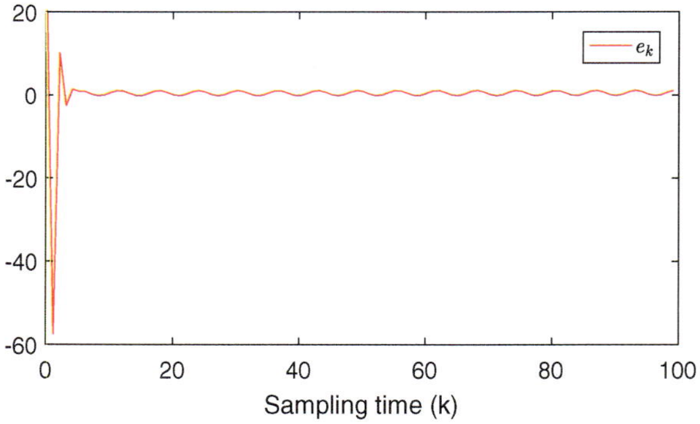

Fig. 3.7 Tracking error e_k (without consider d_k in learning process)

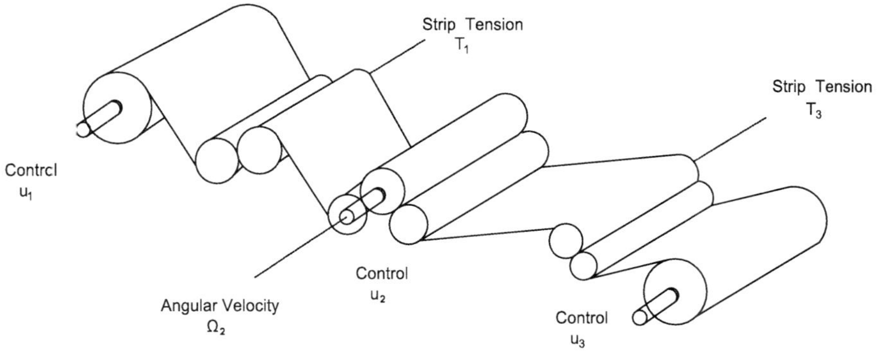

Fig. 3.8 A winding machine

Example 3.2 In this example, we will use a practical example of a winding machine system with three players, as shown in Fig. 3.8, to further verify the effectiveness of the proposed data-driven method. The unwinding reel (strip tension T_1), the traction reel (angular velocity Ω_2) and the rewinding reel (strip tension T_3) could be viewed as three players, whose control inputs are respectively u_1, u_2 and u_3. The linearized system model parameters are shown below

$$\begin{aligned} x_{k+1} &= Ax_k + B_1u_1 + B_2u_2 + B_3u_3 + d_k \\ y_k &= Cx_k, \end{aligned} \tag{3.59}$$

where

$$A = \begin{bmatrix} 0.4126 & 0 & -0.0196 \\ 0.0333 & 0.5207 & -0.0413 \\ -0.0101 & 0 & 0.2571 \end{bmatrix}, \quad B_1 = \begin{bmatrix} -1.7734 \\ 0.0928 \\ -0.0424 \end{bmatrix}$$

$$B_2 = \begin{bmatrix} 0.0696 \\ 0.4658 \\ -0.093 \end{bmatrix}, \quad B_3 = \begin{bmatrix} 0.0734 \\ 0.1051 \\ 2.0752 \end{bmatrix}, \quad C = \begin{bmatrix} 1 & 1 & 1 \end{bmatrix}.$$

The dynamic parameters of d_k and r_{k+1} are

$$S = \begin{bmatrix} \sin(1) & 0 & 0 \\ 0 & \sin(1) & 0 \\ 0 & 0 & \sin(1) \end{bmatrix}, \quad E = \begin{bmatrix} \sin(1) & \cos(1) & \sin(0.1) \\ -\cos(1) & \sin(1) & 0 \\ -\sin(1) & 0 & \sin(1) \end{bmatrix}, \quad F = 1.$$

Choose $Q = \bar{Q} = \text{diag}(1, 1, 1)$, $R_1 = R_2 = R_3 = \bar{R}_1 = \bar{R}_2 = \bar{R}_3 = 10$, $x_0 = [0.35\ 0.55 0.4]^T$, $r_0 = 2.85$ and

$$X_1=\begin{bmatrix}0 & 0 & 0 & 0.3333\\ 0 & 0 & 0 & 0.3333\\ 0 & 0 & 0 & 0.3333\end{bmatrix},\ X_2=\begin{bmatrix}0.3333 & -0.3333 & 0 & 0\\ -0.4553 & 0.6667 & 0 & 0\\ 0.1220 & -0.3333 & 0 & 0\end{bmatrix}$$

$$X_3=\begin{bmatrix}0.3333 & -0.3333 & 0 & 0\\ -0.4553 & -0.3333 & 0 & 0\\ 0.1220 & 0.6667 & 0 & 0\end{bmatrix},\ X_4=\begin{bmatrix}0.3333 & 0 & 0.6667 & 0\\ 0.1220 & 0 & -0.3333 & 0\\ -0.4553 & 0 & -0.3333 & 0\end{bmatrix}$$

$$X_5=\begin{bmatrix}0.3333 & 0 & -0.3333 & 0\\ 0.1220 & 0 & 0.6667 & 0\\ -0.4553 & 0 & -0.3333 & 0\end{bmatrix},\ X_6=\begin{bmatrix}0.3333 & 0 & -0.3333 & 0\\ 0.1220 & 0 & -0.3333 & 0\\ -0.4553 & 0 & 0.6667 & 0\end{bmatrix}$$

$$X_7=\begin{bmatrix}-0.1925 & -0.3849 & 0 & 0.6667\\ 0.2629 & 0.1925 & 0 & -0.3333\\ -0.0704 & 0.1925 & 0 & -0.3333\end{bmatrix},\ X_8=\begin{bmatrix}-0.1925 & -0.3849 & 0 & -0.3333\\ 0.2629 & 0.1925 & 0 & 0.6667\\ -0.0704 & 0.1925 & 0 & -0.3333\end{bmatrix}$$

$$X_9=\begin{bmatrix}-0.1925 & -0.3849 & 0 & -0.3333\\ 0.2629 & 0.1925 & 0 & -0.3333\\ -0.0704 & 0.1925 & 0 & 0.6667\end{bmatrix}.$$

First, assuming the system model parameters are known, we can obtain the optimal transient feedback controller gains (K_1^*, K_2^*, K_3^*), the optimal steady-state controller gains (U_1^*, U_2^*, U_3^*) and the matrix X^* by using Algorithm 3.1 and Lemma 3.2.

$$\begin{gathered}K_1^*=\begin{bmatrix}0.0613 & -0.0031 & -0.0028\end{bmatrix},\ K_2^*=\begin{bmatrix}-0.0054 & -0.0323 & 0.0050\end{bmatrix}\\ K_3^*=\begin{bmatrix}0.00004 & -0.0031 & -0.0382\end{bmatrix},\ U_1^*=\begin{bmatrix}0.0546 & 0.0026 & 0.0097 & -0.1051\end{bmatrix}\\ U_2^*=\begin{bmatrix}0.0093 & -0.1465 & 0.0216 & 0.2802\end{bmatrix}\\ U_3^*=\begin{bmatrix}-0.0093 & -0.0138 & -0.0364 & 0.1436\end{bmatrix}\\ X^*=\begin{bmatrix}-0.0185 & -0.0138 & -0.0406 & 0.3596\\ 0.0480 & 0.0122 & -0.0024 & 0.2765\\ -0.0295 & 0.0015 & 0.0431 & 0.3673\end{bmatrix}.\end{gathered}$$

Then, by Algorithm 3.2, we can get the Q-function matrix H^{j+1}, the transient feedback controller gains (K_1^{j+1}, K_2^{j+1},K_3^{j+1}), the steady-state controller gains (U_1, U_2, U_3) and the matrix X. Figure 3.9 shows the convergence process of H^{j+1}, K_1^{j+1}, K_2^{j+1} and K_3^{j+1} during the learning process of Algorithm 3.2. It can be seen from Fig. 3.9 that the matrices K_1^{j+1}, K_2^{j+1} and K_3^{j+1} converge to the following values after 5 iterations:

$$\begin{gathered}K_1^5=\begin{bmatrix}0.0613 & -0.0031 & -0.0028\end{bmatrix},\ K_2^5=\begin{bmatrix}-0.0054 & -0.0323 & 0.0050\end{bmatrix}\\ K_3^5=\begin{bmatrix}0.00004 & -0.0031 & -0.0382\end{bmatrix}.\end{gathered} \tag{3.60}$$

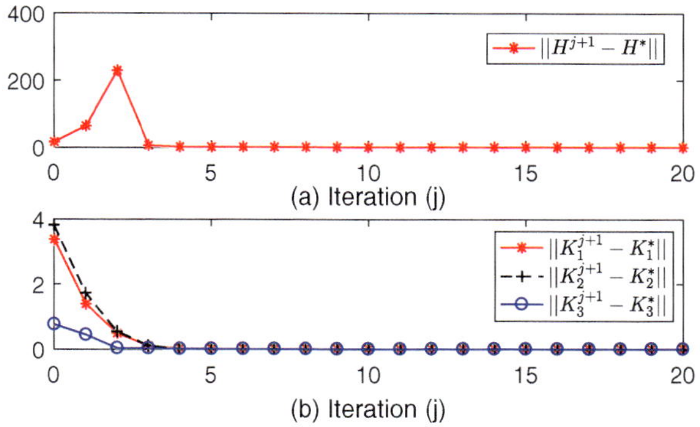

Fig. 3.9 Convergence of the H^{j+1} and $(K_1^{j+1}, K_2^{j+1}, K_3^{j+1})$ by Algorithm 3.2

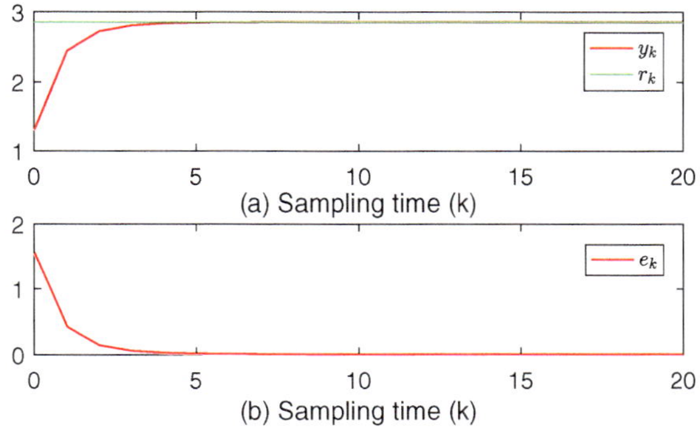

Fig. 3.10 Trajectories of the output y_k, the reference signal r_k, and the tracking error e_k using Algorithm 3.2

The matrices U_1, U_2, U_3 and X are

$$U_1 = \begin{bmatrix} 0.0546 & 0.0026 & 0.0097 & -0.1051 \end{bmatrix}, \ U_2 = \begin{bmatrix} 0.0092 & -0.1465 & 0.0216 & 0.2802 \end{bmatrix}$$
$$U_3 = \begin{bmatrix} -0.0093 & -0.0138 & -0.0364 & 0.1436 \end{bmatrix}$$
$$X = \begin{bmatrix} -0.0185 & -0.0138 & -0.0406 & 0.3596 \\ 0.0479 & 0.0122 & -0.0024 & 0.2765 \\ -0.0295 & 0.0015 & 0.0431 & 0.3673 \end{bmatrix}. \tag{3.61}$$

Figure 3.10 shows the system output y_k, the reference signal r_k along with the corresponding tracking error e_k under the control inputs computed by the proposed algorithm. Figure 3.11 gives the cost performance. From Figs. 3.9 and 3.10, one can see that the transient feedback controller gains can fast converge to the theoretically optimal values by using the proposed algorithm, and the system output can track the reference signal even though there exists external disturbance in the system. Meanwhile, one can find that the consumption of system performance quickly converges to zero from Fig. 3.11.

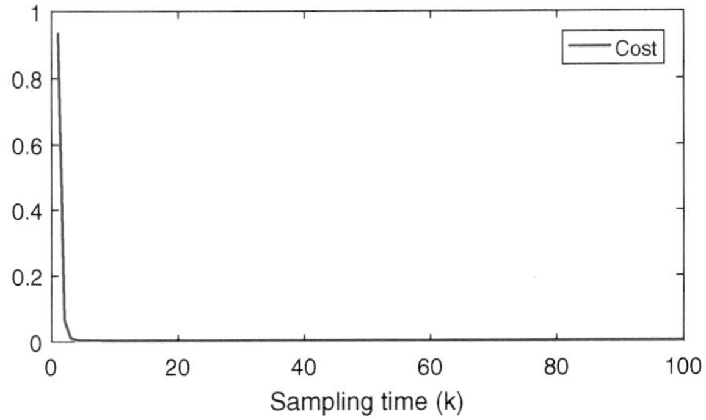

Fig. 3.11 Cost performance using Algorithm 3.2

3.5 Conclusion

In this section, the output regulation method, RL and game theory are integrated for the first time to find the optimal tracking control policies of all players with DT sampling, such that they can struggle together to optimize the common performance and fight the external disturbance without knowing the dynamics of disturbance and reference signal, as well as the input matrices and state matrix of systems. To this end, an OPCG Q-learning algorithm is developed together by providing rigorous proof of the convergence of the proposed algorithm and the unbiasedness of the solution. Finally, illustrative examples are given to show the effectiveness and efficacy of the proposed approach.

References

Adib Yaghmaie F, Hengster Movric K, Lewis FL, Su R, Sebek M (2018) H_∞-output regulation of linear heterogeneous multiagent systems over switching graphs. Int J Robust Nonlinear Control 28(13):3852–3870

Başar T, Olsder GJ (1998) Dynamic noncooperative game theory, 2nd edn. SIAM, PA

ÇImen T, Banks SP (2004) Nonlinear optimal tracking control with application to super-tankers for autopilot design. Automatica 40(11):1845–1863

Francis BA (1977) The linear multivariable regulator problem. SIAM J Control Optim 15(3):486–505

Gao W, Jiang ZP (2016) Adaptive dynamic programming and adaptive optimal output regulation of linear systems. IEEE Trans Autom Control 61(12):4164–4169

Gao W, Jiang Y, Davari M (2019) Data-driven cooperative output regulation of multi-agent systems via robust adaptive dynamic programming. IEEE Trans Circuits Syst II: Express Briefs 66(3):447–451

Gil RA, Johanyák ZC, Kovács T et al (2018) Surrogate model based optimization of traffic lights cycles and green period ratios using microscopic simulation and fuzzy rule interpolation. Int J Artif Intell 16(1):20–40

Huang J (2004) Nonlinear output regulation: theory and applications. SIAM, Philadelphia, PA

Jiang Y, Kiumarsi B, Fan J, Chai T, Li J, Lewis FL (2019) Optimal output regulation of linear discrete-time systems with unknown dynamics using reinforcement learning. IEEE Trans Cybern 50(7):3147–3156

Jiao Q, Modares H, Xu S, Lewis FL, Vamvoudakis KG (2016) Multi-agent zero-sum differential graphical games for disturbance rejection in distributed control. Automatica 69:24–34

Kiumarsi B, Lewis FL, Modares H, Karimpour A, Naghibi-Sistani MB (2014) Reinforcement Q-learning for optimal tracking control of linear discrete-time systems with unknown dynamics. Automatica 50(4):1167–1175

Kiumarsi B, Lewis FL, Jiang Z (2017) H_∞ control of linear discrete-time systems: off-policy reinforcement learning. Automatica 78:144–152

Krener AJ (1992) The construction of optimal linear and nonlinear regulators. In: Systems, models and feedback: theory and applications. Springer, pp 301–322

Kwatny H, Kalnitsky K, Bhatt A (1975) An optimal tracking approach to load-frequency control. IEEE Trans Power Appar Syst 94(5):1635–1643

Li J, Chai T, Lewis FL, Ding Z, Jiang Y (2019) Off-policy interleaved Q-learning: optimal control for affine nonlinear discrete-time systems. IEEE Trans Neural Netw Learn Syst 30(5):1308–1320

Li J, Ding J, Chai T, Lewis FL (2020) Nonzero-sum game reinforcement learning for performance optimization in large-scale industrial processes. IEEE Trans Cybern 50(9):4132–4145

Luo B, Liu D, Huang T, Liu J (2019) Output tracking control based on adaptive dynamic programming with multistep policy evaluation. IEEE Trans Syst Man Cybern Syst 49(10):2155–2165

Luy NT (2018) Distributed cooperative H_∞ optimal tracking control of MIMO nonlinear multi-agent systems in strict-feedback form via adaptive dynamic programming, vol 91. Taylor & Francis

Odekunle A, Gao W, Davari M, Jiang ZP (2020) Reinforcement learning and non-zero-sum game output regulation for multi-player linear uncertain systems. Automatica 112:108672

Preitl S, Precup R, Preitl Z, Vaivoda S, Kilyeni S, Tar JK (2007) Iterative feedback and learning control. Servo systems applications. IFAC Proc Vol 40(8):16–27

Song R, Lewis FL, Wei Q (2016) Off-policy integral reinforcement learning method to solve nonlinear continuous-time multiplayer nonzero-sum games. IEEE Trans Neural Netw Learn Syst 28(3):704–713

Tan K, Zhao S, Xu JX (2007) Online automatic tuning of a proportional integral derivative controller based on an iterative learning control approach. IET Control Theory Appl 1(1):90–96

Vamvoudakis KG, Lewis FL (2011) Multi-player non-zero-sum games: online adaptive learning solution of coupled Hamilton–Jacobi equations. Automatica 47(8):1556–1569

Xiong S, Hou Z, Jin S (2020) Model-free adaptive formation control for unknown multiinput-multioutput nonlinear heterogeneous discrete-time multiagent systems with bounded disturbance. Int J Robust Nonlinear Control 30(15):6330–6350

Ye M, Hu G (2016) Distributed extremum seeking for constrained networked optimization and its application to energy consumption control in smart grid. IEEE Trans Control Syst Technol 24(6):2048–2058

Zhang H, Liang H, Wang Z, Feng T (2015) Optimal output regulation for heterogeneous multiagent systems via adaptive dynamic programming. IEEE Trans Neural Netw Learn Syst 28(1):18–29

Chapter 4
Interleaved Robust Reinforcement Learning

This chapter investigates adaptive robust controller design for DT affine nonlinear systems using an adaptive dynamic programming. A novel adaptive interleaved reinforcement learning algorithm is developed for finding a robust controller of DT affine nonlinear systems subject to matched or unmatched uncertainties. To this end, the robust control problem is converted to the optimal control problem for nominal systems by selecting an appropriate utility function. The performance evaluation and control policy update combined with neural networks approximation are alternately implemented at each time step for solving a simplified HJB equation, such that the UUB stability of DT affine nonlinear systems can be guaranteed, allowing for all realization of unknown bounded uncertainties. The rigorously theoretical proofs of convergence of the proposed interleaved RL algorithm and UUB stability of uncertain systems are provided. Simulation results are given to verify the effectiveness of the proposed method.

It is well known that uncertainty inherently exists in variety of engineering applications caused by inaccurate system identification, disturbances and unmeasurable part of system variables, etc. (Xie et al. 2018; Jiang and Jiang 2014). To remain or enhance performance of systems subject to uncertainty, robust control has become a basic issue in control theory and synthesis. In the past decades, rich achievements about this topic have been published. Petersen (1987) and Petersen and Hollot (1986) started the pioneering research to solve the robust control problem by the usage of optimal control approach for linear systems with time-varying uncertain parameters. In Lin (2000) and Tan et al. (2009), this idea has been extended to linear systems with matched and unmatched uncertainties for finding stabilizing or tracking controller.

ADP or adaptive critic learning (ACL) has the capability of finding the optimal controllers for linear and nonlinear systems (Luo et al. 2014, 2018; Lewis and Vrabie 2009). Fan and Yang (2015) developed an ADP-based integral sliding control method to stabilize a class of nonlinear systems with uncertainty. Jiang and Jiang (2014) and Gao et al. (2016) investigated the robust controller design by integrating the nonlinear

J. Li et al., *Reinforcement Learning*, Advances in Industrial Control,
https://doi.org/10.1007/978-3-031-28394-9_4

small-gain theorem with ADP. Unlike Jiang and Jiang (2014), Fan and Yang (2015) and Gao et al. (2016), where uncertainties need to be measured or estimated, only the information about upper bound of uncertainty was utilized when addressing the robust control problem for continuous-time affine nonlinear systems using ADP methods in Wu et al. (2014), Wang et al. (2014, 2017), Liu et al. (2015) and Zhao et al. (2020). The key strategy of designing robust controller for nonlinear systems proposed in Wu et al. (2014), Wang et al. (2014, 2017), Liu et al. (2015) and Zhao et al. (2020) is to convert the robust control problem into the optimal control problem of nominal nonlinear systems via selecting appropriate utility functions.

Albeit the significant amount of research results in the field of robust control, it is still an interesting issue to develop robust controllers for nonlinear systems from the perspective of discrete-time sampling. Notice that the above-mentioned results on the robust control (Jiang and Jiang 2014; Petersen 1987; Petersen and Hollot 1986; Lin 2000; Tan et al. 2009; Fan and Yang 2015; Gao et al. 2016; Wu et al. 2014; Wang et al. 2014, 2017; Liu et al. 2015; Zhao et al. 2020) are applicable only for continuous-time linear or nonlinear systems. Since discrete-time controllers have the important advantage that they can be directly implemented in digital form using modern-day embedded hardware (Wang et al. 2019, 2015; Chen and Jagannathan 2008; Zhang et al. 2009), and then, an issue is naturally raised, that is how to directly design robust controllers in discrete-time for systems, especially for nonlinear DT systems. The challenge of solving this issue is posed by the natural difference between DT systems and continuous-time systems, and the nonlinearity characteristic would make it more complicated.

The primary advantage of the proposed method in this chapter is that a kind of simplified HJB equation for the first time developed for finding the robust controller for DT affine nonlinear systems, which is more general in the sense of suitability of both unknown matched and unmatched uncertainties. Moreover, the rigorous proofs of convergence of the proposed interleaved RL algorithm and the UUB stability of the closed-loop systems with bounded uncertainty under the designed controller are presented, which is another contribution to this paper.

The organization of this chapter is given as follows. In Sect. 4.1, the robust control problem and the optimal control problem is first formulated and reviewed, and then two sufficient conditions on the robust stability for matched and unmatched uncertain DT affine nonlinear systems are respectively presented. Section 4.2 presents an interleaved RL algorithm integrated with the neural networks approximation to find the robust control policy in the sense of UUB stability. The rigorous proofs of convergence of the developed algorithm and UUB stability of systems under the learned control policy are also provided in Sect. 4.2. Section 4.3 verifies the effectiveness of the proposed method. Conclusions are stated in Sect. 4.4.

4.1 Robust Controller Design and Simplified HJB Equation

In this section, we first devote to formulating the robust control problem for DT affine nonlinear systems with the presence of arbitrary bounded uncertainty. Moreover, some assumptions and definitions are provided for easily following the focused robust control problem. Then, by reviewing the optimal control problem of nominal systems with an augmented utility function, two sufficient conditions on the robust stability of DT affine nonlinear systems are derived respectively allowing for matched and unmatched uncertainties. Moreover, the simplified HJB equations are presented during these derivations. Consider the DT affine nonlinear system described as

$$x_{k+1} = f(x_k) + g(x_k)u_k + \Delta f(x_k), \tag{4.1}$$

where $x_k \in \mathbb{R}^n$ and $u_k \in \mathbb{R}^m$ are the state and the control input, respectively. $\Delta f(x_k)$ denotes the uncertainty caused by unmodeled dynamics, disturbance or inaccurate system identification with $\Delta f(0) = 0$. Here, one kind of general uncertainties with the form of $\Delta f(x_k) = D(x_k)d(\varphi(x_k))$ is focused for system (4.1), where $D(x_k)$ is a bounded function. If $D(x_k) = g(x_k)$, then $\Delta f(x_k)$ is called the matched uncertainty, otherwise it is viewed as the unmatched uncertainty, where $d(\varphi(x_k))$ is an unknown term bounded by $d^T(\varphi(x_k))d(\varphi(x_k)) \leq h^T(x_k)h(x_k)$, and $h(x_k) \in \mathbb{R}^{m\times n}$ is a known functional matrix.

The following general assumption and definitions are presented:

Assumption 4.1 System (4.1) is drift free, i.e., $f(0) = 0$ and $g(0) = 0$. $f(x_k) \in \mathbb{R}^n$ and $g(x_k) \in \mathbb{R}^{n\times m}$ are assumed to be known a priori and bounded, and $f(x_k) + g(x_k)u_k$ is Lipschitz continuous on the compact set $\Omega \subset \mathbb{R}^n$ containing the origin.

Definition 4.1 (Petersen 1987; Petersen and Hollot 1986; Wang et al. 2015) If there exists a control policy u_k under which system (4.1) is asymptotically stable for all uncertainties $\Delta f(x_k)$, then system (4.1) is called robustly stabilizable and the control policy u_k is called the robust controller.

Definition 4.2 (Fan and Yang 2015; Liu et al. 2015) System (4.1) is called to be UUB at the origin, if there exist a bound M ($M > 0$) and a time step N (N is a positive integer) so that $\|x_k\| \leq M$ on the compact set $x_k \in \Omega$ for all $k \geq N$.

The target of this chapter is to develop a novel RL-based robust adaptive algorithm for DT systems with nonlinear dynamics and bounded unknown uncertainties, such that the systems can be guaranteed to be stable in the sense of UUB stability.

Consider the nominal system with respect to system (4.1) described as

$$x_{k+1} = f(x_k) + g(x_k)u_k. \tag{4.2}$$

Suppose that, for system (4.2), one desires to find the control law u_k^* that minimizes a specific performance index below

$$J(x_k) = \sum_{k=0}^{\infty} r(x_k, u_k) \tag{4.3}$$

where $r(x_k, u_k)$ is called the utility function $r(x_k, u_k) = x_k^T Q x_k + u_k^T u_k + \beta(x_k)$, Q is a positive definite matrix and $\beta(x_k) \geq 0$. Notice that the utility function in (4.3) is augmented with $\beta(x_k)$ compared with the traditional expression of $x_k^T Q x_k + u_k^T R u_k$, where R is a positive definite matrix. Notice that, in our defined utility function $r(x_k, u_k)$, the matrix R is set to be an identity matrix for easily designing the robust controller in the sequels. In general, the optimal control policy u_k^* for minimizing performance index (4.3) with respect to system (4.2) can be derived by solving the HJB equation below (Wang et al. 2019, 2015)

$$V^*(x_k) = U(x_k, u_k^*) + \beta(x_k) + V^*\left(f(x_k) + g(x_k)u_k^*\right) \tag{4.4}$$

where $U(x_k, u_k^*) = {x_k}^T Q x_k + (u_k^*)^T u_k^*$ and the optimal control law is derived as

$$u_k^* = -\frac{1}{2} g^T(x) \nabla V^*(x_{k+1}) \tag{4.5}$$

where $x_{k+1} = f(x_k) + g(x_k)u_k^*$ and $V^*(x_k)$ is called the optimal value function with the definition of

$$V^*(x_k) = \min_u \sum_{j=k}^{\infty} \{U(x_k, u_k) + \beta(x_k)\}. \tag{4.6}$$

Inspired by Liu et al. (2015), Wu et al. (2014) and Wang et al. (2014, 2017, 2015), we are going to find the interior relationship between the robust control of system (4.1) and the optimal control of corresponding nominal system (4.2), such that the robust controller for two kinds of uncertain DT affine nonlinear systems can be found via the optimal control approach. The following theorems are derived and clearly show how to design the robust controller for system (4.1).

Theorem 4.1 *For system (4.2), let Assumption 4.1 holds, then the control policy u_k^* in (4.5) ensure system (4.1) with matched uncertainty $\Delta f(x_k) = g(x_k)d(\varphi(x_k))$ to be robustly stable, and there exists a positive definite and twice continuously differentiable function $V^*(x_k)$ on the compact set Ω, if $\beta(x_k)$ is set to be*

$$\begin{aligned}\beta(x_k) &= h^T(x_k)h(x_k) + h^T(x_k)h(x_k)\lambda_{\max}(g^T(x_k)g(x_k)) \\ &\times \left(1 + \frac{1}{4}\lambda_{\max}((\nabla^2 V^*(x_{k+1}))^T \nabla^2 V^*(x_{k+1}))\right)\end{aligned} \tag{4.7}$$

and the following simplified HJB equation holds:

$$V^*(x_k) = x_k^T Q x_k + (u_k^*)^T u_k^* + \beta(x_k) + V^*(f(x_k) + g(x_k)u_k^*) \tag{4.8}$$

where $\nabla^2 V^*(x)$ stands for the Hessian matrix defined as $\nabla^2 V^*(x) = [\partial^2 V^*/\partial x_{ij}]_{n\times n}$.

Proof From the notation of $\beta(x_k)$, it is not hard to know $\beta(x_k) \geq 0$ since $h^T(x_k)h(x_k)$, $g^T(x_k)g(x_k)$ and $(\nabla^2 V^*(x_k))^T \nabla^2 V^*(x_k)$ are nonnegative. Then, the function $V^*(x_k)$ satisfied with (4.8) is the optimal value function compared with HJB equation (4.4) for the optimization problem of nominal system (4.2).

Now we employ the optimal value function $V^*(x_k)$ satisfied with (4.8) and make the Taylor Series Expansion about the operation point x_{k+1} generated along the trajectory of uncertain system (4.1) under the control policy (4.5). Thus, one has

$$\begin{aligned}V^*(x_{k+1}) - V^*(x_k) &= V^*(f(x_k) + g(x_k)u_k^* + g(x_k)d(\varphi(x_k))) - V^*(x_k)\\ &= V^*(f(x_k) + g(x_k)u_k^*) - V^*(x_k) + (g(x_k)d(\varphi(x_k)))^T \nabla V^*(f(x_k) + g(x_k)u_k^*)\\ &+ (g(x_k)d(\varphi(x_k)))^T \nabla^2 V^*(f(x_k) + g(x_k)u_k^*)g(x_k)d(\varphi(x_k)) \end{aligned} \tag{4.9}$$

where $\nabla V^*(x_k)$ denotes the gradient vector defined as $\nabla V^*(x_k) = \partial V^*/\partial x_k$.

Due to $\Delta f(x) = g(x_k)d(\varphi(x_k))$ and (4.5), utilizing the fact of $a^T b \leq a^T a + \frac{1}{4}b^T b$ (Petersen 1987) (a and b are vectors with appropriate dimensions) yields

$$\begin{aligned}&(g(x_k)d(\varphi(x_k))^T \nabla V^*(f(x_k) + g(x_k)u_k^*) \leq d^T(\varphi(x_k))d(\varphi(x_k))\\ &+ \frac{1}{4}(\nabla V^*(f(x_k) + g(x_k)u_k^*))^T g(x_k)g^T(x_k)\nabla V^*(f(x_k) + g(x_k)u_k^*)\\ &\leq h^T(x_k)h(x_k) + (u_k^*)^T u_k^* \end{aligned} \tag{4.10}$$

and

$$\begin{aligned}&(g(x_k)d(\varphi(x_k)))^T \nabla^2 V^*(f(x_k) + g(x_k)u_k^*)g(x_k)d(\varphi(x_k))\\ &\leq d^T(\varphi(x_k))g^T(x_k)g(x_k)d(\varphi(x_k)) + \frac{1}{4}d^T(\varphi(x_k))g^T(x_k)(\nabla^2 V^*(x_{k+1}))^T\\ &\times \nabla^2 V^*(x_{k+1})g(x_k)d(\varphi(x_k))\\ &\leq (1 + \frac{1}{4}\lambda_{\max}((\nabla^2 V^*(x_{k+1}))^T \nabla^2 V^*(x_{k+1}))\\ &\times h^T(x_k)h(x_k)\lambda_{\max}(g^T(x_k)g(x_k)). \end{aligned} \tag{4.11}$$

By (4.9)–(4.11), one has

$$\begin{aligned}V^*(x_{k+1}) - V^*(x_k) &\leq V^*(f(x_k) + g(x_k)u_k^*) - V^*(x_k) + \beta(x_k) + (u_k^*)^T u_k^*\\ &= -x_k^T Q x_k < 0 \end{aligned} \tag{4.12}$$

which indicates $V^*(x_k)$ can be a Lyapunov function candidate for system (4.1), and the control policy u_k^* in (4.5) can make sure system (4.1) to be stable for all possible realization of uncertainty $\Delta f(x_k) = g(x_k)d(\varphi(x_k))$ according to the Lyapunov stability theory. This completes the proof. □

Remark 4.1 Compared (4.7) and (4.8) with (4.4), one can find that (4.8) could be regarded as a special case of HJB equation (4.4) if letting $\beta(x_k)$ to be (4.7). Moreover, different from GHJB obtained in Chen and Jagannathan (2008), Zhang et al. (2009) and Wang et al. (2015) for DT affine nonlinear systems, there is no term of $\nabla V(x_k)$ any longer in (4.8), resulting in less computational load when solving the Lyapunov function $V^*(x_k)$. In this sense, the derived HJB equation (4.8) is called the simplified HJB equation.

Remark 4.2 It is worth emphasizing that the term $\beta(x_k)$ in (4.8) contains the information of the bound of unknown uncertainty $d(\varphi(x_k))$, which makes it possible to stabilize system (4.1) using the control policy u_k^* in (4.5).

Notice that the higher order terms have been ignored in (4.9) like (Chen and Jagannathan 2008; Zhang et al. 2009; Wang et al. 2015) when expanding the optimal value function $V^*(x_{k+1})$ about the operation point x_{k+1} ($x_{k+1} = f(x_k) + g(x_k)u_k^*$) using the Taylor Series Expansion. The error caused by the ignorance of higher order terms can be expressed by $\varepsilon(\|g(x_k)d(\varphi(x_k)\|^2)$, where ε is called infinitesimal with respect to $\|g(x_k)d(\varphi(x_k)\|^2$. That means when $\|g(x_k)d(\varphi(x_k)\|$ is small, then the error is an infinitesimal of a higher order than $\|g(x_k)d(\varphi(x_k)\|^2$. Moreover, if the Lyapunov function $V^*(x_k)$ is quadratic in x_k, which had happened in many cases (Chen and Jagannathan 2008; Zhang et al. 2009; Wang et al. 2015), then (4.9) together with the robust controllers in (4.5) hold without any error. In addition, considering the higher order terms in the Taylor Series Expansion of $V^*(x_k)$ might improve the approximation accuracy in the cost of increasing the computational load, so a tradeoff between accuracy and computational complexity should be made from practical realization perspective.

Now we are in the position to extend the result of Theorem 4.1 into the robust stability of system (4.1) with unmatched uncertainty.

Theorem 4.2 *For system (4.2), let Assumption 4.1 holds, if $\gamma(x_k)$ is set to be*

$$\gamma(x_k) = h^T(x_k)h(x_k)\lambda_{\max}(D^T(x_k)D(x_k))(1 + \frac{1}{4}\lambda_{\max}((\nabla^2 V(x_{k+1}))^T \nabla^2 V(x_{k+1})) + h^T(x_k)h(x_k) \tag{4.13}$$

and there exists a positive definite and twice continuously differentiable function $V(x_k)$ on Ω satisfying the following simplified HJB equation:

$$V(x_k) = x_k^T Q x_k + u_k^T u_k + \gamma(x_k) + V(x_{k+1}) \tag{4.14}$$

then the control policy u_k in (4.15) ensures system (4.1) with unmatched uncertainty $\Delta f(x_k) = D(x_k)d(\varphi(x_k))$ to be robustly stable,

$$u_k = -\frac{1}{2}D^T(x)\nabla V(x_{k+1}). \tag{4.15}$$

Proof For nominal system (4.2), due to (4.14), one has

$$V(x_{k+1}) - V(x_k) = -x_k^T Qx_k - u_k^T u_k - \gamma(x_k) < 0 \tag{4.16}$$

then the control policy u_k in (4.15) is stabilizable for system (4.2). Under the stabilizing control policy u_k in (4.15) and due to $\gamma(x_k) \geq 0$ from (4.13), we can define a cost function below

$$J(x_k, u_k) = \sum_{i=k}^{\infty} \left(x_i^T Qx_i + u_i^T u_i + \gamma(x_i)\right). \tag{4.17}$$

Further, similarly to Wu et al. (2014), one has

$$J(x_{k+1}, u_{k+1}) - J(x_k, u_k) = -x_k^T Qx_k - u_k^T u_k - \gamma(x_k). \tag{4.18}$$

Subtracting both sides of (4.18) from (4.16) yields

$$V(x_{k+1}) - V(x_k) = J(x_{k+1}, u_{k+1}) - J(x_k, u_k). \tag{4.19}$$

If $V(x_0) = J(x_0, u_0)$, then one has $V(x_k) = J(x_k, u_k)$ from (4.19).

Now we start to prove that the control policy u_k in (4.15) can guarantee the stability of uncertain system (4.1). The function $V(x_k)$ derived from (4.14) is considered as a Lyapunov functional candidate and we make Taylor Series Expansion along the trajectory of system (4.1) at the operation point x_{k+1} similar to the steps (4.9)–(4.11) in the proof of Theorem 4.1. Moreover, due to $-\frac{1}{2}D^T(x)\nabla V(x_{k+1}) = u_k$, one has

$$\begin{aligned} V(x_{k+1}) - V(x_k) &\leq V(f(x_k) + g(x_k)u_k) - V(x_k) + \gamma(x_k) + u_k^T u_k \\ &= -x_k^T Qx_k < 0 \end{aligned} \tag{4.20}$$

which indicates system (4.1) is stable for all possible realization of uncertainty limited within the bound. This completes the proof. □

Remark 4.3 It is generally known that finding the Lyapunov function for complex nonlinear systems, especially for uncertain nonlinear systems, is quite difficult. The proofs of Theorems 4.1 and 4.2 have implied that the functions $V^*(x_k)$ and $V(x_k)$ could be Lyapunov functional candidates respectively for matched and unmatched DT affine nonlinear systems, and they can be obtained by solving the simplified HJB equations.

Remark 4.4 Unlike (Liu et al. 2015; Wang et al. 2014, 2017; Zhao et al. 2020) that investigated robust stability of continuous-time affine nonlinear systems, two sufficient conditions for robust controller design of DT affine nonlinear systems are presented. In contrast to Wang et al. (2015), where GHJB equation was employed for solving the robust controller for uncertain DT affine nonlinear systems, there are two differences: (a) a kind of simplified version of GHJB equation, named as the simplified HJB equation in this chapter, is derived for designing the robust controllers; and (b) the robust stabilization of systems with matched uncertainty was investigated in Wang et al. (2015), which cannot work for the case of unmatched uncertain systems. It is worth pointing out that the robust stability conditions of DT nonlinear systems with matched and unmatched uncertainty systems are presented in Theorems 4.1 and 4.2 in this chapter.

If $D(x_k) = g(x_k)$ in uncertainty $\Delta f(x_k)$, then the conclusion of Theorem 4.2 becomes Theorem 4.1. Thus, without loss of generality, we shall investigate how to solve simplified HJB equation (4.14) in Theorem 4.2, such that the robust controller can be obtained. Indeed, it is still hard to get the analytical solution to simplified HJB equation (4.14), since the value function $V(x_k)$ might be nonlinear for the focused nonlinear systems. In the sequel, a numerical solution to simplified HJB equation (4.14) is going to be found by presenting an interleaved RL algorithm combined with the NNs approximation.

4.2 Interleaved RL for Approximating Robust Control

In this section, we devote to developing an interleaved RL algorithm for approximating the robust controller, such that the trajectory of DT affine nonlinear system (4.1) is UUB.

It is well known that the smooth value function $V(x_k)$ and the control policy u_k could be respectively expressed using NNs with errors based on Value Function Approximation (Li et al. 2017) as

$$V(x_k) = (w_c^*)^T \sigma(v_c^T x_k) + \varepsilon_c(x_k) \tag{4.21}$$

and

$$u_k = (w_a^*)^T \sigma(v_a^T x_k) + \varepsilon_a(x_k) \tag{4.22}$$

where $w_c^* \in \mathbb{R}^{N_0}$, $v_c \in \mathbb{R}^{n \times N_0}$, $w_a^* \in \mathbb{R}^{N_1 \times m}$ and $v_a \in \mathbb{R}^{n \times N_1}$ are the weight vectors, $\sigma(\cdot)$ is the NNs activation function and often selected to be $\sigma(\cdot) = \tanh(\cdot)$ for nonlinear functions (Li et al. 2019; Liu et al. 2015). $\varepsilon_c(x_k)$ and $\varepsilon_a(x_k)$ denote the errors generated by the estimation.

Since the interleaved RL has a capability of trading off the convergence speed of algorithm and control updates for systems (Liu et al. 2015; Jiang et al. 2017), then unlike the standard PI and value iteration (VI) RL, here the value function $V(x_k)$ and the control policy u_k respectively defined in (4.21) and (4.22) are alternatively iterated enough number of times at time step k, such that we can get arbitrarily close to the solution of the simplified HJB equation. To be specific, at each iteration i, the critic NN is represented as (Li et al. 2017)

$$\hat{V}^i(x_k) = w_{ci}^T(k)\sigma(v_c^T x_k). \tag{4.23}$$

Thus, one has

$$\hat{V}^i(x_{k+1}) = w_{ci}^T(k)\sigma(v_c^T x_{k+1}), \tag{4.24}$$

and the actor NN u_k^i, at each iteration i, is given by

$$\hat{u}_k^i = w_{ai}^T(k)\sigma(v_a^T x_k) \tag{4.25}$$

where $w_{ci}(k)$ and $w_{ai}(k)$ respectively are the approximations of w_c^* and w_a^*. Here the weights v_c^T and v_a^T are chosen as constant vectors. Training the weight $w_{ci}(k)$ for the critic NN is implemented using the gradient descent algorithm by minimizing the approximation error in terms of (4.14) like (Wang et al. 2017; Zhang et al. 2009; Li et al. 2017)

$$e_{ci}(k) = \rho^i(x_k) + \hat{V}^i(x_{k+1}) - \hat{V}^i(x_k) \tag{4.26}$$

where

$$\begin{aligned}
\rho^i(x_k) &= x_k^T Q x_k + (u_k^i)^T u_k^i + \gamma^i(x_k)\\
\gamma^i(x_k) &= h^T(x_k)h(x_k) + h^T(x_k)h(x_k)\lambda_{\max}(D^T(x_k)D(x_k))\\
&\quad \times (1 + \frac{1}{4}\lambda_{\max}((\nabla^2\hat{V}^i(x_{k+1}))^T \nabla^2\hat{V}^i(x_{k+1}))).
\end{aligned}$$

Thus, the training process of weight $w_{ci}(k)$ with the learning rate l_c can be exactly expressed as

$$w_{c(i+1)}(k) = w_{ci}(k) - l_c\sigma_1(k)e_{ci}^T(k) \tag{4.27}$$

where $\sigma_1(k) = \sigma(v_c^T x_k) - \sigma(v_c^T x_{k+1})$. Similarly, the weight $w_{ai}(k)$ can be trained in the form of

$$w_{a(i+1)}(k) = w_{ai}(k) - l_a\sigma(v_a^T x_k)e_{ai}^T(k) \tag{4.28}$$

where l_a denotes the learning rate and the approximate error $e_{ai}(k)$ is defined as

$$\begin{aligned}
e_{ai}(k) &= \hat{u}_k^i - u_k^i\\
&= w_{ai}^T(k)\sigma(v_a^T x_k) + \frac{1}{2}D^T(x_k)(\nabla\sigma(v_c^T x_{k+1}))^T w_{ci}(k)
\end{aligned} \tag{4.29}$$

where x_{k+1} is generated by using $\hat{u}_{k-1}$ estimated by the last time step $k-1$. The initial weights of critic NN and actor NN are usually set to be random small values that can be adjusted after each training process by the interleaved RL, such that these weights can move in the direction of improving performance. Moreover, if properly increasing the number of neurons of the critic NN and the actor NN and choosing appropriate learning rates, then the estimation errors of NNs can be as small as possible (Lewis and Vrabie 2009; Liu et al. 2015). Now interleaving single-step updates of the weight $w_{ci}(k)$ and the weight $w_{ai}(k)$ at time step k will be employed, thus the following interleaved RL algorithm, Algorithm 4.1, is developed.

Algorithm 4.1 Interleaved RL for uncertain systems

1: Initiation: Set the time step $k = 1$ and the iteration index $i = 0$. Given initial states x_1 and x_2;
2: **Interleaved iteration**
  while do
    Given the weights v_c and v_a, the weights $w_{ci}(k)$ and $w_{ai}(k)$ for the critic NN and the actor NN are initialized, and set the learning rates l_c and l_a;
    Update the weight $w_{ci}(k)$ in terms of (4.27);
    Update the weight $w_{ai}(k)$ in terms of (4.28);
    if $\left\| \hat{V}^{i+1}(x_k) - \hat{V}^i(x_k) \right\| \le \varepsilon$ && $\left\| \hat{u}_k^{i+1} - \hat{u}_k^i \right\| \le \varepsilon$ $(\varepsilon > 0)$ **then**
      break
    else
      $i = i + 1$;
    end if
  end while;
3: Set $w_c(k) = w_{ci}(k)$ and $w_a(k) = w_{ai}(k)$. One can get $\hat{u}_k = w_a^T(k)\sigma(v_a^T x_k)$ which is applied into nominal system (4.2) as the input to generate data x_{k+1} and x_{k+2};
4: **if** $\|w_a(k) - w_a(k-1)\| \le \varepsilon$ && $\|w_c(k) - w_c(k-1)\| \le \varepsilon$ **then**
    go to Step 5;
  else
    set $k = k + 1$ and go back to Step 2;
  end if;
5: Obtain the approximate values $w_c(k)$ and $w_a(k)$ of w_c^* and w_a^*, such that the approximate value $\hat{u}_k = w_a^T(k)\sigma(v_a^T x_k)$ of u_k can be learned.

Remark 4.5 In Algorithm 4.1, the weights $w_{ci}(k)$ and $w_{ai}(k)$ will be kept training in an interleaved way with the iteration index i until convergence at each time step k, and at that time the control action can be obtained and acts system (4.2) for generating new data of next time. In this sense, Algorithm 4.1 is indeed an interleaved RL rather than PI or VI RL. Moreover, from (4.24), (4.26) and (4.27), one can find that it is not hard to implement the training process of weight of critic NN via (4.27) by increasing the iteration index i at each time step k.

Remark 4.6 Note that estimating the control policy u_k using the actor NN can successfully overcome the difficulty caused by the unavailability of the future state x_{k+1} at time step k shown in (4.5) and (4.15).

Remark 4.7 After implementing Algorithm 4.1, the learned approximate value $\hat{u}_k$ of the robust control policy u_k will be adopted and as the control action of system (4.1) to tolerate the uncertainties. To eliminate the negative effect on performance and even stability of systems caused by the dead-zone, backlash, saturation and actuator nonlinearities that are very common in most practical industries, a nonquadratic functional can be added in the utility function in (4.3) in the similar way as Liu et al. (2015) and Zhang et al. (2009), such that these nonlinear control constraints can be dealt with under the ADP framework even for uncertain system (4.1).

The schematic of Algorithm 4.1 can be seen in Fig. 4.1. Actually, there are two differences in Algorithm 4.1 proposed in this paper from the interleaved RL

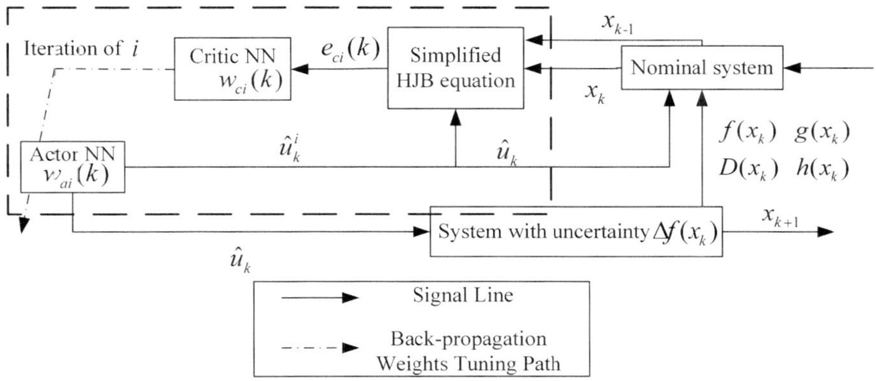

Fig. 4.1 Schematic of the interleaved RL algorithm

algorithms in Li et al. (2019) and Jiang et al. (2017). One is that the controller is learned by Algorithm 4.1 for uncertain systems rather than the optimal controllers for nominal systems which are concerned in Li et al. (2019) and Jiang et al. (2017), then a question will be raised, that is, whether the approximated controller we learned here can stabilize system (4.1) for all uncertainties within the bound or not in the sense of UUB stability. The other one is that the simplified HJB equations are presented in this paper for designing the robust controller of nonlinear DT systems. Although the convergence of the interleaved RL algorithms in Li et al. (2019) and Jiang et al. (2017) has been analyzed, the differences from Li et al. (2019) and Jiang et al. (2017) drive us to prove the convergence of Algorithm 4.1 proposed in this chapter. The next section will deal with the above two issues.

4.2.1 *Theoretical Analysis*

In this subsection, the convergence of Algorithm 4.1 and the UUB stability of uncertain systems are fully analyzed.

Suppose that the real value of iterative value function $V^i(x_k)$ can be exactly represented as

$$V^i(x_k) = (w_{ci}^*(k))^T \sigma(v_c^T x_k) + \varepsilon_{ci}(x_k) \tag{4.30}$$

where $w_{ci}^*(k)$ is the weight vector and $\varepsilon_{ci}(x_k)$ stands for the reconstruction error. The real value of the actor NN u_k^i should have the exact expression below

$$u_k^i = w_{ai}^*(k)\sigma(v_a^T x_k) + \varepsilon_{ai}(x_k) \tag{4.31}$$

where $w_{ai}^*(k)$ is the weight vector and $\varepsilon_{ai}(x_k)$ stands for the reconstruction error.

Another assumption will be made, which is a common technique and has been used in Liu et al. (2015) and Wang et al. (2015). Here we have to point out that one can assume that $\nabla V(x_k)$ and $\nabla V^2(x_k)$ could be bounded on the compact set Ω, since we suppose $V(x_k)$ is twice continuously differential on Ω, and it actually is the value function in (4.17) under the stabilizing control policy u_k in (4.15). In addition, the research results have demonstrated that functions can be approximated as accurately as possible if the proper number of neurons, learning rate and the initial weights for NNs can be selected (Lewis and Vrabie 2009; Liu et al. 2015), which implies $\varepsilon_a(x_k)$, $\varepsilon_c(x_k)$, $\varepsilon_{ci}(x_k)$ and $\varepsilon_{ai}(x_k)$ could be bounded.

Assumption 4.2 The terms $\sigma(v_c^T x_k)$, $\varepsilon_a(x_k)$, $\varepsilon_c(x_k)$, $\varepsilon_{ci}(x_k)$, $\varepsilon_{ai}(x_k)$, $\nabla V(x_k)$ and $\nabla V^2(x_k)$ are bounded on the compact set $\Omega \subset \mathbb{R}^n$.

The following theorem is presented to show the weight $w_c(k)$ and $\hat{u}_k$ learned by Algorithm 4.1 are bounded.

Theorem 4.3 *For system (4.2), the weight $w_{ci}(k)$ is updated in terms of (4.27) and the approximate control policy $\hat{u}_k^i$ is obtained by (4.25) in Algorithm 4.1. Then, when i and k are large enough, there exist $\lambda_w > 0$ and $\lambda_u > 0$ satisfying*

$$\left\|w_c(k) - w_c^*\right\| \leq \lambda_w, \ \left\|\hat{u}_k - u_k\right\| \leq \lambda_u. \tag{4.32}$$

Proof Using the similar way to Li et al. (2017, 2019), one can prove that there exist $\varepsilon_w > 0$ and $\varepsilon_u > 0$, such that $\|w_c(k) - w_{ci}(k)\| \leq \varepsilon_w$ and $\|\hat{u}_k - \hat{u}_k^i\| \leq \varepsilon_u$, if the iteration index i is large enough. The existing PI-based ADP results (Gao et al. 2016; Wang et al. 2014) have proven that when the enough numbers of neurons N_0 and N_1 and approximate learning rates are chosen, one has $\|w_{ci}^*(k) - w_c^*\| \leq \tilde{\varepsilon}_c$ and $\|u_k^i - u_k\| \leq \tilde{\varepsilon}_a$ for small $\tilde{\varepsilon}_c$ and $\tilde{\varepsilon}_a$ ($\tilde{\varepsilon}_c > 0$, $\tilde{\varepsilon}_a > 0$) if the iteration index i is large enough. Moreover, we have

$$\begin{aligned}\left\|w_c(k) - w_c^*\right\| &= \|w_c(k) - w_{ci}(k) + w_{ci}(k) - w_{ci}^*(k) + w_{ci}^*(k) - w_c^*\| \\ &\leq \|w_c(k) - w_{ci}(k)\| + \left\|(w_{ci}(k) - w_{ci}^*(k))\right\| + \left\|(w_{ci}^*(k) - w_c^*)\right\|.\end{aligned} \tag{4.33}$$

Based on (4.23), (4.24), (4.26) and (4.30), one can get

$$(w_{ci}(k) - w_{ci}^*(k))^T(\sigma(v_c^T x_k) - \sigma(v_c^T x_{k+1})) = \varepsilon_{ci}(x_k). \tag{4.34}$$

It follows

$$\left\|(w_{ci}(k) - w_{ci}^*(k))\right\| \leq \tilde{\varepsilon}_w \tag{4.35}$$

where $\tilde{\varepsilon}_w = \|\varepsilon_{ci}(x_k)/\sigma_1(k)\|$. Then, under Assumption 4.2 and based on (4.32)–(4.34), it naturally follows that there exists $\lambda_w = \varepsilon_w + \tilde{\varepsilon}_w + \tilde{\varepsilon}_c$ satisfying

$$\left\|w_{ci}(k) - w_c^*\right\| \leq \lambda_w. \tag{4.36}$$

In the similar way, one has

$$\|\hat{u}_k - u_k\| = \|\hat{u}_k - \hat{u}_k^i + \hat{u}_k^i - u_k^i + u_k^i - u_k\|. \tag{4.37}$$

Based on (4.25) and (4.31), one has $\left\|\hat{u}_k^i - u_k^i\right\| \leq \|\varepsilon_{ai}(x_k)\|$. Since $\left\|u_k^i - u_k\right\| \leq \tilde{\varepsilon}_a$ holds, based on (4.37), then there exists $\lambda_u = \varepsilon_u + \|\varepsilon_{ai}(x_k)\| + \tilde{\varepsilon}_a$ satisfying

$$\left\|\hat{u}_k - u_k\right\| \leq \lambda_u. \tag{4.38}$$

This completes the proof. □

From Theorem 4.3, a control policy $\hat{u}_k$ and an approximate value function $\hat{V}(x_k)$ can be obtained after the iterations of i and k by implementing Algorithm 4.1. It has to point out that the learned control policy $\hat{u}_k$ is an approximated value of u_k and the bound λ_u depends on the number of neurons, learning rate and the initial weights for the critic and actor NNs. In the next theorem, we will prove that $\hat{u}_k$ can guarantee the UUB stability of uncertain system (4.1).

Theorem 4.4 *Under the learned control policy $\hat{u}_k$ by Algorithm 4.1, the dynamics of the uncertain closed-loop system (4.1) is UUB.*

Proof The dynamics of system (4.1) under the learned control policy $\hat{u}_k$ becomes

$$x_{k+1} = f(x_k) + g(x_k)\hat{u}_k + \Delta f(x_k). \tag{4.39}$$

We choose the solution $V(x_k)$ of simplified HJB equation (4.14) as a Lyapunov function candidate, and along the trajectory of (4.39), we have

$$V(x_{k+1}) - V(x_k) = V(f(x_k) + g(x_k)\hat{u}_k + \Delta f(x_k)) - V(x_k). \tag{4.40}$$

Due to (4.38), one has $\hat{u}_k - u_k = \hat{\varepsilon}_u$ ($\left\|\hat{\varepsilon}_u\right\| \leq \lambda_u$). Further, (4.40) is rewritten as using the Taylor Series Expansion

$$\begin{aligned} V(x_{k+1}) - V(x_k) &= V(f(x_k) + g(x_k)u_k + g(x_k)\hat{\varepsilon}_u + \Delta f(x_k)) - V(x_k) \\ &= V(w_{k+1}) - V(x_k) + (g(x_k)\hat{\varepsilon}_u + \Delta f(x_k))^T \nabla V(w_{k+1}) \\ &+ (g(x_k)\hat{\varepsilon}_u + \Delta f(x_k))^T \nabla^2 V(w_{k+1})(g(x_k)\hat{\varepsilon}_u + \Delta f(x_k)) \end{aligned} \tag{4.41}$$

where $w_{k+1} = f(x_k) + g(x_k)u_k$. Similarly to (4.10) and (4.11), one has

$$\Delta f(x_k)^T \nabla V(w_{k+1}) \leq h^T(x_k)h(x_k) + (u_k)^T u_k \tag{4.42}$$

and

$$\Delta f(x_k)^T \nabla^2 V(w_{k+1}) \Delta f(x_k) \leq \gamma(x_k) - h^T(x_k)h^T(x_k). \tag{4.43}$$

Utilizing simplified HJB equation (4.14), (4.42) and (4.43), then (4.41) becomes

$$\begin{aligned} V(x_{k+1}) - V(x_k) &\leq -x_k^T Q x_k + (g(x_k)\hat{\varepsilon}_u)^T \nabla V(w_{k+1}) \\ &+ (g(x_k)\hat{\varepsilon}_u)^T \nabla V^2(w_{k+1}) g(x_k)\hat{\varepsilon}_u + 2(g(x_k)\hat{\varepsilon}_u)^T \nabla^2 V(w_{k+1}) \Delta f(x_k) \end{aligned} \tag{4.44}$$

and using the similar way to (4.10), it follows

$$\begin{aligned} 2(g(x_k)\hat{\varepsilon}_u)^T \nabla^2 V(w_{k+1}) \Delta f(x_k) &\leq h^T(x_k) h(x_k) \lambda_{\max}(D^T(x_k) D(x_k)) \\ &+ \left\| g(x_k)\hat{\varepsilon}_u \right\|^2 \lambda_{\max}(\nabla^2 V(w_{k+1})(\nabla^2 V(w_{k+1}))^T). \end{aligned} \tag{4.45}$$

Due to Assumptions 4.1, 4.2 and the boundedness of $D(x_k)$, the following form must be bounded on the compact set Ω. Thus, there must exist a positive number κ satisfying

$$\begin{aligned} h^T(x) h(x) \lambda_{\max}(D^T(x_k) D(x_k)) &+ \left\| g(x_k)\hat{\varepsilon}_u \right\|^2 \lambda_{\max}(\nabla^2 V(w_{k+1})(\nabla^2 V(w_{k+1}))^T) \\ &+ (g(x_k)\hat{\varepsilon}_u)^T \nabla V(w_{k+1}) + (g(x_k)\hat{\varepsilon}_u)^T \nabla V^2(w_{k+1}) g(x_k)\hat{\varepsilon}_u \leq \kappa \end{aligned} \tag{4.46}$$

then, one has

$$V(x_{k+1}) - V(x_k) \leq -x_k^T Q x_k + \kappa \leq -(\lambda_{\min}(Q)\|x_k\|^2 - \kappa). \tag{4.47}$$

Thus, if $\|x_k\| > \sqrt{\frac{\kappa}{\lambda_{\min}(Q)}}$, then $V(x_{k+1}) - V(x_k) < 0$. This shows the control policy $\hat{u}_k$ learned by Algorithm 4.1 can guarantee the trajectory of system (4.1) to be UUB. This completes the proof. □

4.3 Illustrative Examples

In this section, the effectiveness and efficacy of the proposed interleaved RL algorithm are illustrated in three representative examples.

Example 4.1 In our first example, we examine the performance of the developed Algorithm 4.1 by using a torsional pendulum system subject to matched uncertainty. The dynamics of the torsional pendulum is described below (Wei et al. 2014)

$$\begin{cases} \frac{d\theta}{dt} = \omega \\ J\frac{d\omega}{dt} = u - Mgl\sin\theta - f_d\frac{d\theta}{dt} \end{cases} \tag{4.48}$$

where the current angle θ and the angular velocity ω are viewed as the system states. The parameters J, M, l, f_d respectively denote the rotary inertia, the mass, length of the pendulum bar and frictional factor, and they are set to be $J = 4/3Ml^2$, $M = 1/3\,\text{kg}$, $l = 2/3\,\text{m}$ and $f_d = 2$, respectively. $g = 9.8\,\text{m/s}^2$ is the gravity. Using Euler and trapezoidal methods (Wei et al. 2014), System (4.48) can be discretized as the following form if the sampling period is chosen as $\Delta t = 0.1\,\text{s}$.

$$x_{k+1} = f(x_k) + g(x_k)(u_k + p\sin(x_1(k)) + x_2(k)) \tag{4.49}$$

where $p \in [-10, 10]$ is an unknown bounded parameter,

$$f(x_k) = \begin{bmatrix} x_1(k) + 0.1x_2(k) \\ -0.49\sin(x_1(k)) + 0.98x_2(k) \end{bmatrix}, \; g(x_k) = \begin{bmatrix} 0 \\ 0.1 \end{bmatrix}.$$

System (4.48) indeed has matched uncertainty caused by the input disturbance and measurement error represented by $p\sin(x_1(k)) + x_2(k)$. Observing system (4.48), one can find that $d(\varphi(x_k) = p\sin(x_1(k)) + x_2(k)$ that indicates $h^T(x_k)h(x_k) = 10^2 x^T(k)x(k)$.

Before implementing Algorithm 4.1, we set the structures of the critic and actor NNs to be 2-8-1 and 2-2-1, respectively. Choose $Q = \text{diag}(1, 1)$ and the NNs activation function $\sigma(\cdot) = \tanh(\cdot)$. Let the learning rates of the critic and action networks respectively be 0.1 and 0.3. In this example, we find that the satisfactory results can be obtained when selecting eight neuros and two neuros in the hidden layers for the critic NN and the actor NN during computer simulations.

The weights from the input layer to the hidden layer of the critic NN and the actor NN are respectively chosen as the following vector and an unit vector:

$$v_c = \begin{bmatrix} 1.0690 & -2.8982 & -3.1954 & -0.2263 & -1.5441 & -1.2247 & -4.5784 & -2.9014 \\ -0.9853 & -0.8763 & -0.7539 & -0.8104 & -1.7055 & -0.1988 & -0.8437 & -0.2369 \end{bmatrix}.$$

The initial weights from the hidden layer to output layer for the critic NN and the actor NN are respectively chosen below

$$w_c^T = \begin{bmatrix} 0.1703 & -0.1050 & -1.1158 & -0.3483 & 0.0425 & 0.0429 & 0.0292 & -0.1289 \end{bmatrix}$$
$$w_a^T = \begin{bmatrix} -0.4140 & -0.1992 \end{bmatrix}.$$

The probing noise is chosen as $e_k = 0.2\text{rand}(1, 1)$ for satisfying the PE condition. Implementing Algorithm 4.1 yields the training results of weights of the critic and actor NNs in Fig. 4.2a, b, which indicate the convergence of these weights. For verifying the efficacy of the learned control policy, we suppose $p = 10$. Choose the initial state $x_0 = [0.3 \; -0.3]^T$, Fig. 4.3a shows the system state trajectories under the learned control policy shown in Fig. 4.3b. This shows the control policy learned by Algorithm 4.1 can guarantee system (4.48) to be stable even though there exists uncertainty in the system.

Example 4.2 Now we consider another affine nonlinear system with matched uncertainty

$$\begin{bmatrix} x_1(k+1) \\ x_2(k+1) \end{bmatrix} = \begin{bmatrix} -0.5x_1(k) + 0.2x_2(k) \\ \sin(x_1(k)) + 0.3x_2(k) \end{bmatrix} + \begin{bmatrix} 0 \\ 0.1 \end{bmatrix} (u_k + p\sin^2(x_1(k))x_2(k)) \tag{4.50}$$

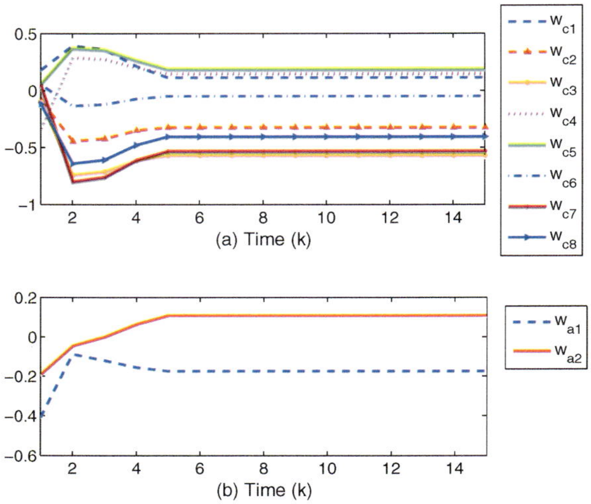

Fig. 4.2 Evolution of weights of the critic NN and the actor NN

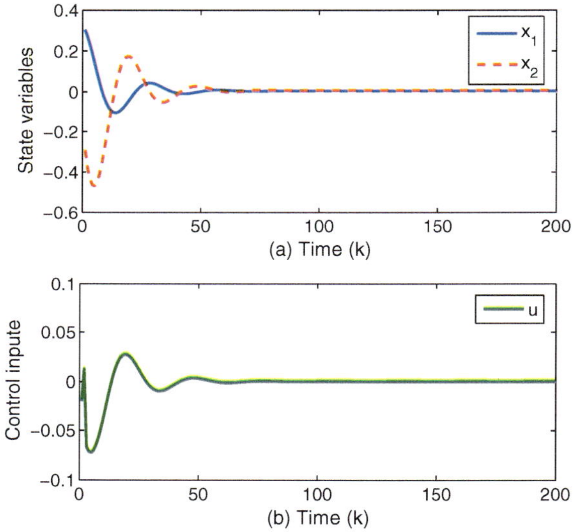

Fig. 4.3 States of system with $p = 10$ under the learned control policy

where $p \in [-10, 10]$ is an unknown bounded parameter. Observing system (4.49), we know $d(\varphi(x_k) = p\sin^2(x_1(k))x_2(k)$ indicating $h^T(x_k)h(x_k) = 10^2 x^T(k)x(k)$.

Like Example 4.1, we let the structures of the critic and actor NNs be 2-8-1 and 2-2-1, respectively. Choose $Q = \text{diag}(1, 1)$ and the NNs activation function $\sigma(\cdot) = \tanh(\cdot)$. The learning rates of the critic and action networks are set to be 0.1 and 0.3, respectively.

The weights from the input layer to the hidden layer of the critic NN and the actor NN are respectively chosen as the following vector and an unit vector.

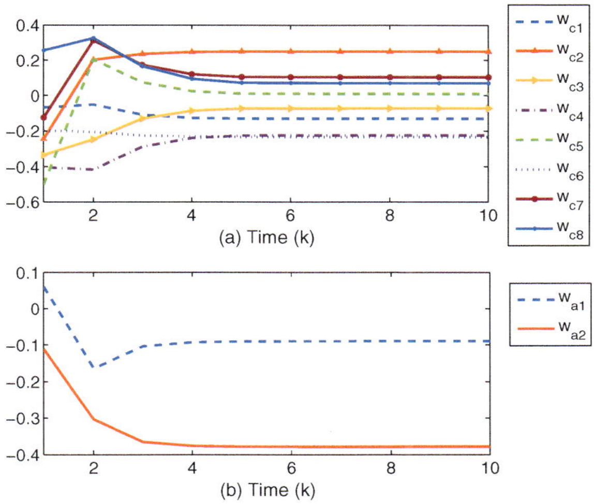

Fig. 4.4 Convergence of weights of the critic NN and the actor NN

$$v_c = \begin{bmatrix} -1.4842 & 0.2545 & -1.9152 & -0.9952 & 0.3234 & -0.9286 & -0.5006 & 5.0454 \\ -2.9931 & 1.1506 & 1.8105 & 3.2322 & -4.0321 & -1.4386 & -5.3543 & -1.9947 \end{bmatrix}.$$

The initial weights from the hidden layer to output layer for the critic NN and the actor NN are respectively chosen below

$$w_c^T = \begin{bmatrix} -0.0675 & -0.2452 & -0.3355 & -0.4031 & -0.5023 & -0.1985 & -0.1247 & 0.2559 \end{bmatrix}$$
$$w_a^T = \begin{bmatrix} 0.0605 & -0.1115 \end{bmatrix}.$$

The probing noise is chosen as $e_k = 0.2\text{rand}(1, 1)$ for satisfying the PE condition. Implementing Algorithm 4.1, Fig. 4.4a, b plot the training results of weights of the critic and actor NNs. For verifying the efficacy of the learned control policy, we suppose $p = 5$ from time step 0 to time step 20, $p = (-1)^k 10$ from time step 21 to time step 40 and $p = 5$ for other time steps. Figure 4.5a shows the evolution of system (4.49) from the initial state $x_0 = [0.3 \; -0.3]^T$ under the learned control policy shown in Fig. 4.5b. Figure 4.5 shows that the learned control policy can make the uncertain system stable.

Example 4.3 We shall show the efficacy of the developed algorithm using the following affine nonlinear system with unmatched uncertainty:

$$\begin{aligned} \begin{bmatrix} x_1(k+1) \\ x_2(k+1) \end{bmatrix} &= \begin{bmatrix} -\sin(0.5x_2(k)) \\ -\cos(1.4x_2(k))\sin(0.9x_1(k)) \end{bmatrix} \\ &+ \begin{bmatrix} 0 \\ 1 \end{bmatrix} u_k + \begin{bmatrix} 1 \\ (x_1(k))^2 \end{bmatrix} p\sin^2(x_1(k))x_2(k)) \end{aligned} \tag{4.51}$$

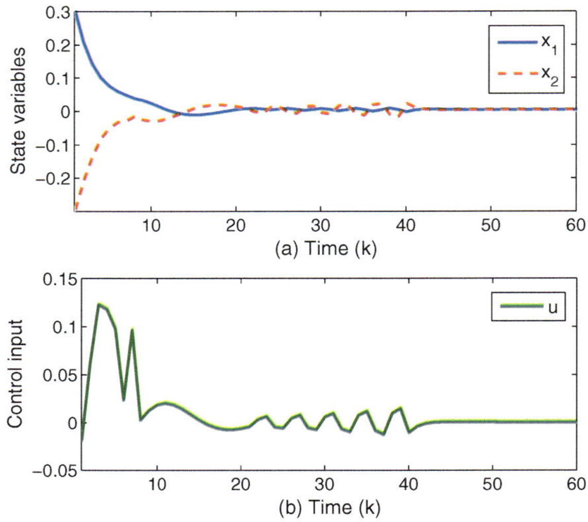

Fig. 4.5 States of system under the learned control policy

where $p \in [-0.5,\ 0.5]$ is an unknown bounded parameter. Observing system (4.50), one has

$$D(x_k) = \begin{bmatrix} 1 \\ (x_1(k))^2 \end{bmatrix},\ d(\varphi(x_k)) = p\sin(x_1(k))\sin(x_2(k)).$$

Then, one has $d^T(x_k)d(x_k) \leq 0.25\ \|x_k\|^2$. The NNs structure of critic and actor are respectively set to be 3-10-1 and 2-2-1, and the learning rates of these two networks are chosen as 0.5 and 0.1, respectively, by computer simulations. The probing noise is chosen as $e_k = 0.2\text{rand}(1, 1)$ for satisfying the PE condition. The weights from the input layer to the hidden layer of the critic NN and the actor NN are respectively chosen as the following vector and an unit vector:

$$v_c = \begin{bmatrix} -0.1639 & 0.0357 & -0.0725 & -0.0036 & 0.0306 \\ -0.0170 & -0.0357 & -0.1965 & -0.2281 & -0.0820 \end{bmatrix.$$
$$\begin{matrix} -0.0161 & -0.2250 & -0.1256 & 0.0633 & 0.0865 \\ -0.0458 & -0.1797 & -0.0677 & 0.1502 & -0.2402 \end{matrix}\Bigg].$$

The initial weights from the hidden layer to output layer for the critic NN and the actor NN are respectively chosen below

$$\begin{aligned} w_c^T &= \big[-0.0413\ 0.0596\ -0.0892\ -0.1391\ 0.1073 \\ &\quad 0.0304\ -0.0068\ -0.0472\ 0.0308\ -0.0907 \big] \\ w_a^T &= \big[-0.0259\ -0.0802 \big]. \end{aligned}$$

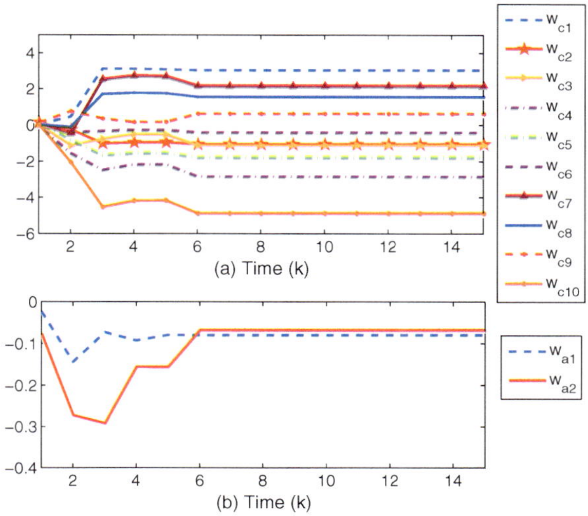

Fig. 4.6 Evolution of weights of the critic NN

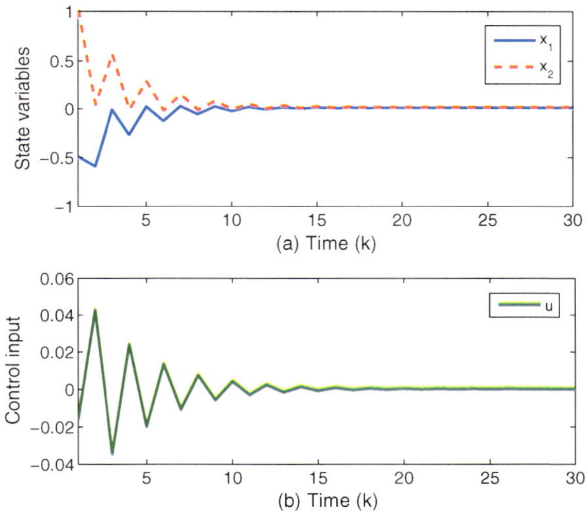

Fig. 4.7 States of system under the learned control policy

Let the initial state vector $x_0 = [-0.5\ 1]^T$ and the initial control $u^0 = -0.05$, implementing Algorithm 4.1 yields the weights from the hidden layer to output layer of the critic NN and the actor NN, which are plotted in Fig. 4.6a, b. Figure 4.7a shows the state trajectory of system with time-varying parameter $p \in [-0.5, 0.5]$ under the control policy learned by Algorithm 4.1 as given in Fig. 4.7b. The satisfactory performance can be seen in Fig. 4.7 by using the proposed robust controller design method.

4.4 Conclusion

In this section, the simplified HJB equations are presented by integrating dynamic programming, Lyapunov theory and Taylor Series Expansion, such that solving it renders the controllers that guarantee the UUB stability of affine nonlinear DT systems subject to unknown matched and unmatched uncertainties. Interleavedly implementing value function approximation and control policy updating over and over again with the iteration index at each time step for successively approximating the solution to the simplified HJB equations yields the controllers, under which DT affine nonlinear systems with matched and unmatched uncertainties can be UUB. The systematic derivation and analysis of the proposed robust control method for general DT affine nonlinear systems are presented. For linear quadratic regulation without consideration of uncertainty, the proposed interleaved RL algorithm could learn the approximately optimal controller, otherwise it is a controller with the ability of guaranteeing DT affine nonlinear systems to be UUB. Simulation results have demonstrated the effectiveness of the proposed method.

References

Chen Z, Jagannathan S (2008) Generalized Hamilton–Jacobi–Bellman formulation-based neural network control of affine nonlinear discrete-time systems. IEEE Trans Neural Netw 19(1):90–106

Fan QY, Yang GH (2015) Adaptive actor–critic design-based integral sliding-mode control for partially unknown nonlinear systems with input disturbances. IEEE Trans Neural Netw Learn Syst 27(1):165–177

Gao W, Jiang Y, Jiang ZP, Chai T (2016) Output-feedback adaptive optimal control of interconnected systems based on robust adaptive dynamic programming. Automatica 72:37–45

Jiang Y, Jiang ZP (2014) Robust adaptive dynamic programming and feedback stabilization of nonlinear systems. IEEE Trans Neural Netw Learn Syst 25(5):882–893

Jiang Y, Fan J, Chai T, Li J, Lewis FL (2017) Data-driven flotation industrial process operational optimal control based on reinforcement learning. IEEE Trans Ind Inform 14(5):1974–1989

Lewis FL, Vrabie D (2009) Reinforcement learning and adaptive dynamic programming for feedback control. IEEE Circuits Syst Mag 9(3):32–50

Li J, Modares H, Chai T, Lewis FL, Xie L (2017) Off-policy reinforcement learning for synchronization in multiagent graphical games. IEEE Trans Neural Netw Learn Syst 28(10):2434–2445

Li J, Chai T, Lewis FL, Ding Z, Jiang Y (2019) Off-policy interleaved Q-learning: optimal control for affine nonlinear discrete-time systems. IEEE Trans Neural Netw Learn Syst 30(5):1308–1320

Lin F (2000) An optimal control approach to robust control design. Int J Control 73(3):177–186

Liu D, Yang X, Wang D, Wei Q (2015) Reinforcement-learning-based robust controller design for continuous-time uncertain nonlinear systems subject to input constraints. IEEE Trans Cybern 45(7):1372–1385

Luo B, Wu HN, Li HX (2014) Adaptive optimal control of highly dissipative nonlinear spatially distributed processes with neuro-dynamic programming. IEEE Trans Neural Netw Learn Syst 26(4):684–696

Luo B, Yang Y, Liu D (2018) Adaptive Q-learning for data-based optimal output regulation with experience replay. IEEE Trans Cybern 48(12):3337–3348

Petersen IR (1987) A stabilization algorithm for a class of uncertain linear systems. Syst Control Lett 8(4):351–357

Petersen IR, Hollot CV (1986) A Riccati equation approach to the stabilization of uncertain linear systems. Automatica 22(4):397–411

Tan H, Shu S, Lin F (2009) An optimal control approach to robust tracking of linear systems. Int J Control 82(3):525–540

Wang D, Liu D, Li H (2014) Policy iteration algorithm for online design of robust control for a class of continuous-time nonlinear systems. IEEE Trans Autom Sci Eng 11(2):627–632

Wang D, Liu D, Li H, Luo B, Ma H (2015) An approximate optimal control approach for robust stabilization of a class of discrete-time nonlinear systems with uncertainties. IEEE Trans Syst Man Cybern Syst 46(5):713–717

Wang D, Liu D, Mu C, Zhang Y (2017) Neural network learning and robust stabilization of nonlinear systems with dynamic uncertainties. IEEE Trans Neural Netw Learn Syst 29(4):1342–1351

Wang D, Ha M, Qiao J (2019) Self-learning optimal regulation for discrete-time nonlinear systems under event-driven formulation. IEEE Trans Autom Control 65(3):1272–1279

Wei Q, Wang FY, Liu D, Yang X (2014) Finite-approximation-error-based discrete-time iterative adaptive dynamic programming. IEEE Trans Cybern 44(12):2820–2833

Wu HN, Li MM, Guo L (2014) Finite-horizon approximate optimal guaranteed cost control of uncertain nonlinear systems with application to mars entry guidance. IEEE Trans Neural Netw Learn Syst 26(7):1456–1467

Xie S, Xie Y, Huang T, Gui W, Yang C (2018) Coordinated optimization for the descent gradient of technical index in the iron removal process. IEEE Trans Cybern 48(12):3313–3322

Zhang H, Luo Y, Liu D (2009) Neural-network-based near-optimal control for a class of discrete-time affine nonlinear systems with control constraints. IEEE Trans Neural Netw 20(9):1490–1503

Zhao J, Na J, Gao G (2020) Adaptive dynamic programming based robust control of nonlinear systems with unmatched uncertainties. Neurocomputing 395:56–65

Chapter 5
Optimal Networked Controller and Observer Design

In this chapter, we mainly discuss the optimal control of linear discrete-time networked single controller systems and networked multi-player systems in the presence of network-induced delays. In the view of unknown system model parameters, unmeasurable system states, and the existence of network-induced delays, we design learning algorithms to implement the network controller and observer of the system and the Nash equilibrium of the multi-player system by a data-driven approach. To this end, we first estimate the system state and compensate for the network-induced delay by adding an observer and a Smith predictor and make the separation law hold so that the observer and the network controller can be designed separately. Then, the optimal observer gain and networked control policy are learned by a model-free off-policy Q-learning algorithm. In addition, we give rigorous proof of the theoretical derivation and the convergence of the algorithm. Simulation results verify the effectiveness of our proposed method.

5.1 Off-Policy Q-Learning for Single-Player Networked Control Systems

It is well known that networked control systems have attracted much attention by researchers and have been applied into wide variety of practical applications, such as industries, unmanned vehicles, medical treatment, etc., over the past couples of decades (Zhang et al. 2001; Wang and Han 2018; Wang et al. 2018b). This is because the basic characteristics that information and control signals are transmitted via wire or wireless networks among sensors, controllers and actuators brings the advantages to control systems, such as low cost, reduced weight, easy installation and maintenance. However, the negative impact brought by network-induced delays and packet losses on performance of control systems inherently exist in networked control sys-

J. Li et al., *Reinforcement Learning*, Advances in Industrial Control,
https://doi.org/10.1007/978-3-031-28394-9_5

tems (Zhang et al. 2001; Wang and Han 2018; Wang et al. 2018b; Li et al. 2018b; Liu et al. 2017).

For alleviating the above-mentioned negative impact and optimizing the control performance of systems, rich achievements have been reported in the fields of control and communication, such as handling network-induced delay, packet losses, bandwidth, coding and decoding (Zhang et al. 2001; Wang and Han 2018; Wang et al. 2018b; Li et al. 2018b; Liu et al. 2016, 2017; Quevedo et al. 2009), etc. It can be noticed that all the above-mentioned approaches have something in common, that is they all require the system dynamics to be accurately known a prior.

Consider the following networked control system

$$\begin{aligned} x_{k+1} &= Ax_k + Bu_k \\ y_k &= Cx_k \end{aligned} \tag{5.1}$$

and the networked controller

$$u_k = Kx_{k\text{-}d_k} \tag{5.2}$$

where $x_k = x(k) \in \mathbb{R}^n$, $u_k = u(k) \in \mathbb{R}^m$ and $y_k = y(k) \in \mathbb{R}^p$ are state, control input and output, respectively. k $(k = 0, 1, 2, \ldots)$ is the sampling time instant. A, B and C are matrices with appropriate dimensions. $d_k = d(k)$ denote the bounded network-induced delays including the transmission delays from the sensor to the controller and the computing time occurred in the controller. Without loss of generality, d_k is a nonnegative integer with $0 \leq d_k \leq \delta_{\max}$, and $\delta_{\max}$ is some positive integer (Zhang et al. 2001; Wang and Han 2018).

For the case that the system matrices A, B and C cannot be modeled accurately even if they are completely unknown, the RL has been widely used to learn the optimal controller $u_k = Kx_k$ for stabilizing and optimizing the control performance of system (5.1) (Li et al. 2017, 2018a; Kiumarsi et al. 2014; Luo et al. 2014; Wei et al. 2014). However, network-induced delays and unmeasurable state were not taken into account in these results. Since the states x_k might be unmeasurable or the cost used for measuring them is very high in most of practical control system applications (Kiumarsi et al. 2015; Pham et al. 2015; Wang et al. 2017b), such as unmanned vehicles and smart grid, etc., and network-induced delays inevitably occur when sending information from the sensor to the controller via a network in the above-mentioned practical industries (Wang and Han 2018; Wang et al. 2018b; Kiumarsi et al. 2015; Chen and Guo 2017), then these two problems have to be considered even though it would pose challenge on designing controller for system (5.1) subject to inaccurate system model.

To simultaneously handle these challenging problems, inaccurate system matrices, unmeasured state and network-induced delays, we will present a novel off-policy Q-learning algorithm to make networked control system (5.1) be optimum by introducing a Smith predictor, designing an observer state-based optimal controller and doing some appropriate mathematical manipulation. That is, we will develop a novel off-policy Q-learning method to find the optimal observer gain and the optimal controller for achieving optimality of network communication-based linear DT systems using only measured data in this part. An optimization problem for networked control systems composed of a plant, a state observer and a Smith predictor is formulated first. The Smith predictor is employed to not only compensate network-induced delays, but also make the separation principle hold, thus the observer and controller can be designed separately. Then, the off-policy Q-learning is implemented for learning the optimal observer gain and the optimal controller combined with the Smith predictor, such that a novel off-policy Q-learning algorithm is derived using only input, output and delayed estimated state of systems, not the inaccurate system matrices. The convergences of the iterative observer gain and the iterative controller gain are rigorously proven. Finally, simulation results are given to verify the effectiveness of the proposed method.

The structure of this section is organized as follows. Section 5.1.1 formulates an optimal control problem for networked control systems with a state observer and a Smith predictor. Section 5.1.2 devotes to the optimal observer design using the off-policy Q-learning approach. Section 5.1.3 presents a novel off-policy Q-learning algorithm to find the optimal controller. The results of simulations are given in Sect. 5.1.4.

5.1.1 Problem Formulation

In this subsection, an optimal control problem of networked control system with the compensation of network-induced delays and the state observer is formulated.

As shown in Fig. 5.1, a state observer is put in system (5.1) to estimate the unmeasured state of the plant. Since there exist network-induced delays from the sensor of the plant to the controller, then the estimated states will be delayed when they arrive at the controller via the network. A Smith predictor similar to that in Jiang et al. (2017) is presented for compensating the network-induced delays and thus the control input is computed using the predicted observer state to enable the closed-loop plant stable and high performance.

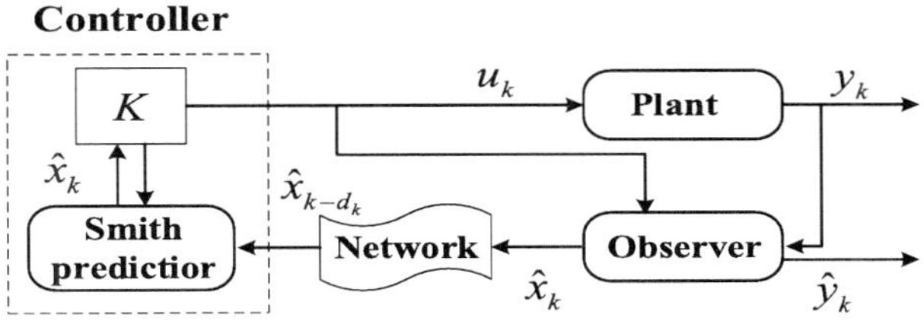

Fig. 5.1 The Architecture of networked control system with observer and predictor

The dynamics of the added state observer is given below

$$\begin{aligned}\hat{x}_{k+1} &= A\hat{x}_k + Bu_k + L\left(y_k - \hat{y}_k\right)\\ \hat{y}_k &= C\hat{x}_k\end{aligned} \tag{5.3}$$

where $\hat{x}_k$ and $\hat{y}_k$ are states and outputs of the observer, respectively. L is a gain matrix of the observer. Assume that the pair (A, B) is controllable and the pair (A, C) is observable. The controllability and observability of systems can generally be identified in practical applications even though system matrices A, B and C are inaccurate, such as some industrial control processes, unmanned aerial vehicles, etc. (Wang and Han 2018; Wang et al. 2017b, 2018b; Pham et al. 2015; Li et al. 2018a). This class of systems is what we focused on in this part.

Let $e_k = x_k - \hat{x}_k$, combining (5.1) and (5.3) yields the following dynamics of observer error:

$$e_{k+1} = (A - LC)e_k. \tag{5.4}$$

Remark 5.1 From (5.4), one can find that the error of observer is only dominated by the observer gain L. Compared with Kiumarsi et al. (2015), the observer states instead of the past inputs and outputs are used to approximate the states of system (5.1).

From Fig. 5.1, one can see that the observer state $\hat{x}_k$ is unavailable at the time instant k due to the network-induced delays, while $\hat{x}_{k-d_k}$ is available. By (5.3), one has

$$\begin{aligned}\hat{x}_{k-d_k+1} &= A\hat{x}_{k-d_k} + Bu_{k-d_k} + L(y_{k-d_k} - \hat{y}_{k-d_k})\\ \hat{x}_{k-d_k+2} &= A^2\hat{x}_{k-d_k} + Bu_{k-d_k+1} + ABu_{k-d_k} + L(y_{k-d_k+1} - \hat{y}_{k-d_k+1})\\ &\quad + AL(y_{k-d_k} - \hat{y}_{k-d_k})\\ &\vdots\\ \hat{x}_k &= A^{d_k}\hat{x}_{k-d_k} + \sum_{i=1}^{d_k} A^{i-1}Bu_{k-i} + \sum_{i=1}^{d_k} A^{i-1}L(y_{k-i} - \hat{y}_{k-i}).\end{aligned} \tag{5.5}$$

Thus, $\hat{x}_k$ can be predicted according to the most recent observer states by using the novel Smith predictor (5.5).

A static feedback controller (Li et al. 2018c) based on the predicted observer state is chosen as

$$u_k = K\hat{x}_k. \tag{5.6}$$

The target of this paper is to find the controller (5.6) to stabilize system (5.1) and minimize the following prescribed performance index without use of system matrices A, B and C.

$$J=\frac{1}{2}\sum_{k=0}^{\infty}y_k^TQ_py_k+u_k^TR_pu_k \tag{5.7}$$

where Q_p and R_p are respectively positive semi-definite matrix and positive definite matrix. Thus, we express the concerned optimization problem in this part as

Problem 5.1

$$\min_{u_k}\frac{1}{2}\sum_{k=0}^{\infty}y_k^TQ_py_k+u_k^TR_pu_k \tag{5.8}$$

s.t. (5.1), (5.4), (5.5) and (5.6)

since the observer state $\hat{x}_k$ can be available at the time instant k due to the usage of Smith predictor (5.5), the closed-loop system composed by (5.1), (5.4) and (5.6) can be written as an augmented form

$$\xi_{k+1}=\begin{bmatrix}A+BK & -BK\\ 0 & A-LC\end{bmatrix}\xi_k \tag{5.9}$$

where $\xi_k=\begin{bmatrix}x_k & e_k\end{bmatrix}^T$.

Obviously, the eigenvalues of the closed-loop (5.9) are those of $A+BK$ and those of $A-LC$. Thus the stability of the observer (5.3) and the stability and cost (5.8) of the system (5.1) are independent, which is the Separation Principal. Thus, when solving Problem 5.1, L and K can be separately designed by decomposing Problem 5.1 into two subproblems below

Subproblem 5.1

$$\min_{u_k}\frac{1}{2}\sum_{k=0}^{\infty}y_k^TQ_py_k+u_k^TR_pu_k$$

s.t.

$$\begin{aligned}x_{k+1}&=Ax_k+Bu_k\\ y_k&=Cx_k\\ u_k&=Kx_k.\end{aligned} \tag{5.10}$$

Subproblem 5.2 find the observer gain L for (5.4) to make $\lim_{k\to\infty}e_k=0$.

Remark 5.2 All of the pairs (K, L) that can ensure $\lim_{k\to\infty}x_k=0$ and $\lim_{k\to\infty}e_k=0$ compose the feasible solutions to the focused optimization problem (5.8). Since $y_k=Cx_k$, then the performance index in (5.10) can be transformed as $\min\limits_{u_k}\frac{1}{2}\sum_{k=0}^{\infty}x_k^T\tilde{Q}_px_k+u_k^TR_pu_k$, where $\tilde{Q}=CQ_pC$. In this sense, Subproblem 5.1 becomes a standard linear quadratic regulation problem, then there exists unique one solution u_k to Problem 5.1 since it is decomposed into separately finding optimal controller u_k and observer.

Note that Subproblem 5.1 and Subproblem 5.2 can be easily solved if system parameters A, B and C are accurately known. But if these system parameters cannot be accurately identified, then the traditional model-based observer and controller design method as well as the Smith predictor proposed in this paper cannot work. What we are interested in here is to find a data-driven method to solve the above two subproblems and implement the proposed Smith predictor using only measured data.

Before solving Subproblem 5.1, designing the observer gain L is going to be the first thing to do.

5.1.2 Optimal Observer Design

This subsection is devoted to finding the optimal observer gain L which minimizes the accumulated observer error. The idea of off-policy Q-Learning is used here to present a learning algorithm with no deed of accurate system matrices A, B and C, so that the optimal observer gain L can be found using only measured data.

Define an observer policy

$$w_k = L(y_k - \hat{y}_k) \tag{5.11}$$

then the dynamics of observer error (5.4) can be rewritten as

$$e_{k+1} = Ae_k - w_k. \tag{5.12}$$

To obtain the optimal observer, Subproblem 5.2 is changed into the optimization problem below

Problem 5.2

$$\min_{u_k} \frac{1}{2} \sum_{k=0}^{\infty} (y_k - \hat{y}_k)^T Q_o (y_k - \hat{y}_k) + w_k^T R_o w_k \tag{5.13}$$

s.t. (5.11) and (5.12)

where Q_o and R_o are respectively positive semi-definite matrix and positive definite matrix. If we try to find the optimal observer gain L for Problem 5.2, then unique one solution can be found since it is a standard linear quadratic regulation problem if $y_k - \hat{y}_k$ is replaced by Ce_k.

The model-based Q-learning algorithm is first given, then an off-policy Q-learning algorithm is derived to find the optimal observer gain using only measured data.

5.1.2.1 Model-Based Optimal Observer Design

Let $w_k = L(y_k - \hat{y}_k)$ be an admissible policy, a value function and an action-dependent Q-function can be respectively defined as (Wang and Han 2018; Al-Tamimi et al. 2007; Wang et al. 2017a)

$$V_o(x_k) = \frac{1}{2}\sum_{i=k}^{\infty} e_i^T \tilde{Q}_o e_i + w_i^T R_o w_i \tag{5.14}$$

and

$$Q_{ob}(e_k, w_k) = \frac{1}{2}(e_k^T \tilde{Q}_o e_k + w_k^T R_o w_k) + V_o(e_{k+1}) \tag{5.15}$$

where $\tilde{Q}_o = C Q_o C$. Thus, the following relationship holds:

$$\begin{aligned} V^*(e_k) &= \min_{w_k} \frac{1}{2}\sum_{i=k}^{\infty} e_i^T \tilde{Q}_o e_i + w_i^T R_o w_i \\ &= \min_{w_k} Q_{ob}(e_k, w_k) \\ &= Q_{ob}(e_k, w_k^*). \end{aligned} \tag{5.16}$$

According to the dynamic programming theory, the Q-function-based ARE can be derived below

$$\begin{aligned} Q_{ob}\left(e_k, w_k^*\right) &= \min_{w_k} \left\{\frac{1}{2}\left(e_k^T \tilde{Q}_o e_k + w_k^T R_o w_k\right) + V_o^*\left(e_{k+1}\right)\right\} \\ &= \frac{1}{2}\left(e_k^T \tilde{Q}_o e_k + \left(w_k^*\right)^T R_o w_k^*\right) + Q_{ob}\left(e_{k+1}, w_{k+1}^*\right). \end{aligned} \tag{5.17}$$

From (5.16), it follows that a unique optimal observer policy should be in the form of

$$w_k^* = \arg\min_{w_k} Q_{ob}\left(e_k, w_k\right). \tag{5.18}$$

The following lemma is useful to find the optimal observer policy.

Lemma 5.1 (Al-Tamimi et al. 2007; Modares et al. 2015) *For Problem 5.2, under the admissible policy (5.11), the value function and the Q-function have the quadratic forms of*

$$V_o\left(e_k\right) = \frac{1}{2} e_k^T P_o e_k \tag{5.19}$$

and

$$Q_{ob}(e_k, w_k) = \frac{1}{2}\begin{bmatrix} e_k \\ w_k \end{bmatrix}^T H_1 \begin{bmatrix} e_k \\ w_k \end{bmatrix} \tag{5.20}$$

where

$$H_1 = \begin{bmatrix} H_{1,ee} & H_{1,ew} \\ * & H_{1,ww} \end{bmatrix} = \begin{bmatrix} A^T P_o A + \tilde{Q}_o & -A^T P_o \\ * & R_o + P_o \end{bmatrix} \tag{5.21}$$

$$P_o = \begin{bmatrix} I \\ LC \end{bmatrix}^T H_1 \begin{bmatrix} I \\ LC \end{bmatrix} \tag{5.22}$$

where the matrix I denotes an identity matrix with approximate dimensions.

Substituting (5.20) into ARE (5.17), one has

$$\begin{aligned} \begin{bmatrix} e_k \\ w_k^* \end{bmatrix}^T H_1 \begin{bmatrix} e_k \\ w_k^* \end{bmatrix} &= e_k^T \tilde{Q}_o e_k + (w_k^*)^T R_o w_k^* + \begin{bmatrix} e_{k+1} \\ w_{k+1}^* \end{bmatrix}^T H_1 \begin{bmatrix} e_{k+1} \\ w_{k+1}^* \end{bmatrix} \\ &= e_k^T \tilde{Q}_o e_k + (w_k^*)^T R_o w_k^* \\ &+ \begin{bmatrix} e_k \\ w_k^* \end{bmatrix}^T \left(\begin{bmatrix} I \\ L^*C \end{bmatrix} \begin{bmatrix} A & -I \end{bmatrix} \right)^T H_1 \left(\begin{bmatrix} I \\ L^*C \end{bmatrix} \begin{bmatrix} A & -I \end{bmatrix} \right) \begin{bmatrix} e_k \\ w_k^* \end{bmatrix}. \end{aligned} \tag{5.23}$$

Based on the necessary condition of optimality (Kiumarsi et al. 2014; Wei et al. 2014; Li et al. 2018a), implementing $\frac{\partial Q_{ob}(e_k, w_k)}{\partial w_k}$ yields

$$w_k^* = -H_{1,ww}^{-1} H_{1,ew}^T e_k. \tag{5.24}$$

Compared with (5.11) and (5.24), one has

$$L^*C = -H_{1,ww}^{-1} H_{1,ew}^T. \tag{5.25}$$

It means that the optimal observer policy w_k^* can be obtained if the Q-function matrix H_1 is solved from (5.23). To learn H_1, Algorithm 5.1 is provided.

Remark 5.3 Al-Tamimi et al. (2007) and Modares et al. (2015) have proven that $\lim_{j\to\infty} H_1^{j+1} = H_1^*$ and $\lim_{j\to\infty} w_k^{j+1} = w_k^*$. One can find that calculating H_1^{j+1} requires the system matrices A and C to be accurately known when implementing Algorithm 5.1, while it is natural not to accurately model the system matrices A, B and C in real applications.

Next we will focus on investigating an off-policy Q-learning algorithm to get optimal observer for system (5.1) with inaccurate system matrices A, B and C.

Algorithm 5.1 Model-based policy iteration algorithm

1: Initialization: Given an admissible observer gain L^0, and let $j = 0$, where j denotes the iteration index;

2: Policy evaluation: Calculate H_1^{j+1} with

$$H_1^{j+1} = \begin{bmatrix} \tilde{Q}_o & 0 \\ 0 & R_o \end{bmatrix} + \left(\begin{bmatrix} I \\ L^j C \end{bmatrix} \begin{bmatrix} A & -I \end{bmatrix} \right)^T H_1^{j+1} \left(\begin{bmatrix} I \\ L^j C \end{bmatrix} \begin{bmatrix} A & -I \end{bmatrix} \right). \tag{5.26}$$

Iterative ARE equation (5.26) can be derived from (5.23) by deleting $\begin{bmatrix} e_k^T & (w_k^*)^T \end{bmatrix}^T$ and its transpose from the both sides of (5.23);

3: Policy update:

$$w_k^{j+1} = -(H_{1,ww}^{j+1})^{-1} (H_{1,ew}^{j+1})^T e_k \tag{5.27}$$

or

$$L^{j+1} = -(H_{1,ww}^{j+1})^{-1} (H_{1,ew}^{j+1})^T C^T (CC^T)^{-1} \tag{5.28}$$

where CC^T is invertible;

4: Stop when $\left\| H_1^{j+1} - H_1^j \right\| \leq \epsilon$ with a small constant ϵ ($\epsilon > 0$); Otherwise, $j = j + 1$ and go back to Step 2.

5.1.2.2 Data-Driven Off-Policy Q-Learning for the Optimal Observer

Here, we shall introduce two manipulations for reaching the goal of getting the optimal observer policy w_k^* using the data-driven off-policy learning approach. One manipulation is to define a virtual Q-function matrix $\bar{H}_1$ satisfied with

$$H_1 = \begin{bmatrix} C & 0 \\ 0 & I \end{bmatrix}^T \bar{H}_1 \begin{bmatrix} C & 0 \\ 0 & I \end{bmatrix}. \tag{5.29}$$

The other manipulation is to introduce an auxiliary variable $w_k^j = L^j C e_k$ into system (5.4), which yields

$$\begin{aligned} e_{k+1} &= Ae_k - w_k - w_k^j + w_k^j \\ &= \left(Ae_k - w_k^j \right) + \left(w_k^j - w_k \right) \end{aligned} \tag{5.30}$$

where w_k is called the behavior policy to generate data and w_k^j is viewed as the target policy needed to be learned.

By Lemma 5.1 and (5.29), one has

$$H_1 = \begin{bmatrix} A^T P_o A + \tilde{Q}_o & -A^T P_o \\ * & R_o + P_o \end{bmatrix} = \begin{bmatrix} C & 0 \\ 0 & I \end{bmatrix}^T \bar{H}_1 \begin{bmatrix} C & 0 \\ 0 & I \end{bmatrix}$$

$$= \begin{bmatrix} C & 0 \\ 0 & I \end{bmatrix}^T \begin{bmatrix} \bar{H}_{1,11} & \bar{H}_{1,12} \\ * & \bar{H}_{1,22} \end{bmatrix} \begin{bmatrix} C & 0 \\ 0 & I \end{bmatrix}$$
$$= \begin{bmatrix} C^T \bar{H}_{1,11} C & C^T \bar{H}_{1,12} \\ * & \bar{H}_{1,22} \end{bmatrix}. \tag{5.31}$$

Along the trajectory of (5.30) and refer to (5.26), one has

$$\begin{aligned}
&\begin{bmatrix} e_k \\ w_k^j \end{bmatrix}^T H_1^{j+1} \begin{bmatrix} e_k \\ w_k^j \end{bmatrix} - (y_{k+1} - \hat{y}_{k+1}) \begin{bmatrix} I \\ L^j \end{bmatrix}^T \bar{H}_1^{j+1} \begin{bmatrix} I \\ L^j \end{bmatrix} (y_{k+1} - \hat{y}_{k+1}) \\
&= \begin{bmatrix} e_k \\ w_k^j \end{bmatrix}^T H_1^{j+1} \begin{bmatrix} e_k \\ w_k^j \end{bmatrix} - \left((Ae_k - w_k^j) + w_k^j - w_k\right)^T \begin{bmatrix} C \\ L^j C \end{bmatrix} \\
&\times \bar{H}_1^{j+1} \begin{bmatrix} C \\ L^j C \end{bmatrix} \left((Ae_k - w_k^j) + w_k^j - w_k\right) \\
&= \begin{bmatrix} e_k \\ w_k^j \end{bmatrix}^T H_1^{j+1} \begin{bmatrix} e_k \\ w_k^j \end{bmatrix} - \begin{bmatrix} e_k \\ w_k^j \end{bmatrix}^T \left(\begin{bmatrix} I \\ L^j C \end{bmatrix} [A \ -I]\right)^T H_1^{j+1} \left(\begin{bmatrix} I \\ L^j C \end{bmatrix} [A \ -I]\right) \\
&\times \begin{bmatrix} e_k \\ w_k^j \end{bmatrix} - 2(Ae_k - w_k^j) \begin{bmatrix} C \\ L^j C \end{bmatrix}^T \bar{H}_1^{j+1} \begin{bmatrix} C \\ L^j C \end{bmatrix} (w_k^j - w_k) \\
&- (w_k^j - w_k)^T \begin{bmatrix} C \\ L^j C \end{bmatrix}^T \bar{H}_1^{j+1} \begin{bmatrix} C \\ L^j C \end{bmatrix} (w_k^j - w_k) \\
&= (y_k - \hat{y}_k)^T Q_o (y_k - \hat{y}_k) + (w_k^j)^T R_o w_k^j \\
&- 2(Ae_k - w_k^j) \begin{bmatrix} C \\ L^j C \end{bmatrix}^T \bar{H}_1^{j+1} \begin{bmatrix} C \\ L^j C \end{bmatrix} (w_k^j - w_k) \\
&- (w_k^j - w_k)^T \begin{bmatrix} C \\ L^j C \end{bmatrix}^T \bar{H}_1^{j+1} \begin{bmatrix} C \\ L^j C \end{bmatrix} (w_k^j - w_k).
\end{aligned} \tag{5.32}$$

Due to (5.22), (5.29) and (5.31), (5.32) becomes

$$\begin{aligned}
&\begin{bmatrix} y_k - \hat{y}_k \\ w_k \end{bmatrix} \bar{H}_1^{j+1} \begin{bmatrix} y_k - \hat{y}_k \\ w_k \end{bmatrix} - (y_{k+1} - \hat{y}_{k+1}) \begin{bmatrix} I \\ L^j \end{bmatrix}^T \bar{H}_1^{j+1} \begin{bmatrix} I \\ L^j \end{bmatrix} (y_{k+1} - \hat{y}_{k+1}) \\
&= (y_k - \hat{y}_k)^T Q_o (y_k - \hat{y}_k) + (w_k^j)^T R_o w_k^j - 2e_k^T A^T P_o^{j+1} (w_k^j - w_k) \\
&+ 2w_k^T A^T P_o^{j+1} (w_k^j - w_k) - (w_k^j - w_k)^T P_o^{j+1} (w_k^j - w_k) \\
&= (y_k - \hat{y}_k)^T Q_o (y_k - \hat{y}_k) + (w_k^j)^T R_o w_k^j + 2(y_k - \hat{y}_k)^T \bar{H}_{1,12}^{j+1} (w_k^j - w_k) \\
&+ (w_k^j + w_k)^T (\bar{H}_{1,22}^{j+1} - R_o)(w_k^j - w_k).
\end{aligned} \tag{5.33}$$

Due to (5.31), one has

$$H_{1,ww}^{j+1} = \bar{H}_{1,22}^{j+1}, \ H_{1,ew}^{j+1} = C^T \bar{H}_{1,12}^{j+1}. \tag{5.34}$$

Thus, iterative observer policy (5.27) can be rewritten as

$$w_k^{j+1} = -\left(\bar{H}_{1,22}^{j+1}\right)^{-1}\left(\bar{H}_{1,12}^{j+1}\right)^T(y_k - \hat{y}_k) \tag{5.35}$$

which indicates

$$L^{j+1} = -\left(\bar{H}_{1,22}^{j+1}\right)^{-1}\left(\bar{H}_{1,12}^{j+1}\right)^T. \tag{5.36}$$

Theorem 5.1 is developed to prove iterative observer policy w_k^{j+1} (5.35) is going to converge to the optimal observer policy w_k^*, i.e., $\lim_{j\to\infty} w_k^{j+1} = w_k^*$.

Theorem 5.1 *If the matrix CC^T is invertible, then there exists unique matrix $\bar{H}_1^{j+1}$, which satisfied with*

$$H_1^{j+1} = \begin{bmatrix} C & 0 \\ 0 & I \end{bmatrix}^T \bar{H}_1^{j+1} \begin{bmatrix} C & 0 \\ 0 & I \end{bmatrix} \tag{5.37}$$

and (5.33), such that (5.35) converges to the optimal observer policy, i.e., $\lim_{j\to x} w_k^{j+1} = w_k^*$.

Proof First, we shall prove that if $\bar{H}_1^{j+1}$ is the solution of (5.33), then the matrix H_1^{j+1} derived by (5.37) is the solution of (5.26). We know $y_k - \hat{y}_k = Ce_k$ and the dynamics of e_k is (5.30). If $\bar{H}_1^{j+1}$ is the solution of (5.33), then $\bar{H}_1^{j+1}$ will satisfy with the form below

$$\begin{aligned}
&\begin{bmatrix} y_k - \hat{y}_k \\ w_k \end{bmatrix} \bar{H}_1^{j+1} \begin{bmatrix} y_k - \hat{y}_k \\ w_k \end{bmatrix} - \left((Ae_k - w_k^j) + w_k^j - w_k\right)^T \begin{bmatrix} C \\ L^j C \end{bmatrix} \\
&\quad \times \bar{H}_1^{j+1} \begin{bmatrix} C \\ L^j C \end{bmatrix} \left((Ae_k - w_k^j) + w_k^j - w_k\right) \\
&\quad = (y_k - \hat{y}_k)^T Q_o (y_k - \hat{y}_k) + (w_k^j)^T R_o w_k^j - 2e_k^T A^T P_o^{j+1}(w_k^j - w_k) \\
&\quad + 2(w_k^j)^T A^T P_o^{j+1}(w_k^j - w_k) - (w_k^j - w_k)^T P_o^{j+1}(w_k^j - w_k).
\end{aligned} \tag{5.38}$$

By (5.22) of Lemma 5.1 and from (5.38), H_1^{j+1} defined in (5.37) makes (5.26) hold. And then, we shall prove that there is only one solution $\bar{H}_1^{j+1}$ satisfied with (5.33). If there are two different solutions $\bar{H}_1^{j+1}$ and $\bar{W}_1^{j+1}$ both of which make (5.33) hold, then we get H_1^{j+1} computed by (5.37) and W_1^{j+1} computed by

$$W_1^{j+1} = \begin{bmatrix} C & 0 \\ 0 & I \end{bmatrix}^T \bar{W}_1^{j+1} \begin{bmatrix} C & 0 \\ 0 & I \end{bmatrix}. \tag{5.39}$$

Since the matrix CC^T is invertible, we have

$$\begin{bmatrix}(CC^T)^{-1} & 0\\ 0 & I\end{bmatrix}\begin{bmatrix}C & 0\\ 0 & I\end{bmatrix}W_1^{j+1}\begin{bmatrix}C^T & 0\\ 0 & I\end{bmatrix}\begin{bmatrix}(CC^T)^{-1} & 0\\ 0 & I\end{bmatrix}=\hat{W}_1^{j+1} \tag{5.40}$$

and

$$\begin{bmatrix}(CC^T)^{-1} & 0\\ 0 & I\end{bmatrix}\begin{bmatrix}C & 0\\ 0 & I\end{bmatrix}H_1^{j+1}\begin{bmatrix}C^T & 0\\ 0 & I\end{bmatrix}\begin{bmatrix}(CC^T)^{-1} & 0\\ 0 & I\end{bmatrix}=\hat{H}_1^{j+1}. \tag{5.41}$$

If H_1^{j+1} and W_1^{j+1} are the same, then $\bar{H}_1^{j+1}$ and $\bar{W}_1^{j+1}$ are equal. So the distinct H_1^{j+1} and W_1^{j+1} satisfy (5.26). However, there is only one solution of (5.26) for Problem 5.2. By contradiction, there is only one solution of (5.33). Due to (5.34), one has

$$\begin{aligned} w_k^{j+1} &= -\left(\bar{H}_{1,22}^{j+1}\right)^{-1}\left(\bar{H}_{1,12}^{j+1}\right)^T(y_k-\hat{y}_k)\\ &= -\left(H_{1,ww}^{j+1}\right)^{-1}\left(H_{1,ew}^{j+1}\right)^T e_k. \end{aligned} \tag{5.42}$$

Since $\lim_{j\to\infty} w_k^{j+1} = \lim_{j\to\infty} -\left(H_{1,ww}^{j+1}\right)^{-1}\left(H_{1,ew}^{j+1}\right)^T e_k = w_k^*$, then iterative observer policy w_k^{j+1} in (5.35) converges to the optimal observer policy w_k^*. This completes the proof. □

Remark 5.4 Theorem 5.1 shows that learning w_k^* by solving $\bar{H}_1^{j+1}$ from (5.33) requires the reversibility of CC^T. If the matrix C is full row rank then CC^T is invertible. Here, we require the matrix C can be identified is invertible or not based on the practical application, even if it is inaccurate.

Let $e_{y_k} = y_k - \hat{y}_k$, (5.33) can be further rewritten as

$$\theta^j(k)h_o^{j+1} = \rho_k^j \tag{5.43}$$

where

$$\begin{aligned} \rho_k^j &= e_{yk}^T Q_o e_{yk} + w_k^T R_o w_k\\ h_o^{j+1} &= \left[\left(\text{vec}(\bar{H}_{1,11}^{j+1})\right)^T \left(\text{vec}(\bar{H}_{1,12}^{j+1})\right)^T \left(\text{vec}(\bar{H}_{1,22}^{j+1})\right)^T\right]^T\\ \theta^j(k) &= \left[\theta_1^j\ \theta_2^j\ \theta_3^j\right]\\ \theta_1^j &= e_{yk}^T \otimes e_{yk}^T - e_{yk+1}^T \otimes e_{yk+1}^T\\ \theta_2^j &= 4e_{yk}^T \otimes w_k^T - 2e_{yk+1}^T \otimes (L^j e_{yk+1})^T - 2e_{yk}^T \otimes (L^j e_{yk})^T\\ \theta_3^j &= 2w_k^T \otimes w_k^T - (L^j e_{yk+1})^T \otimes (L^j e_{yk+1})^T - (L^j e_{yk})^T \otimes (L^j e_{yk})^T. \end{aligned} \tag{5.44}$$

Algorithm 5.2 is presented below to learn $\bar{H}_1^{j+1}$.

Algorithm 5.2 Off-policy Q-learning algorithm

1: Data collection: Choose an arbitrary admissible observer policy as a behavior policy w_k, system data e_{y_k} are collected and stored in the sample sets $\theta^j(k)$ and ρ_k^j;
2: Initiation: Choose an initial gain L^0, such that $w_k^0 = L^0 e_{y_k}$ makes (5.4) stable. Let $j = 0$;
3: Implementing off-policy Q-learning: By using least squares methods for (5.43), $\bar{H}_1^{j+1}$ can be estimated using the collected data in Step 1, and then L^{j+1} can be updated in terms of (5.36);
4: If $\|L^{j+1} - L^j\| \leq \epsilon$ ($\epsilon > 0$ is some small positive numbers), then stop the iteration and the optimal control policy has been obtained; Otherwise, let $j = j + 1$ and go back to Step 3.

Remark 5.5 Different from Kiumarsi et al. (2015) where the past inputs and outputs are used to estimate the states of systems, a state observer is employed in this paper, such that the unmeasured states can be approximated by the observer states. Introducing of the matrix $\bar{H}_1$ makes it possible to learn the optimal observer gain using only measured data. It is worth pointing out that it is the first time to design optimal observer by using a way that is independent of the system matrices A, B and C, except for requiring CC^T to be invertible.

Remark 5.6 If a proportional and integral observer is employed instead of proportional observer (5.3) for estimating the state of discrete-time linear system (5.1), then it is potential that the optimal observer policy with proportional gain and integral gain can be learned by using the proposed data-driven off-policy Q-learning method since the separation principle still holds for the case of proportional and integral observer state-based feedback control for linear systems (Beale and Shafai 1989).

5.1.3 Optimal Controller Design

In this subsection, we shall solve Subproblem 5.1 to find the optimal control policy using only measured data.

5.1.3.1 On-Policy Q-Learning for the Optimal Controller

Let $u_k = Kx_k$ be an admissible control policy. For solving Subproblem 5.1, a value function and an action-dependent Q-function can be respectively defined as (Wang and Han 2018; Al-Tamimi et al. 2007)

$$V(x_k) = \frac{1}{2}\sum_{i=k}^{\infty} y_i^T Q_p y_i + u_i^T R_p u_i \tag{5.45}$$

and

$$Q(x_k, u_k) = \frac{1}{2}\left(y_k^T Q_p y_k + u_k^T R_p u_k\right) + V(x_{k+1}). \tag{5.46}$$

Thus, the following relation holds:

$$\begin{aligned} V^*(x_k) &= \min_{u_k} \frac{1}{2} \sum_{i=k}^{\infty} y_i^T Q_p y_i + u_i^T R_p u_i \\ &= \min_{u_k} Q(x_k, u_k) \\ &= Q(x_k, u_k^*). \end{aligned} \tag{5.47}$$

According to the dynamic programming theory, the Q-function-based ARE can be derived below

$$\begin{aligned} Q\left(x_k, u_k^*\right) &= \min_{u_k} \left\{ \frac{1}{2} \left(y_k^T Q_p y_k + u_k^T R_p u_k \right) + V^*\left(x_{k+1}\right) \right\} \\ &= \frac{1}{2} \left(y_k^T Q_p y_k + \left(u_k^*\right)^T R_p u_k^* \right) + Q\left(x_{k+1}, u_{k+1}^*\right). \end{aligned} \tag{5.48}$$

From (5.47), it follows that the optimal control policy should be with the form of

$$u_k^* = \arg\min_{u_k} Q\left(x_k, u_k\right). \tag{5.49}$$

The following lemma is useful to find the optimal control policy.

Lemma 5.2 (Al-Tamimi et al. 2007) *For Subproblem 5.1, under the admissible policy* $u_k = K x_k$*, the value function and the Q-function have the quadratic forms of*

$$V\left(x_k\right) = \frac{1}{2} x_k^T P_2 x_k \tag{5.50}$$

and

$$Q\left(x_k, u_k\right) = \frac{1}{2} \begin{bmatrix} x_k \\ u_k \end{bmatrix}^T H_2 \begin{bmatrix} x_k \\ u_k \end{bmatrix} \tag{5.51}$$

where

$$H_2 = \begin{bmatrix} H_{2,xx} & H_{2,xu} \\ * & H_{2,uu} \end{bmatrix} = \begin{bmatrix} A^T P_2 A + \tilde{Q}_p & A^T P_2 B \\ * & R_p + B^T P_2 B \end{bmatrix} \tag{5.52}$$

$$P_2 = \begin{bmatrix} I \\ K \end{bmatrix}^T H_2 \begin{bmatrix} I \\ K \end{bmatrix} \tag{5.53}$$

where $P_2 = P_2^T > 0$ *and* $\tilde{Q}_p = C^T Q_p C$.

Substituting (5.51) into ARE (5.48), one has

$$\begin{bmatrix} x_k \\ u_k^* \end{bmatrix}^T H_2^* \begin{bmatrix} x_k \\ u_k^* \end{bmatrix} = y_k^T Q_p y_k + (u_k^*)^T R_p u_k^* + \begin{bmatrix} x_{k+1} \\ u_{k+1}^* \end{bmatrix}^T H_2^* \begin{bmatrix} x_{k+1} \\ u_{k+1}^* \end{bmatrix}. \tag{5.54}$$

Based on the necessary condition of optimality, implementing $\frac{\partial Q(x_k,u_k)}{\partial u_k} = 0$ yields

$$u_k^* = -H_{2,uu}^{-1} H_{2,ux} x_k. \tag{5.55}$$

Since $u_k = Kx_k$, one has

$$K^* = -H_{2,uu}^{-1} H_{2,ux}. \tag{5.56}$$

Theorem 5.2 *If $K^* = -H_{2,uu}^{-1} H_{2,ux}$ and H_2 makes ARE (5.54) hold, then $u_k = K^* \hat{x}_k$ is the optimal control policy for Problem 5.1.*

Proof Note that $K^* = -H_{2,uu}^{-1} H_{2,ux}$ is the optimal controller gain of Subproblem 5.1. From (5.9), the performance index (5.8) of system (5.1) is only affected by controller gain K which is independent of the observer gain L in terms of Separation Principle. And the performance index in Problem 5.1 is the same as that in Subproblem 5.1. Therefore, $u_k = K^* \hat{x}_k$ can minimize the performance index in Problem 5.1. This completes the proof. □

The optimal controller gain K^* can be obtained if the Q-function matrix H_2 is solved from (5.54). To learn H_2, Algorithm 5.3 is provided by referring to Li et al. (2019), Kiumarsi et al. (2014), Al-Tamimi et al. (2007) and Modares et al. (2015).

Algorithm 5.3 On-policy iteration algorithm

1: Initialization: Given stabilizing controller gain K^0, and let $j = 0$, where j denotes iteration index;

2: Policy evaluation by solving Q-function matrix H_2^{j+1}:

$$\begin{bmatrix} x_k \\ K^j x_k \end{bmatrix}^T H_2^{j+1} \begin{bmatrix} x_k \\ K^j x_k \end{bmatrix} = y_k^T Q_p y_k + (K^j x_k)^T R_p K^j x_k + \begin{bmatrix} x_{k+1} \\ K^j x_{k+1} \end{bmatrix}^T H_2 \begin{bmatrix} x_{k+1} \\ K^j x_{k+1} \end{bmatrix} \tag{5.57}$$

where $x_{k+1} = Ax_k + BK^j x_k$;

3: Policy update:

$$K^{j+1} = -(H_{2,uu}^{j+1})^{-1} H_{2,ux}^{j+1}; \tag{5.58}$$

4: Stop when $\left\| H_2^{j+1} - H_2^j \right\| \leq \epsilon$ with a small constant ϵ ($\epsilon > 0$); Otherwise, $j = j + 1$ and go back to Step 2.

Remark 5.7 Al-Tamimi et al. (2007), Modares et al. (2015) have proven $\lim_{j\to\infty} H_2^{j+1} = H_2^*$ and $\lim_{j\to\infty} K^{j+1} = K^*$ if online implementing Algorithm

5.3. It is worth pointing out that Algorithm 5.3 does not need to know the system matrices A, B and C, but the states x_k should be measurable when solving H_2^{j+1} in (5.57). However, the network-induced delay and indirectly measured state make state x_k unavailable. Even though state x_k is actually replaced by the predicted $\hat{x}_k$ at the time instant k, it is impossible to predict the observer states $\hat{x}_k$ in (5.5) without knowing the accurate system matrices A and B. Therefore, Algorithm 5.3 cannot work for learning the optimal controller gain K^* at this situation.

For overcoming these obstacles, we shall develop an off-policy Q-learning algorithm. First, let us analyze the controller in the networked control system in Fig. 5.1.

According to the different cases of network-induced delays, a new variable $\bar{z}(k)$ is defined correspondingly using the similar way to Jiang et al. (2017).

- If $d_k = 0$,

$$\bar{z}_k = \Big[\underbrace{\hat{x}_k^T\ 0\ \ldots\ 0}_{\delta_{\max}+1}\ \underbrace{0\ \ldots\ 0}_{\delta_{\max}}\ \underbrace{0\ \ldots\ 0}_{\delta_{\max}}\Big]^T. \tag{5.59}$$

- If $d_k = 1$,

$$\bar{z}_k = \Big[\underbrace{0\ \hat{x}_k^T\ 0\ \ldots\ 0}_{\delta_{\max}+1}\ \underbrace{u_{k-1}^T\ 0\ \ldots\ 0}_{\delta_{\max}}\ \underbrace{LCe_{k-1}^T\ 0\ \ldots\ 0}_{\delta_{\max}}\Big]^T. \tag{5.60}$$

- ...
- If $d_k = \delta_{\max}$,

$$\bar{z}_k = \Big[\underbrace{0\ \ldots\ 0\ \hat{x}_k^T}_{\delta_{\max}+1}\ \underbrace{u_{k-1}^T\ \ldots\ u_{k-\delta_{\max}}^T}_{\delta_{\max}}\ \underbrace{LCe_{k-1}^T\ \ldots\ LCe_{k-\delta_{\max}}^T}_{\delta_{\max}}\Big]^T. \tag{5.61}$$

Referring to Smith predictor (5.5), the predicted observer state can be rewritten as

$$\hat{x}_k = M\bar{z}_k \tag{5.62}$$

where

$$M = \begin{bmatrix} M_1 & A^{\delta_{\max}} M_2 & M_1 \end{bmatrix}$$
$$M_1 = \begin{bmatrix} I & A & \ldots & A^{\delta_{\max}-1} \end{bmatrix},\ M_2 = \begin{bmatrix} B & AB & \ldots & A^{\delta_{\max}-1}B \end{bmatrix}.$$

Then, the predicted observer state-based feedback controller in Fig. 5.1 is in fact

$$u_k = K\hat{x}_k = \bar{K}\bar{z}_k \tag{5.63}$$

where $\bar{K} = KM$.

Now, we are in the position of finding $\bar{K}^*$ which satisfies $\bar{K}^* = K^*M$. Here, another new variable $\hat{z}(k)$ is defined as

- If $d_k = 0$,

$$\bar{z}_k = \begin{bmatrix} \underbrace{x_k^T\ 0\ \ldots\ 0}_{\delta_{\max}+1} & \underbrace{0\ \ldots\ 0}_{\delta_{\max}} \end{bmatrix}^T. \tag{5.64}$$

- If $d_k = 1$,

$$\bar{z}_k = \begin{bmatrix} \underbrace{0\ x_k^T\ 0\ \ldots\ 0}_{\delta_{\max}+1} & \underbrace{u_{k-1}^T\ 0\ \ldots\ 0}_{\delta_{\max}} \end{bmatrix}^T. \tag{5.65}$$

- ...
- If $d_k = \delta_{\max}$,

$$\bar{z}_k = \begin{bmatrix} \underbrace{0\ \ldots\ 0\ x_k^T}_{\delta_{\max}+1} & \underbrace{u_{k-1}^T\ \ldots\ u_{k-\delta_{\max}}^T}_{\delta_{\max}} \end{bmatrix}^T. \tag{5.66}$$

Similar to the derivation of (5.5), one has $x_k = \hat{M}\hat{z}_k$, where $\hat{M} = [M_1\ A^{\delta_{\max}}\ M_2]$. In terms of the definitions of M and $\hat{M}$, one can find

$$M = \hat{M}\hat{I}_1\hat{I}_2 + [\hat{M}\ 0] \tag{5.67}$$

where $\hat{I}_1 = [I\ 0\ 0]^T$ and $\hat{I}_2 = [0\ 0\ 0\ I]$. Let $\hat{K}^* = K^*\hat{M}$, then

$$\bar{K}^* = K^*M = \hat{K}^*\hat{I}_1\hat{I}_2 + [\hat{K}^*\ 0]. \tag{5.68}$$

So, if $\hat{K}^*$ can be learned even though the system matrices A, B and C are inaccurate or unknown, then $\bar{K}^*$ can be obtained for systems with the inaccurate system matrices A, B and C. Next, an off-policy Q-learning algorithm is present to learn $\hat{K}^*$.

5.1.3.2 Off-Policy Q-Learning for the Optimal Controller

Introducing an auxiliary variable BK^jx_k into system (5.1) yields

$$\begin{aligned} x_{k+1} &= Ax_k + BK^jx_k + B\left(u_k - K^jx_k\right) \\ &= \left(A + BK^j\right)\hat{M}\hat{z}_k + B\left(u_k - K^j\hat{M}\hat{z}_k\right) \end{aligned} \tag{5.69}$$

where u_k is called as the behavior policy to generate data and $K^jx_k = \hat{K}^j\hat{z}_k$ ($\hat{K}^j = K^j\hat{M}$) is viewed as the target policy needed to be learned.

Actually, (5.57) is equivalent to the form below

$$\begin{bmatrix} I \\ K^j \end{bmatrix}^T H_2^{j+1} \begin{bmatrix} I \\ K^j \end{bmatrix} = C^T Q_p C_k + (K^j)^T R_p K^j + (A + BK^j)^T \begin{bmatrix} I \\ K^j \end{bmatrix}^T H_2^{j+1} \begin{bmatrix} I \\ K^j \end{bmatrix} (A + BK^j). \tag{5.70}$$

Along the trajectory of (5.69), one has

$$\begin{aligned} & x_k^T \begin{bmatrix} I \\ K^j \end{bmatrix}^T H_2^{j+1} \begin{bmatrix} I \\ K^j \end{bmatrix} x_k - x_{k+1}^T \begin{bmatrix} I \\ K^j \end{bmatrix}^T H_2^{j+1} \begin{bmatrix} I \\ K^j \end{bmatrix} x_{k+1} \\ & = x_k^T \begin{bmatrix} I \\ K^j \end{bmatrix}^T H_2^{j+1} \begin{bmatrix} I \\ K^j \end{bmatrix} x_k - \left(Ax_k + BK^j x_k + B\left(u_k - K^j x_k\right)\right)^T \begin{bmatrix} I \\ K^j \end{bmatrix}^T \\ & \times H_2^{j+1} \begin{bmatrix} I \\ K^j \end{bmatrix} \left(Ax_k + BK^j x_k + B\left(u_k - K^j x_k\right)\right). \end{aligned} \tag{5.71}$$

Due to (5.70) and $x_k = \hat{M}\hat{z}_k$, (5.71) becomes

$$\begin{aligned} & \begin{bmatrix} \hat{z}_k \\ \hat{K}^j \hat{z}_k \end{bmatrix}^T \bar{H}_2^{j+1} \begin{bmatrix} \hat{z}_k \\ \hat{K}^j \hat{z}_k \end{bmatrix} - \begin{bmatrix} \hat{z}_{k+1} \\ \hat{K}^{j+1} \hat{z}_{k+1} \end{bmatrix}^T \bar{H}_2^{j+1} \begin{bmatrix} \hat{z}_{k+1} \\ \hat{K}^{j+1} \hat{z}_{k+1} \end{bmatrix} \\ & = \begin{bmatrix} \hat{z}_k \\ \hat{K}^j \hat{z}_k \end{bmatrix}^T \begin{bmatrix} M^T \tilde{Q}_p M & 0 \\ 0 & R_p \end{bmatrix} \begin{bmatrix} \hat{z}_k \\ \hat{K}^j \hat{z}_k \end{bmatrix} \\ & - 2x_k^T (A + BK^j)^T \begin{bmatrix} I \\ K^j \end{bmatrix}^T H_2^{j+1} \begin{bmatrix} I \\ K^j \end{bmatrix} B(u_k - \hat{K}^j \hat{z}_k) \\ & - (u_k - \hat{K}^j \hat{z}_k)^T B^T \begin{bmatrix} I \\ K^j \end{bmatrix}^T H_2^{j+1} \begin{bmatrix} I \\ K^j \end{bmatrix} B(u_k - \hat{K}^j \hat{z}_k) \end{aligned} \tag{5.72}$$

where

$$\bar{H}_2^{j+1} = \begin{bmatrix} \hat{M} & 0 \\ 0 & I \end{bmatrix}^T H_2^{j+1} \begin{bmatrix} \hat{M} & 0 \\ 0 & I \end{bmatrix}. \tag{5.73}$$

By Lemma 5.2, one has

$$\begin{aligned} \bar{H}_2^{j+1} & = \begin{bmatrix} \hat{M} & 0 \\ 0 & I \end{bmatrix}^T \begin{bmatrix} A^T P_2^{j+1} A + \tilde{Q}_p & A^T P_2^{j+1} B \\ B^T P_2^{j+1} A & B^T P_2^{j+1} B + R_p \end{bmatrix} \begin{bmatrix} \hat{M} & 0 \\ 0 & I \end{bmatrix} \\ & = \begin{bmatrix} \hat{M}^T (A^T P_2^{j+1} A + \tilde{Q}_p)\hat{M} & \hat{M}^T A^T P_2^{j+1} B \\ * & B^T P_2^{j+1} B + R_p \end{bmatrix} = \begin{bmatrix} \bar{H}_{2,zz}^{j+1} & \bar{H}_{2,zu}^{j+1} \\ * & \bar{H}_{2,uu}^{j+1} \end{bmatrix} \end{aligned} \tag{5.74}$$

and

$$
\begin{aligned}
&-2x_k^T(A+BK^j)^T\begin{bmatrix} I \\ K^j \end{bmatrix}^T H_2^{j+1}\begin{bmatrix} I \\ K^j \end{bmatrix} B(u_k-\hat{K}^j\hat{z}_k)\\
&\quad=-2\hat{z}_k^T(A\hat{M}+BK^j\hat{M})^T P^{j+1}B(u_k-\hat{K}^j\hat{z}_k)\\
&\quad=-2\hat{z}_k^T\bar{H}_{2,zu}^{j+1}(u_k-\hat{K}^j\hat{z}_k)-2u_k^j(\bar{H}_{2,uu}^{j+1}-R_p)(u_k-\hat{K}^j\hat{z}_k)
\end{aligned}
\tag{5.75}
$$

and

$$
\begin{aligned}
&-(u_k-\hat{K}^j\hat{z}_k)^T B^T\begin{bmatrix} I \\ K^j \end{bmatrix}^T H_2^{j+1}\begin{bmatrix} I \\ K^j \end{bmatrix} B(u_k-\hat{K}^j\hat{z}_k)\\
&\quad=-(u_k-\hat{K}^j\hat{z}_k)^T B^T P_2^{j+1}B(u_k-\hat{K}^j\hat{z}_k)\\
&\quad=-(u_k-\hat{K}^j\hat{z}_k)^T(\bar{H}_{2,uu}^{j+1}-R_p)(u_k-\hat{K}^j\hat{z}_k).
\end{aligned}
\tag{5.76}
$$

Thus, (5.72) becomes

$$
\begin{aligned}
&\begin{bmatrix} \hat{z}_k \\ \hat{K}^j\hat{z}_k \end{bmatrix}^T \bar{H}_2^{j+1}\begin{bmatrix} \hat{z}_k \\ \hat{K}^j\hat{z}_k \end{bmatrix}-\begin{bmatrix} \hat{z}_{k+1} \\ \hat{K}^j\hat{z}_{k+1} \end{bmatrix}^T \bar{H}_2^{j+1}\begin{bmatrix} \hat{z}_{k+1} \\ \hat{K}^j\hat{z}_{k+1} \end{bmatrix}\\
&\quad=\begin{bmatrix} \hat{z}_k \\ \hat{K}^j\hat{z}_k \end{bmatrix}^T\begin{bmatrix} \hat{M}^T\tilde{Q}\hat{M} & 0 \\ 0 & R_p \end{bmatrix}\begin{bmatrix} \hat{z}_k \\ \hat{K}^j\hat{z}_k \end{bmatrix}-2\hat{z}_k^T\bar{H}_{2,zu}^{j+1}(u_k-\hat{K}^j\hat{z}_k)\\
&\quad+(\hat{K}^j\hat{z}_k)^T(\bar{H}_{2,uu}^{j+1}-R_p)(\hat{K}^j\hat{z}_k)-u_k^T(\bar{H}_{2,uu}^{j+1}-R_p)u_k.
\end{aligned}
\tag{5.77}
$$

Further, (5.77) can be rewritten as

$$
\varphi^j(k)h_c^{j+1}=\beta_k^j \tag{5.78}
$$

where

$$
\begin{aligned}
\beta_k^j&=y_k^TQ_py_k+u_k^TR_pu_k\\
h_c^{j+1}&=\left[(\mathrm{vec}(\bar{H}_{2,zz}^{j+1}))^T\ (\mathrm{vec}(\bar{H}_{2,zu}^{j+1}))^T\ (\mathrm{vec}(\bar{H}_{2,uu}^{j+1}))^T\right]^T\\
\varphi^j(k)&=\left[\varphi_1^j(k)\ \varphi_2^j(k)\ \varphi_3^j(k)\right]\\
\varphi_1^j(k)&=\hat{z}_k^T\otimes\hat{z}_k^T-\hat{z}_{k+1}^T\otimes\hat{z}_{k+1}^T\\
\varphi_2^j(k)&=2\left(\hat{z}_k^T\otimes\left(\hat{K}^j\hat{z}_k\right)^T-\hat{z}_{k+1}^T\otimes\left(\hat{K}^j\hat{z}_{k+1}\right)^T\right)+2\hat{z}_k^T\otimes\left(u_k-\hat{K}^j\hat{z}_k\right)^T\\
\varphi_3^j(k)&=u_k^T\otimes u_k^T-\left(\hat{K}^j\hat{z}_{k+1}\right)^T\otimes\left(\hat{K}^j\hat{z}_{k+1}\right)^T.
\end{aligned}
\tag{5.79}
$$

Theorem 5.3 *There exists a unique matrix* $\bar{H}_2^{j+1}$ *which satisfied with (5.78), and* $\hat{K}^j$ *converges to* $\hat{K}^*$ *as* $j\to\infty$.

Proof Because there exists a unique solution H_2^{j+1} to (5.70), there exists a matrix $\bar{H}_2^{j+1}$ that makes (5.78) by the derivation of (5.78) from (5.70). Now, the uniqueness of solution to (5.78) is going to be proved. Assume there are two distinct solutions $\bar{H}_2^{j+1}$ and $\bar{W}_2^{j+1}$ to (5.78), from (5.73), one has two distinct matrices H_2^{j+1} and W_2^{j+1}. By contradiction, there is only one solution $\bar{H}_2^{j+1}$ to (5.78).

By (5.58) and (5.68), one has

$$\hat{K}^{j+1} = K^{j+1}\hat{M} = -\left(H_{2,uu}^{j+1}\right)^{-1} H_{2,ux}^{j+1}\hat{M}. \tag{5.80}$$

By Lemma 5.2, one has

$$H_2^{j+1} = \begin{bmatrix} H_{2,xx}^{j+1} & H_{2,xu}^{j+1} \\ * & H_{2,uu}^{j+1} \end{bmatrix} = \begin{bmatrix} A^T P_2^{j+1} A + \tilde{Q}_p & A^T P_2^{j+1} B \\ B^T P_2^{j+1} A & B^T P_2^{j+1} B + R_p \end{bmatrix}. \tag{5.81}$$

Compared (5.74) with (5.81), it is not difficult to find

$$\bar{H}_{2,zu}^{j+1} = \hat{M} H_{2,xu}^{j+1}, \ \bar{H}_{2,uu}^{j+1} = H_{2,uu}^{j+1}. \tag{5.82}$$

Then (5.80) becomes

$$\hat{K}^{j+1} = -\left(\bar{H}_{2,uu}^{j+1}\right)^{-1} \bar{H}_{2,u\bar{z}}^{j+1} \tag{5.83}$$

and

$$\begin{aligned} \lim_{j\to\infty} \hat{K}^{j+1} &= \lim_{j\to\infty}\left(-\left(\bar{H}_{2,uu}^{j+1}\right)^{-1} \bar{H}_{2,u\bar{z}}^{j+1}\right) \\ &= \lim_{j\to\infty}\left(-\left(H_{2,uu}^{j+1}\right)^{-1} H_{2,ux}^{j+1}\hat{M}\right) = K^*\hat{M} = \hat{K}^*. \end{aligned} \tag{5.84}$$

This completes the proof. □

If $\hat{K}^*$ can be found by learning $\bar{H}_2$, then $\bar{K}^* = \hat{K}^*\hat{I}_1\hat{I}_2 + [\hat{K}^*\ 0]$ can be calculated. The approximate optimal control law $u_k^* = \bar{K}^*\bar{z}_k = K^*\hat{x}_k$ can be derived for Problem 5.1. To learn $\bar{H}_2^{j+1}$, Algorithm 5.4 is presented as follows.

Remark 5.8 As shown by the proof of Theorem 5.2 and the derivation of (5.78) and (5.83), Algorithm 5.4 used for learning $\hat{K}^*$ is proposed, such that $\bar{K}^*$ can be finally calculated just with the help of (5.57) and (5.58) in Algorithm 5.3.

Remark 5.9 For system (5.1) subject to inaccurate system matrices, network-induced delays and unmeasured states, the optimal observer gain and the optimal controller gain can be found by implementing Algorithm 5.4 using only measured inputs, outputs and observer errors, not system matrices A, B and C. This is different from Li et al. (2017, 2019), Kiumarsi et al. (2014), Luo et al. (2014), Wei

Algorithm 5.4 Off-policy Q-learning algorithm

1: Data collection: Choose a behavior policy u_k and a behavior observer policy w_k to act the plant and the observer in Fig. 5.1. When $\hat{y}_{k-d_k}$ approaches the output y_{k-d_k} of the system as close as it can, x_{k-d_k} in $\hat{z}_k$ is replaced by $\hat{x}_{k-d_k}$ and store $\hat{z}_k$ and u_k in the sample sets φ^j and β_k^j. The system data e_{y_k} are collected and stored in the sample sets $\theta^j(k)$ and ρ_k^j;
2: Learning the optimal observer gain: Implementing step 2-step 4 in Algorithm 5.2 to find the approximately optimal observer gain;
3: Learning $\hat{K}^*$:

 3.1: Choose the initial stabilizing gain $\hat{K}^0$ and let $j = 0$;
 3.2: By using the least squares method, $\bar{H}_2^{j+1}$ in (5.78) can be estimated using the collected data in Step 1, and then $\hat{K}^{j+1}$ can be updated in terms of (5.83);
 3.3: If $\left\| \bar{H}_2^{j+1} - \bar{H}_2^j \right\| \leq \varepsilon$ $(\varepsilon > 0)$, then stop the iteration and $\hat{K}^*$ can be estimated by $\hat{K}^{j+1}$, thus $\bar{K}^*$ can further be calculated (5.68); Otherwise, let $j = j + 1$ and go back to Step 3.2.

et al. (2014), Wang et al. (2017a), Al-Tamimi et al. (2007) and Modares et al. (2015) without consideration of network-induced delays and unmeasurable states, and Jiang et al. (2017) and Zhang et al. (2018b), where states of systems are assumed to be measurable. Moreover, it is worth pointing out that, until now, it has not been reported about off-policy Q-learning algorithm used for solving optimal controller for networked control systems with accurate model parameters, network-induced delay and together with unavailable sates information.

Remark 5.10 Notice that the number of unknown elements in the iterative matrix $\bar{H}_2^{j+1}$ is $(m(\delta_{\max}+1)+n(\delta_{\max}+1)+1) \times (m(\delta_{\max}+1)+n(\delta_{\max}+1))/2$ in (5.78). If no network-induced delay is considered, then it is $(m+n+1) \times (m+n)/2$. The number of unknown elements in the iterative matrix $\bar{P}^{j+1}$ in Liu et al. (2014a) without network-induced delay is $(Nm+Np) \times (Nm+Np+1)/2\,(Np \geq n,\ N \geq 1)$. So, the computational complexity is less than the method in Kiumarsi et al. (2015).

5.1.4 Simulation Results

In this subsection, the proposed off-policy Q-learning algorithm for predicted observer state-based feedback control of system (5.1) is verified respectively for systems with and without network-induced delays. Moreover, simulations show the efficiency of the designed observer and predictor for alleviating the negative influence of network-induced delays on the control performance of systems.

Example 5.1 Consider the following discrete-time model of F-16 aircraft autopilot (Al-Tamimi et al. 2007; Modares et al. 2015):

$$
\begin{aligned}
x_{k+1} &= \begin{bmatrix} 0.906488 & 0.0816012 & -0.0005 \\ 0.074349 & 0.90121 & -0.000708383 \\ 0 & 0 & 0.132655 \end{bmatrix} x_k + \begin{bmatrix} -0.00150808 \\ -0.0096 \\ 0.867345 \end{bmatrix} u_k \\
y_k &= \begin{bmatrix} 1 & 0 & 0 \\ 2 & 1 & 1 \\ 1 & 0 & 3 \end{bmatrix} x_k .
\end{aligned}
\tag{5.85}
$$

Choose $Q_1 = Q_2 = \text{diag}(10, 10, 10)$, $R_1 = \text{diag}(1, 1, 1)$ and $R_2 = 1$.

5.1.4.1 Model-Based the Optimal Observer and the Optimal Controller

Using the command "dare" in MATLAB yields the optimal Q-function matrix H_1^* and the optimal observer gain L^*C in terms of (5.21) and (5.24). Further, $\bar{H}_1^*$ can be calculated by (5.29).

$$
\bar{H}_1^* = \begin{bmatrix} 21.4051 & 1.2230 & 0.4671 & -15.8013 & 0.0737 & -19.8418 \\ 1.2230 & 22.0225 & -3.4352 & -23.1629 & -11.2431 & -13.0895 \\ 0.4671 & -3.4352 & 11.1489 & 5.5249 & 3.3115 & -0.0488 \\ -15.8013 & -23.1629 & 5.5249 & 61.7846 & 20.1982 & 50.0020 \\ 0.0737 & -11.2431 & 3.3115 & 20.1982 & 11.6467 & 9.9968 \\ -19.8418 & -13.0895 & -0.0488 & 50.0020 & 9.9968 & 101.0173 \end{bmatrix} \tag{5.86}
$$

and the optimal observer gain

$$
L^*C = \begin{bmatrix} 0.8539 & 0.1610 & 0.0025 \\ 0.1426 & 0.7030 & -0.0038 \\ 0.0193 & -0.0197 & 0.1301 \end{bmatrix} . \tag{5.87}
$$

For Subproblem 5.1, using the command "dare" in MATLAB yields the optimal Q-function matrix H_2^* and the gain K^*.

$$
H_2^* = \begin{bmatrix} 761.9176 & 630.3343 & 55.3593 & 33.2776 \\ 630.3343 & 569.9991 & 11.0594 & 5.2257 \\ 55.3593 & 11.0594 & 101.7525 & 11.4653 \\ 33.2776 & 5.2257 & 11.4653 & 76.0158 \end{bmatrix} \tag{5.88}
$$

$$
K^* = \begin{bmatrix} -0.4378 & -0.0687 & -0.1508 \end{bmatrix} . \tag{5.89}
$$

5.1.4.2 Learning the Optimal Observer and the Optimal Controller

We assume that the system matrices A, B and C of system (5.85) are not accurately known. First, the network-induced delay is not taken into account. In this case, the

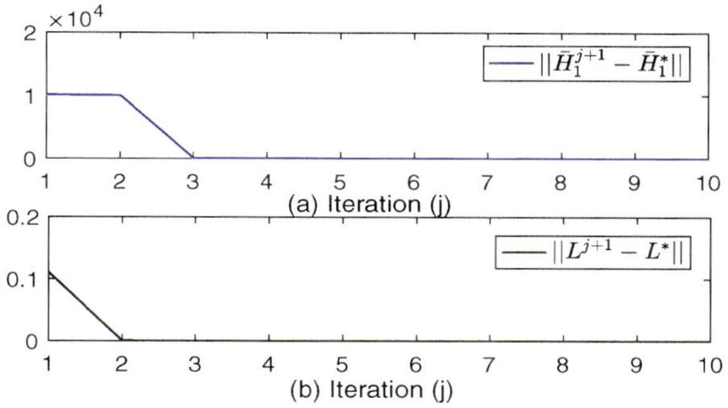

Fig. 5.2 Convergence of matrices $\bar{H}_1^{j+1}$ and L^{j+1}

maximum delay upper bound $\delta_{\max} = 0$, $\hat{z}_k = x_k$ and $M = I$. Here x_k will be replaced by $\hat{x}_k$ when e_k is very close to zero. By Algorithm 5.4, Fig. 5.2 plots the results of convergence of $\bar{H}_1^{j+1}$ and L^{j+1} when implementing the off-policy Q-learning after 5 iterations, and we get

$$L^5 = \begin{bmatrix} 0.8294 & 0.0755 & -0.0005 \\ 0.0689 & 0.8246 & -0.0006 \\ 0 & 0 & 0.1206 \end{bmatrix} \tag{5.90}$$

and

$$L^5 C = \begin{bmatrix} 0.8294 & 0.0755 & -0.0005 \\ 0.0689 & 0.8246 & -0.0006 \\ 0 & 0 & 0.1206 \end{bmatrix} = L^* C \tag{5.91}$$

which means that the optimal observer gain L^* has been learned using Algorithm 5.4 using only available data.

Moreover, implementing Algorithm 5.4 also yields

$$\bar{H}_2^5 = \begin{bmatrix} 761.9176 & 630.3343 & 55.3593 & 33.2776 \\ 630.3343 & 569.9991 & 11.0594 & 5.2257 \\ 55.3593 & 11.0594 & 101.7525 & 11.4653 \\ 33.2776 & 5.2257 & 11.4653 & 76.0158 \end{bmatrix} \tag{5.92}$$

$$\hat{K}^5 = \begin{bmatrix} -0.4378 & -0.0687 & -0.1508 \end{bmatrix}. \tag{5.93}$$

When no network-induced delay occurred, one can notice that $\bar{H}_2^5$ and $\hat{K}^5$ respectively converge to H_2^* and K^*.

Second, if the maximum network-induced delay bound $\delta_{\max} = 1$, then $\hat{M} = [I \ A \ B]$ and $M = [I \ A \ B \ I]$. Implementing Algorithm 5.4 yields

$$\hat{K}^6 = \begin{bmatrix} -0.4378 & -0.068 & 0.1508 & -0.4019 & -0.0977 & 0.0197 & 0.1295 \end{bmatrix} \tag{5.94}$$

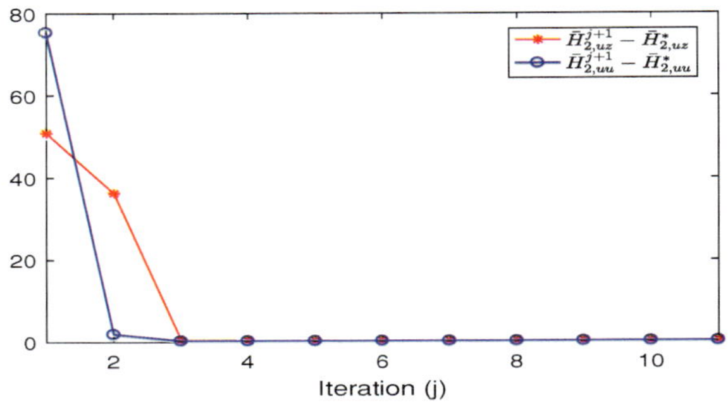

Fig. 5.3 Convergence of matrix $\bar{H}_2^{j+1}$

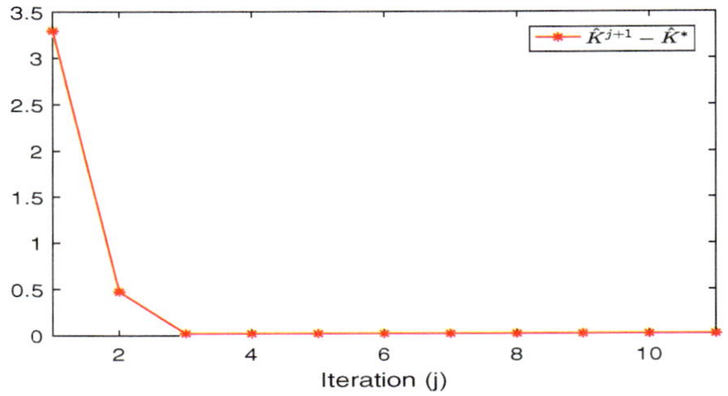

Fig. 5.4 Convergence of matrix $\hat{K}^{j+1}$

and $L^6 = L^*$. One can find that $\hat{K}^6$ is equal to $K^*\hat{M}$. Further, we can calculate

$$\begin{aligned}\bar{K}^6 &= \hat{K}^6 \hat{I}_1 \hat{I}_2 + [\hat{K}^6 \; 0] \\ &= [-0.4378 \;\; -0.0687 \;\; -0.1508 \;\; -0.4019 \;\; -0.0977 \\ &\quad 0.0197 \; 0.1295 \;\; -0.4378 \;\; -0.068 \; 0.1508]\end{aligned} \tag{5.95}$$

where

$$\hat{I}_1 = \begin{bmatrix} I_{3\times 3} \\ 0 \\ 0 \end{bmatrix}, \; \hat{I}_2 = \begin{bmatrix} 0 \; 0 \; 0 \; I_{3\times 3} \end{bmatrix}, \; I_{3\times 3} = \begin{bmatrix} 1 \; 0 \; 0 \\ 0 \; 1 \; 0 \\ 0 \; 0 \; 1 \end{bmatrix} \tag{5.96}$$

which shows $\bar{K}^6 = \bar{K}^* = K^* M$. Then $u_k^* = \bar{K}^* \bar{z}_k$ can be obtained. Figures 5.3 and 5.4 present the detailed convergence results.

5.1.4.3 Simulations and Comparisons

Choose the initial states of the system and its observer $x_0 = \begin{bmatrix} -1 \; 2 \; 1 \end{bmatrix}^T$ and $\hat{x}_0 = \begin{bmatrix} 0 \; 1 \; -1 \end{bmatrix}^T$. For the case of absence of network-induced delays, Fig. 5.5 is given to

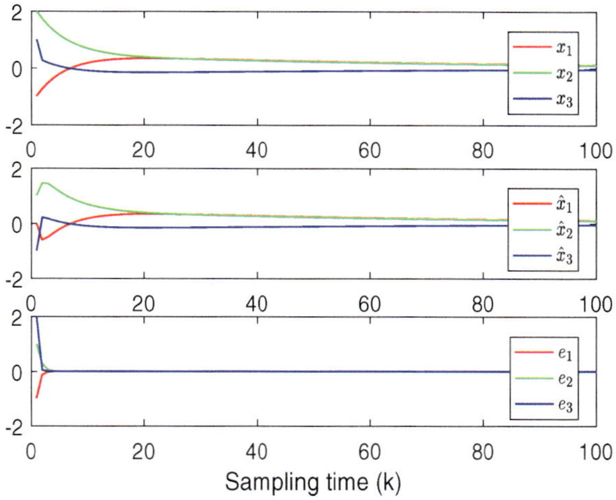

Fig. 5.5 Trajectories of states, estimated states and estimation errors of control system without network-induced delay

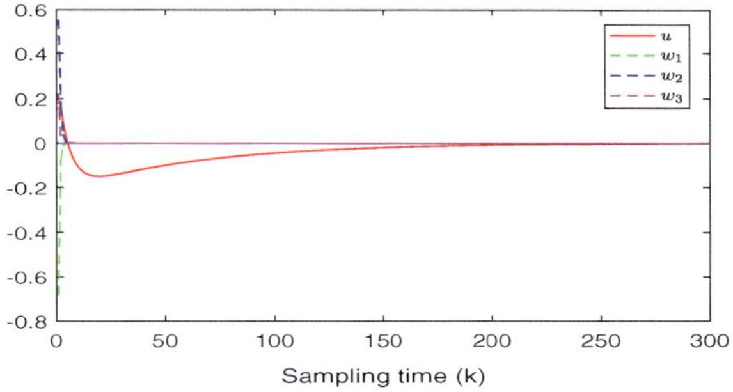

Fig. 5.6 Curves of the optimal control policy and the optimal observer policy

show the system sate trajectories and estimated system state trajectories, as well as the estimated errors using the optimal controller gain (5.93) and the optimal observer gain (5.90) (see Fig. 5.6).

For the case, there exist network-induced delays plotted in Fig. 5.7, we choose the initial states of the system and its observer $x_0 = \begin{bmatrix}-1 & 2 & 1\end{bmatrix}^T$, $\hat{x}_0 = \begin{bmatrix}0 & 1 & -1\end{bmatrix}^T$, $x_{-1} = \begin{bmatrix}1.5 & 0 & -1\end{bmatrix}^T$ and $\hat{x}_{-1} = \begin{bmatrix}-1 & -1 & 0\end{bmatrix}^T$, and initial control input $u_0 = 1$. Figure 5.8 shows the system state trajectories and estimated system state trajectories, as well as the estimated errors under the approximate optimal control policy (5.95) and using the approximate optimal observer (5.90) (see Fig. 5.9). By comparing Fig. 5.8 with Fig. 5.5, even though there exist network-induced delays, the good control performance (the same convergence speed and overshoot as those under the case of absence of network-induced delays) can be obtained by using the proposed data-drive learning algorithm by combining prediction control and state observer estimation.

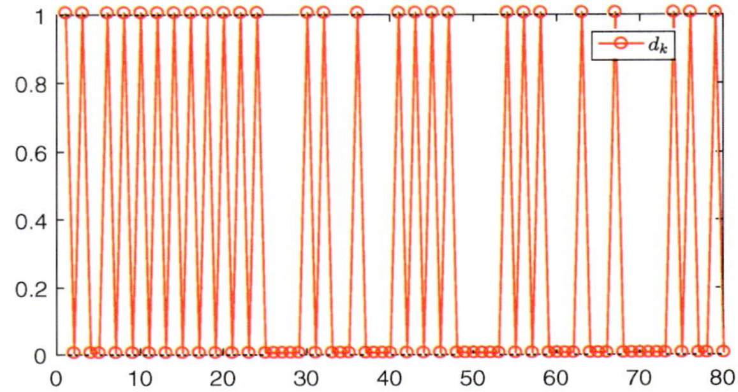

Fig. 5.7 Network-induced delays

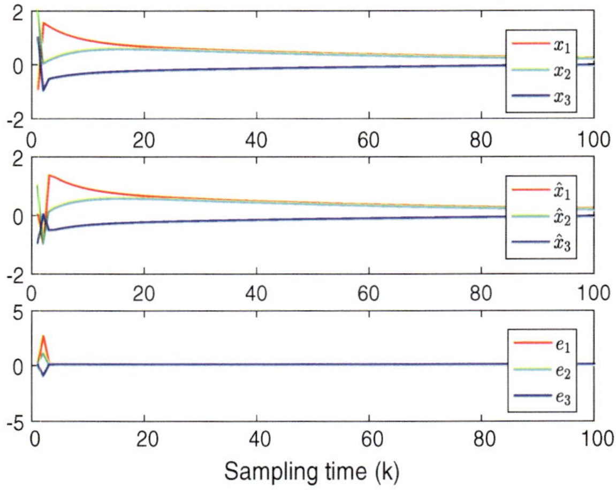

Fig. 5.8 Trajectories of states, estimated states and estimation errors of networked control system

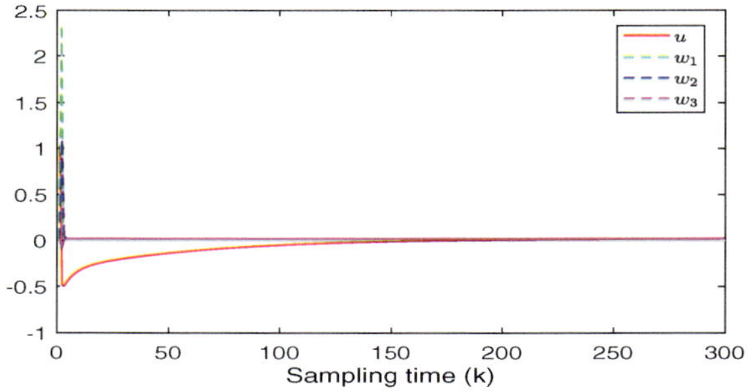

Fig. 5.9 Curves of the optimal control policy and the optimal observer policy under time-varying network-induced delays

5.2 Off-Policy Q-Learning for Multi-player Networked Control Systems

In Sect. 5.1, we have discussed the linear DT networked system with single controller. However, From the point of view of practical applications, network communication-based control systems are more preferred even though the unavoidable existence of network-induced delay and packet dropout would cause negative impacts on control systems, such as unmanned marine vehicles (Wang et al. 2018a), industrial process control (Wang et al. 2015) and Wireless HART (Maass et al. 2019), because of the advantages including remote manipulation, easy maintenance and less cost (Linsenmayer et al. 2019; Liu and Shi 2019; Xu and Jagannathan 2014). Given these advantages of network communication-based control systems, this section considers a class of multi-player games with transmitting the common system state from the plant to all players via networks.

The most relevant works focused on model-free multi-player or multi-agent control using ADPRL approaches are presented since multi-player systems could be taken as a special case of multi-agent systems if replacing the dynamics of each agent by the shared state of the entire systems. They are Li et al. (2020), Jiao et al. (2016), Lv and Ren (2019), Odekunle et al. (2020), Gao et al. (2018) and Jiang et al. (2020). Indeed, the above-mentioned literature have played a remarkable role in investigating data-driven optimal control problems of multi-player or multi-agent systems via solving zero-sum games (Lv and Ren 2019; Jiao et al. 2016) and nonzero-sum games (Li et al. 2020; Odekunle et al. 2020; Jiang et al. 2020). Particularly, Qin et al. (2019) and Zhang et al. (2018a) presented the off-policy ADPRL methods for achieving the optimal synchronization of multi-agent systems. However, they cannot be directly used to cope with the Nash equilibrium solution of multi-player or multi-agent games where network communication is employed for transmitting signals with the occurrence of network-induced delay. On the other hand, the developed model-free optimal control methods with the consideration of packet dropout occurred in network communication (Jiang et al. 2017) cannot work out the faced challenges when solving the coupled HJB equation derived in multi-player games. Thus, one may wonder if a novel model-free ADPRL method could be developed for networked multi-player systems suffering from network-induced delay, unknown model parameters and even unmeasured system state. This section will present an affirmative answer.

Q-learning, one of RL methods, has the capability of finding the optimal control policy for optimization problems, without the knowledge of system dynamics (Al-Tamimi et al. 2007). Compared with the on-policy RL, the off-policy RL offers some merits including the guarantee of unbiased solution even though adding probing noises for satisfying PE and full exploration by introducing arbitrary admissible behavior policies (Kiumarsi et al. 2017). Therefore, this section dedicates to developing a novel off-policy Q-learning algorithm under the framework of ADPRL, such that the Nash equilibrium solution to nonzero-sum games for networked multiple players is found without requiring the system matrices to be known and measurable system state. Model-free solving the multiple coupled AREs with the obstacle caused by network-induced delay would increase the difficulty without doubts when learning the Nash equilibrium solution. The main contributions of this section are as follows. Firstly, a novel methodology for solving the nonzero-sum game of networked multi-player systems is presented, which differs from those in Zhang et al. (2016), Song et al. (2016), Liu et al. (2014b), Lv et al. (2020), Zhao et al. (2015), Johnson et al. (2014), Li et al. (2020), Jiao et al. (2016), Lv and Ren (2019), Odekunle et al. (2020) and Gao et al. (2018) applied for multi-player or multi-agent systems without the consideration of network-induced delay and its compensation, and those in Jiang et al. (2017) and Li et al. (2020) for single-player systems in the presence of network-induced delay or packet dropout. Secondly, the proposed methodology can successfully handle the network-induced delay, unknown model parameter and unmeasurable system state via an off-policy Q-learning approach, which is achieved under the help of a virtual Smith predictor and state observer. Thus, the Nash equilibrium solution with compensation capability of network-induced delay can be learned using only data. Finally, the theoretical proof of convergence of the proposed algorithm is provided.

This section is organized as follows. In Sect. 5.2.1, we first formulate the nonzero-sum multi-player game problem, with adding a state observer and a virtual Smith predictor. Section 5.2.2 develops a novel off-policy Q-learning algorithm for finding the solution to the nonzero-sum game of networked multi-player systems, together with a theoretical proof of convergence of the proposed algorithm. In Sect. 5.2.3, an illustrative example is given to verify the effectiveness of the developed method.

5.2.1 Problem Formulation

Consider a class of linear discrete-time systems with multiple players described by

$$\begin{aligned} x_{k+1} &= Ax_k + \sum\nolimits_{i=1}^{s} B_i u_{ik} \\ y_k &= Cx_k \end{aligned} \tag{5.97}$$

where $x_k \in \mathbb{R}^{n_x}$ is the system state with initial state x_0 and $y_k \in \mathbb{R}^{n_y}$ is system output. Define a set $\mathbb{S} = \{1, 2, \ldots, s\}$ composed by all players, and for any $i \in \mathbb{S}$, $u_{ik} = u_i(k) \in \mathbb{R}^{n_{u_i}}$ is control input of player i, s is a positive integer and k is the sampling time. A, B_i and C are unknown system matrices with appropriate dimensions.

Assumption 5.1 The pair (A, B_i) is stabilizable, $\forall i \in \mathbb{S}$. The pair (A, C) is observable.

The nonzero-sum game problem of the multi-player system (5.97) is formulated as follows.

Problem 5.3 Each player i makes efforts to find the control policy u_{ik}^* by

$$\min_{u_{ik}} \sum_{k=0}^{\infty} \left(y_k^T Q_i y_k + \sum_{i=1}^{s} u_{ik}^T R_i u_{ik} \right) \tag{5.98}$$

subject to (5.97) and $u_{ik} = -K_i x_k$ where $Q_i = Q_i^T \geq 0$ and $R_i = R_i^T > 0$. The pairs $(\sqrt{Q_i}, A)$ are detectable.

Definition 5.1 (Başar and Olsder 1998) Given an s-tuple of control policies $\{u_1^*, u_2^*, \ldots, u_s^*\}$ satisfying the following s inequalities,

$$J_i^* \equiv J_i(u_1^*, \ldots, u_i^*, \ldots, u_s^*) \leq J_i(u_1^*, \ldots, u_i, \ldots, u_s^*) \tag{5.99}$$

then the s-tuple of control policies is called a Nash equilibrium solution to the nonzero-sum game and the s-tuple of quantities $\{J_1^*, J_2^*, \ldots, J_s^*\}$ is viewed as a Nash equilibrium outcome of the nonzero-sum game.

The objective of this subsection is to find the unique Nash equilibrium solution to the nonzero-sum game Problem 5.3. However, there are three obstacles that we have to face. One is how to eliminate the negative impact caused by the network-induced delay when transmitting the signals from the sensor to the controllers via a communication network. The second one is how to handle the case of unmeasurable system state when designing state-feedback control policies since the state of systems usually cannot be directly measured in practice. The last one is how to find the Nash equilibrium solution if the system matrices of system (5.97) are completely unknown. To successfully fix out these difficulties, as shown in Fig. 5.10, a state observer and a virtual Smith predictor have been added for respectively estimating the system state and compensating the network-induced delay. Moreover, a data-driven learning algorithm will be developed by utilizing ADP combined with Q-learning in the sequential subsection.

Virtual Smith Predictor: A system state-based virtual Smith predictor is given by

$$x_k = A^{d_k} x_{k-d_k} + \sum_{v=1}^{d_k} A^{v-1} \sum_{i=1}^{s} B_i u_{i(k-v)} \tag{5.100}$$

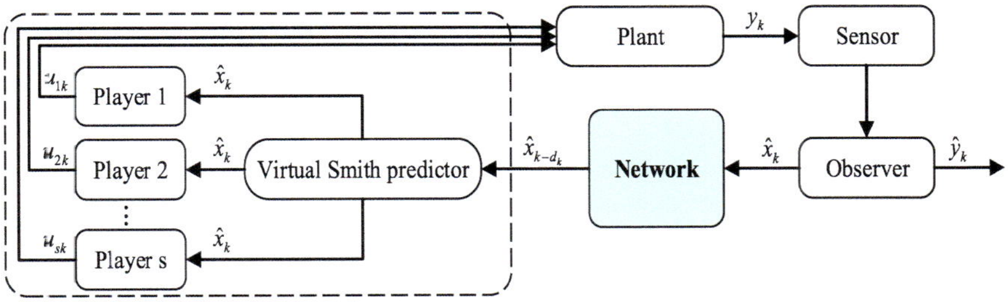

Fig. 5.10 A nonzero-sum game of networked multi-player system

where $d_k = d(k)$ denotes the bounded network-induced delay with $d_k \in [0, \delta_{\max}]$, where $\delta_{\max}$ is some positive value.

Remark 5.11 Equation (5.100) can be easily derived by (5.97). However, due to the system matrices are unknown and the system state x_{k-d_k} are unmeasurable, the state x_k at the time k cannot be predicted by the Smith predictor, that is why it is called a virtual Smith predictor in this subsection.

To achieve our goal of finding the Nash equilibrium solution to Problem 5.3, we shall follow two steps. Firstly, an observer should be designed with the outcome of $\lim_{k\to\infty}(\hat{x}_k - x_k) = 0$ using data including the output and input of system (5.97) since system matrices of (5.97) are unknown. Then, the Nash equilibrium solution will be learned with the help of the virtual Smith predictor and the replacement for x_{k-d_k} using $\hat{x}_{k-d_k}$. Now, we are going to start the first step to design the observer referring to Ding et al. (2011) and Alipouri et al. (2020), where the state of systems can be observed through a Luenberger type observer below

$$\begin{aligned} \hat{y}_{k+1} &= G\hat{y}_k + \sum\nolimits_{i=1}^{s} D_i u_{ik} + L y_k \\ T\hat{x}_k &= \hat{y}_k \\ r_k &= g_y y_k - c_y \hat{y}_k \end{aligned} \tag{5.101}$$

where $\hat{x}_k$ and $\hat{y}_k$ are the state and output of the observer, r_k is the residual signal. G (should be stable), D_i, L, T, g_y and c_y are matrices with appropriate dimensions. By Algorithm D2PS and Algorithm PS2DO in Ding et al. (2011), the parameters of observer (5.101) can be designed using only input and output data from system (5.97).

Remark 5.12 Collecting the historic data consisted of inputs and outputs of system (5.97) yields sufficient data with the satisfaction of PE condition, thus the asymptotic stability of $\hat{x}_k - x_k$ can surely be guaranteed if the selected parameters in (5.101) satisfying the Luenberger conditions (Ding et al. 2011; Thomas 2002).
Now we are in the position to find controller gains K_i in Problem 5.3 with the constructed virtual Smith predictor and estimated state $\hat{x}_{k-d_k}$ from (5.101).

5.2.2 *Main Results*

In this subsection, a novel off-policy Q-learning algorithm is proposed for finding the Nash equilibrium solution for the multi-player game Problem 5.3 with unknown system matrices by using the observer state and control inputs.

According to the performance index in Problem 5.3, the value function and the Q-function can be defined as

$$V_i(x_k) = \sum_{q=k}^{\infty} \left(y_q^T Q_i y_q + \sum_{i=1}^{s} u_{iq}^T R_i u_{iq} \right) \tag{5.102}$$

$$Q_i(x_k, u_{1k}, \ldots, u_{sk}) = y_k^T Q_i y_k + \sum_{i=1}^{s} u_{ik}^T R_i u_{ik} + V_i(x_{k+1}) \tag{5.103}$$

and have quadratic forms (Kiumarsi et al. 2015; Vamvoudakis 2015) as $V_i(x_k) = x_k^T P_i x_k$ and $Q_i(x_k, u_{1k}, \ldots, u_{sk}) = Z_k^T H_i Z_k$, where H_i $(i \in \mathbb{S})$ are the Q-function matrices, $Z_k = [x_k^T \ u_{1k}^T \ \ldots \ u_{sk}^T]^T$, $P_i = M^T H_i M$, $M = [I \ -K_1^T \ \ldots \ -K_s^T]^T$, $P_i = P_i^T > 0$, and

$$H_i = \begin{bmatrix} H_{i,xx} & H_{i,xu_1} & H_{i,xu_2} & \cdots & H_{i,xu_s} \\ H_{i,xu_1}^T & H_{i,u_1u_1} & H_{i,u_1u_2} & \cdots & H_{i,u_1u_s} \\ \vdots & \vdots & \vdots & \ddots & \vdots \\ H_{i,xu_s}^T & H_{i,u_1u_s}^T & H_{i,u_2u_s}^T & \cdots & H_{i,u_su_s} \end{bmatrix}$$
$$= \begin{bmatrix} A^T P_i A + C^T Q_i C & A^T P_i B_1 & \cdots & A^T P_i B_s \\ (A^T P_i B_1)^T & B_1^T P_i B_1 + R_1 & \cdots & B_1^T P_i B_s \\ \vdots & \vdots & \ddots & \vdots \\ (A^T P_i B_s)^T & (B_1^T P_i B_s)^T & \cdots & B_s^T P_i B_s + R_s \end{bmatrix}. \tag{5.104}$$

Utilizing the dynamic programming method and implementing the manipulation $\partial Q_i / \partial u_{ik} = 0$ (Wang et al. 2019; Na et al. 2020), we derive the Q-function-based game AREs,

$$Z_k^T H_i^* Z_k = y_k^T Q_i y_k + \sum_{i=1}^{s} (u_{ik}^*)^T R_i u_{ik}^* + Z_{k+1}^T H_i^* Z_{k+1} \tag{5.105}$$

and the optimal control policies $u_{ik}^* = -K_i^* x_k$, where

$$K_i^* = (H_{i,u_iu_i}^*)^{-1} [(H_{i,xu_i}^*)^T - \sum_{t=1, t\neq i}^{s} H_{i,u_iu_t}^* K_t^*]. \tag{5.106}$$

Based on the nonzero-sum game theory (Song et al. 2016; Jiang et al. 2018; Vamvoudakis 2015), the s-tuple of $\{u_{1k}^*, u_{2k}^*, \ldots, u_{sk}^*\}$ constitutes the unique Nash equilibrium solution to nonzero-sum game Problem 5.3.

If we follow the existing methods like Song et al. (2016) and Al-Tamimi et al. (2007). Algorithm 5.5 can be presented below for fixing out the nonlinear coupling of matrices H_i $(i = 1, 2, \ldots, s)$ in (5.105) via an iterative approach, resulting in finding the optimal controller gains K_i^*.

Remark 5.13 In Algorithm 5.5, it can be found that $H_i^{j+1} \to H_i^*$ and $K_i^{j+1} \to K_i^*$ as $j \to \infty$, resulting in the convergence of $\{u_{1k}^j, u_{2k}^j, \ldots, u_{sk}^j\}$ to the Nash equilibrium solution. This has been proven in Song et al. (2016) and Al-Tamimi et al. (2007).

Algorithm 5.5 Policy iteration Q-learning algorithm

1: Let K_i^0 $(i \in \mathbb{S})$ be any admissible control gains and set the iteration index j to be zero;

2: Policy evaluation: using the least squares method solves H_i^{j+1} with $\{u_{1k}^j, u_{2k}^j, \ldots, u_{sk}^j\}$ in terms of

$$Z_k^T H_i^{j+1} Z_k = y_k^T Q_i y_k + \sum_{i=1}^{s} (u_{ik}^j)^T R_i u_{ik}^j + Z_{k+1}^T H_i^{j+1} Z_{k+1}$$
$$(i = 1, 2, \ldots, s); \tag{5.107}$$

3: Policy updating: given $\{H_1^{j+1}, H_2^{j+1}, \ldots, H_s^{j+1}\}$, we update K_i^{j+1} $(i = 1, 2, \ldots, s)$ by

$$K_i^{j+1} = (H_{i,u_i u_i}^{j+1})^{-1}[(H_{i,xu_i}^{j+1})^T - \sum\nolimits_{t=1,t\neq i}^{s} H_{i,u_i u_t}^{j+1} K_t^j]; \tag{5.108}$$

4: If $\left\| K_i^{j+1} - K_i^j \right\| \leq \epsilon$ $(\epsilon > 0)$ for all $i \in \mathbb{S}$, stop. Otherwise, $j = j + 1$ and go back to Step 2.

Remark 5.14 It is worthwhile pointing out that one of the intractable obstacles when implementing Algorithm 5.5 is that system state x_k is unavailable at time k because the virtual Smith predictor cannot work when system matrices A, B_i and C are unknown, such that the control gains K_i^* cannot be learned. In this case, the existing model-based predictor design methods, such as Yang et al. (2013), cannot be directly adopted to fix out the above-mentioned difficulty. Additionally, the existing model-free ADPRL methods for game problems (Al-Tamimi et al. 2007; Li et al. 2020) cannot directly work either, since they took no consideration of the network-induced delay.

Next, some significant mathematical manipulations are made by analyzing the added virtual Smith predictor in our scheme shown in Fig. 5.10, such that the control policies $\{u_{1k}^*, u_{2k}^*, \ldots, u_{sk}^*\}$ can be derived by proposing a novel Q-learning algorithm using available observer states.

5.2.2.1 Analysis of Observer State-Based Virtual Smith Predictor

According to the network-induced delay, we define two new variables $\bar{z}_k$ and $\hat{z}_k$.

• If $d_k = 0$, $\bar{z}_k = [\underbrace{\hat{x}_k^T \; 0 \; \ldots \; 0}_{\delta_{\max}+1} \underbrace{\tilde{u}_{k-1} \; \ldots \; \tilde{u}_{k-\delta_{\max}}}_{\delta_{\max}}]^T$, $\hat{z}_k = [\underbrace{x_k^T \; 0 \; \ldots \; 0}_{\delta_{\max}+1} \underbrace{\tilde{u}_{k-1} \; \ldots \; \tilde{u}_{k-\delta_{\max}}}_{\delta_{\max}}]^T$,

where $\tilde{u}_{k-d} = [0 \; \ldots \; 0]$ and $d = 1, 2, \ldots, \delta_{\max}$.

• If $d_k = 1$, $\bar{z}_k = [\underbrace{0 \; \hat{x}_{k-1}^T \; 0 \; \ldots \; 0}_{\delta_{\max}+1} \underbrace{\tilde{u}_{k-1} \; \ldots \; \tilde{u}_{k-\delta_{\max}}}_{\delta_{\max}}]^T$, $\hat{z}_k = [\underbrace{0 \; x_{k-1}^T \; 0 \; \ldots \; 0}_{\delta_{\max}+1} \underbrace{\tilde{u}_{k-1} \; \ldots \; \tilde{u}_{k-\delta_{\max}}}_{\delta_{\max}}]^T$,

where $\tilde{u}_{k-1} = [u_{1(k-1)}^T \; \ldots \; u_{s(k-1)}^T]$, $\tilde{u}_{k-d} = [0 \; \ldots \; 0]$ and $d = 2, 3, \ldots, \delta_{\max}$.

• …

- If $d_k = \delta_{\max}$, $\bar{z}_k = [\underbrace{0\ \ldots\ 0\ \hat{x}^T_{k-\delta_{\max}}}_{\delta_{\max}+1}\ \underbrace{\tilde{u}_{k-1}\ \ldots\ \tilde{u}_{k-\delta_{\max}}}_{\delta_{\max}}]^T$, $\hat{z}_k = [\underbrace{0\ \ldots\ 0\ x^T_{k-\delta_{\max}}}_{\delta_{\max}+1}\ \underbrace{\tilde{u}_{k-1}\ \ldots\ \tilde{u}_{k-\delta_{\max}}}_{\delta_{\max}}]^T$,

where $\tilde{u}_{k-d} = [u^T_{1(k-d)}\ \ldots\ u^T_{s(k-d)}]$ and $d = 1, 2, \ldots, \delta_{\max}$.

In fact, $\bar{z}_k$ represents the received signals of the virtual Smith predictor via the communication network from the state observer unlike $\hat{z}_k$, which contains the delayed system states. The predicted system state referring to (5.100) at time k can be expressed as $x_k = \hat{W}\hat{z}_k$, where $\hat{W} = [W_1\ A^{\delta_{\max}}\ W_2]$, $W_1 = [I\ A\ A^2 \ldots\ A^{\delta_{\max}-1}]$, $W_2 = [W_3\ AW_3\ A^2W_3\ \ldots\ A^{\delta_{\max}-1}W_3]$, $W_3 = [B_1\ B_2\ \ldots\ B_s]$. Therefore, the feedback controller of each player based on the virtual Smith predictor can be translated as $u_{ik} = -K_i x_k = -\hat{K}_i \hat{z}_k$, where $\hat{K}_i = K_i \hat{W}$.

Next, we shall try to design an algorithm for finding $\hat{K}^*_i$ instead of calculating K^*_i using available data.

5.2.2.2 Off-Policy Game Q-Learning Algorithm Design

Adding the auxiliary variables $u^j_{ik} = -K^j_i x_k$ into system (5.97) yields

$$\begin{aligned} x_{k+1} &= Ax_k - \sum\nolimits_{i=1}^{s} B_i K^j_i x_k + \sum\nolimits_{i=1}^{s} B_i (u_{ik} + K^j_i x_k) \\ &= \left(A - \sum\nolimits_{i=1}^{s} B_i K^j_i\right) \hat{W}\hat{z}_k + \sum\nolimits_{i=1}^{s} B_i (u_{ik} + K^j_i \hat{W}\hat{z}_k) \end{aligned} \tag{5.109}$$

where u^j_{ik} are called the target policies that are going to be learned and updated by the proposed algorithm. u_{ik} are the behavior control policies actually used to generate data for learning. Along the trajectory of system (5.109) and due to $x_k = \hat{W}\hat{z}_k$ and (5.107), one has

$$\begin{aligned} &\hat{Z}^T_k \bar{H}^{j+1}_i \hat{Z}_k - \hat{Z}^T_{k+1} \bar{H}^{j+1}_i \hat{Z}_{k+1} \\ &= \hat{Z}^T_k \Lambda_{2i} \hat{Z}_k - 2x^T_k \left(A - \sum\nolimits_{i=1}^{s} B_i K^j_i\right)^T (M^j)^T H^{j+1}_i \\ &\times M^j \sum\nolimits_{i=1}^{s} B_i (u_{ik} + \hat{K}^j_i \hat{z}_k) + \sum\nolimits_{i=1}^{s} (u_{ik} + \hat{K}^j_i \hat{z}_k)^T B^T_i \\ &\times (M^j)^T H^{j+1}_i M^j \sum\nolimits_{i=1}^{s} B_i (u_{ik} + \hat{K}^j_i \hat{z}_k) \end{aligned} \tag{5.110}$$

where $\hat{Z}_k = [\hat{z}^T_k\ -(\hat{K}^j_1 \hat{z}_k)^T\ \ldots\ -(\hat{K}^j_s \hat{z}_k)^T]^T$, $M^j = [I\ -(K^j_1)^T\ \ldots\ -(K^j_s)^T]^T$, $\Lambda_{2i} = \mathrm{diag}(\hat{W}^T(C^T Q_i C)\hat{W}, R_1, \ldots, R_s)$, and

$$\bar{H}^{j+1}_i = \Pi^T_1 H^{j+1}_i \Pi_1 = \begin{bmatrix} \bar{H}^{j+1}_{i,zz} & G^{j+1}_{i1} \\ (G^{j+1}_{i1})^T & G^{j+1}_{i2} \end{bmatrix}$$

where $\Pi_1 = \mathrm{diag}(\hat{W},\ I)$, $G^{j+1}_{i1} = [\bar{H}^{j+1}_{i,zu_1}\ \bar{H}^{j+1}_{i,zu_1}\ \ldots\ \bar{H}^{j+1}_{i,zu_s}]$, $G^{j+1}_{i2} = [\bar{H}^{j+1}_{i,u_a u_r}]$ $(a, r \in \mathbb{S})$, $\bar{H}^{j+1}_{i,u_a u_r}$ is the block matrix located in the ath row and the rth column in symmetric matrix G_{i2}^{j+1}. The dimension of matrix $\bar{H}_i^{j+1}$ is $\mu \times \mu$, where $\mu = (\delta_{\max}+1)\times(n_x+\sum_{i=1}^{s} n_{u_i})$. Besides,

$$
\begin{aligned}
&-2x_k^T(A-\sum_{i=1}^{s} B_i K_i^j)^T (M^j)^T H_i^{j+1} M^j \sum_{i=1}^{s} B_i(u_{ik}+\hat{K}_i^j \hat{z}_k)\\
&= -2\hat{z}_k^T G_{i1}^{j+1}(u_{ik}+\hat{K}_i^j \hat{z}_k) + 2u_{ik}^T G_{i3}^{j+1}(u_{ik}+\hat{K}_i^j \hat{z}_k)\\
&\sum_{i=1}^{s}(u_{ik}+\hat{K}_i^j \hat{z}_k)^T B_i^T (M^j)^T H_i^{j+1} M^j \sum_{i=1}^{s} B_i(u_{ik}+\hat{K}_i^j \hat{z}_k)\\
&= \sum_{i=1}^{s}(u_{ik}+\hat{K}_i^j \hat{z}_k)^T G_{i3}^{j+1} \sum_{i=1}^{s}(u_{ik}+\hat{K}_i^j \hat{z}_k)
\end{aligned}
\tag{5.111}
$$

where $G_{i3}^{j+1} = G_{i2}^{j+1} - \mathrm{diag}(R_1,\ldots,R_s)$. Thus, (5.110) is written as

$$
\hat{\theta}_i^j(k)\hat{L}_i^{j+1} = \hat{\rho}_{i,k}^j \tag{5.112}
$$

where

$$
\begin{aligned}
\hat{\rho}_{i,k}^j &= y_k^T Q_i y_k + \sum_{i=1}^{s} u_{ik}^T R_i u_{ik}\\
\hat{L}_i^{j+1} &= \Big[(\mathrm{vec}(\bar{H}_{i,zz}^{j+1}))^T\ (\mathrm{vec}(\bar{H}_{i,zu_1}^{j+1}))^T \ldots (\mathrm{vec}(\bar{H}_{i,zu_s}^{j+1}))^T\\
&\quad (\mathrm{vec}(\bar{H}_{i,u_1u_1}^{j+1}))^T \ldots (\mathrm{vec}(\bar{H}_{i,u_1u_s}^{j+1}))^T\ (\mathrm{vec}(\bar{H}_{i,u_2u_2}^{j+1}))^T\\
&\quad \ldots (\mathrm{vec}(\bar{H}_{i,u_2u_s}^{j+1}))^T \ldots (\mathrm{vec}(\bar{H}_{i,u_su_s}^{j+1}))^T\Big]^T\\
\hat{\theta}_i^j(k) &= \Big[\hat{\theta}_{i,zz}^j\ \hat{\theta}_{i,zu_1}^j \ldots \hat{\theta}_{i,zu_s}^j\ \hat{\theta}_{i,u_1u_1}^j \ldots \hat{\theta}_{i,u_1u_s}^j\\
&\quad \hat{\theta}_{i,u_2u_2}^j \ldots \hat{\theta}_{i,u_2u_s}^j \ldots \hat{\theta}_{i,u_su_s}^j\Big]\\
\hat{\theta}_{i,zz}^j &= \hat{z}_k^T \otimes \hat{z}_k^T - \hat{z}_{k+1}^T \otimes \hat{z}_{k+1}^T\\
\hat{\theta}_{i,zu_l}^j &= 2\hat{z}_k^T \otimes u_l^T + 2\hat{z}_{k+1}^T \otimes (\hat{K}_l^j \hat{z}_{k+1})^T\\
\hat{\theta}_{i,u_lu_l}^j &= u_l^T \otimes u_l^T - (\hat{K}_l^j \hat{z}_{k+1})^T \otimes (\hat{K}_l^j \hat{z}_{k+1})^T\\
\hat{\theta}_{i,u_lu_t}^j &= 2u_l^T \otimes u_t^T - 2(\hat{K}_l^j \hat{z}_{k+1})^T \otimes (\hat{K}_t^j \hat{z}_{k+1})^T\\
&(l = 1,2,\ldots,s,\ t = l+1, l+2,\ldots,s).
\end{aligned}
$$

Moreover, one can find $\bar{H}_{i,zu_l}^{j+1} = \hat{W}^T H_{i,xu_l}^{j+1}$, $\bar{H}_{i,u_lu_l}^{j+1} = H_{i,u_lu_l}^{j+1}$ and $\bar{H}_{i,u_lu_t}^{j+1} = H_{i,u_lu_t}^{j+1}$ from (5.111). Therefore, according to $\hat{K}_i^{j+1} = K_i^{j+1}\hat{W}$ and (5.108), one obtains

$$
\hat{K}_i^{j+1} = (\bar{H}_{i,u_iu_i}^{j+1})^{-1}\Big[(\bar{H}_{i,zu_i}^{j+1})^T - \sum\nolimits_{t=1,t\neq i}^{s} \bar{H}_{i,u_iu_t}^{j+1}\hat{K}_t^j\Big]. \tag{5.113}
$$

Therefore, correctly solving (5.112) and using its solution $\hat{L}_i^{j+1}$, the matrices $\hat{K}_i^{j+1}$ can be finally calculated in terms of (5.113). The detailed procedure is given in Algorithm 5.6 for finding the matrices $\hat{K}_i^*$.

Algorithm 5.6 Off-policy game Q-learning algorithm

1: Choose admissible behavior policies u_{ik} with the added probing noises, which are acted as control inputs of the networked multi-player systems (5.97);
2: Given the initial admissible controller gains $\hat{K}_i^0$ ($i \in \mathbb{S}$) and set the iteration index j to be zero;
3: Policy evaluation: solve $\hat{L}_i^{j+1}$ of (5.112) using the iterative controller gains $\{\hat{K}_1^j, \hat{K}_2^j, \ldots, \hat{K}_s^j\}$;
4: Policy updating: calculate matrices $\hat{K}_i^{j+1}$ in terms of (5.113);
5: If $\left\|\hat{K}_i^{j+1} - \hat{K}_i^j\right\| \le \epsilon$ ($\epsilon > 0$) for all $i \in \mathbb{S}$, stop. Otherwise, $j = j + 1$ and go back to Step 3.

Theorem 5.4 is given below to prove that the matrices $\hat{K}_i^{j+1}$ in Algorithm 5.6 finally converge to matrices $\hat{K}_i^*$.

Theorem 5.4 *There exist matrices* $\hat{K}_i^* = K_i^* \hat{W}$, *such that* $\lim_{j\to\infty} \hat{K}_i^{j+1} = \hat{K}_i^*$.

Proof From (5.110) to (5.113), it concludes that there exists a matrix $\bar{H}_i^{j+1}$ satisfied with (5.112) for player i, because there exists a unique solution H_i^{j+1} to (5.107). Now, we are going to prove the uniqueness of solution to (5.112). Suppose there are two distinct solutions $\bar{H}_i^{j+1}$ and $\bar{M}_i^{j+1}$ to (5.112), one has two distinct solutions H_i^{j+1} and M_i^{j+1} to (5.107) due to (5.111) with the row full rank matrix Π_1. By contradiction, the solution $\bar{H}_i^{j+1}$ is unique to (5.112) for player i. Thus, it follows $\lim_{j\to\infty} \bar{H}_i^{j+1} = \Pi_1^T H_i^* \Pi_1$ and $\lim_{j\to\infty} \hat{K}_i^{j+1} = K_i^* \hat{W}$ due to (5.111), $\hat{K}_i^{j+1} = K_i^{j+1} \hat{W}$, $\lim_{j\to\infty} H_i^{j+1} = H_i^*$ and $\lim_{j\to\infty} K_i^{j+1} = K_i^*$ explained in Remark 5.13. □

Remark 5.15 If $\hat{K}_i^*$ can be found by implementing Algorithm 5.6, then the Nash equilibrium solution to the multi-player game can be finally derived by $\lim_{j\to\infty} u_{ik}^{j+1} = \lim_{j\to\infty} -\hat{K}_i^{j+1}\hat{z}_k = -\hat{K}_i^*\hat{z}_k = -K_i^* x_k$, i.e., $u_{ik}^* = -\hat{K}_i^*\hat{z}_k$.

In Algorithm 5.6, the admissible behavior policies u_{ik} and the admissible control gains $\hat{K}_i^0$ are generally chosen empirically according to the practical applications or utilizing the robust control methods (Modares and Lewis 2014). As claimed in Al-Tamimi et al. (2007), Jiao et al. (2016) and Kiumarsi et al. (2015), the least squares method together with probing noises added into behavior control policies could be employed for ensuring the PE condition of system (5.97), such that the unique solution $\hat{L}_i^{j+1}$ (resulting in getting $\bar{H}_i^{j+1}$) to (5.112) can be obtained. In addition, the unbiased solution of $\bar{H}_i^{j+1}$ can be guaranteed by the off-policy learning approach even though adding probing noises, usually white noises, as proved in Kiumarsi et al. (2017).

Remark 5.16 Notice that solving (5.112) needs data $\hat{z}_k$, in which the state x_{k-d_k} are unmeasurable. The available data $\bar{z}_k$ will be utilized instead of $\hat{z}_k$ with the arbitrarily small error of $\bar{z}_k - \hat{z}_k$ promised by the Luenberger conditions, which is the similar idea to Modares et al. (2016).

Remark 5.17 Different form Li et al. (2020) and Kiumarsi et al. (2015) with the consideration of unmeasurable system state for single-player systems and Liang et al. (2020) with the developed model-based reduced-order state observer, we introduced the auxiliary variables $(\bar{z}_k, \hat{z}_k)$, multiple matrices $\hat{K}_i^{j+1}$ by some mathematical manipulations based on the virtual Smith predictor and the observer, so that the Nash equilibrium solution $u_{ik}^* = -K_i^* x_k$ $(i = 1, 2, \ldots, s)$ can be derived by the form of $u_{ik}^* = -\hat{K}_i^* \bar{z}_k$ using only measured data. These manipulations overcome the technical challenges of solving the coupled AREs for multi-player games with unknown system matrices, network-induced delay and unmeasurable system state.

Remark 5.18 Following the idea of Sect. 5.2.2, the proposed off-policy game Q-learning algorithm for linear multi-player systems can be extended to the case of nonlinear multi-player systems, while even more challenges could be brought out than that for linear systems. This is because the Q-functions could be complex nonlinear rather than quadratic form and the virtual Smith predictor could be nonlinear as well caused by the natural complexity of nonlinear dynamics of systems (Jiang et al. 2018; Johnson et al. 2014; Li et al. 2020; Poveda et al. 2019). Besides, the coupling of control policies and unmeasurable system state certainly set the obstacles for solving Nash equilibrium of nonlinear multi-player systems via model-free Q-learning approaches.

5.2.3 *Illustrative Example*

To verify the effectiveness of the proposed data-driven Algorithm 5.6, a three-player nonzero-sum game problem characterized by winding machine system (5.114) is taken into account. The three players are the unwinding reel, the traction reel and the rewinding reel and their control inputs respectively are u_1, u_2 and u_3.

$$\begin{aligned} x_{k+1} &= Ax_k + B_1 u_{1k} + B_2 u_{2k} + B_3 u_{3k} \\ y_k &= Cx_k \end{aligned} \tag{5.114}$$

where $B_1 = [-1.7734\ 0.0928\ -0.0424]^T$, $B_2 = [0.0696\ 0.4\ 658\ -0.093]^T$, $B_3 = [0.0734\ 0.1051\ 2.0752]^T$ and

$$A = \begin{bmatrix} 0.4126 & 0 & -0.0196 \\ 0.0333 & 0.5207 & -0.0413 \\ -0.0101 & 0 & 0.2571 \end{bmatrix}, \quad C = \begin{bmatrix} 1 & 0 & -1 \\ -2 & 1 & 0 \\ 1 & 1 & 1 \end{bmatrix}.$$

We choose $Q_1 = \text{diag}(5, 5, 5)$, $Q_2 = \text{diag}(6, 6, 6)$, $Q_3 = \text{diag}(7, 7, 7)$ and $R_1 = R_2 = R_3 = 1$, and set the initial states of the system and observer to be $x_0 = [-3\ -1\ 3]^T$, $\hat{x}_0 = [1\ -1\ 0.5]^T$. Based on the system model, one has

$$\begin{aligned}
\hat{K}_1^* = [\,&0.2255\ -0.0482\ -0.0108\ 0.0915\ -0.0251\\
&-0.0052\ 0.0370\ -0.0131\ -0.0021\ -0.4039\\
&-0.0057\ -0.0109\ -0.1644\ -0.0048\ -0.0067\,]\\
\hat{K}_2^* = [\,&-0.0759\ -0.7165\ 0.0742\ -0.0559\ -0.3731\\
&0.0501\ -0.0360\ -0.1943\ 0.0294\ 0.0649\\
&-0.3549\ 0.0730\ 0.0624\ -0.1823\ 0.0607\,]\\
\hat{K}_3^* = [\,&0.0010\ -0.0764\ -0.1139\ 0.0009\ -0.0398\\
&-0.0261\ -0.0015\ -0.0207\ -0.0051\ -0.0041\\
&-0.0249\ -0.2443\ -0.0008\ -0.0162\ -0.0585\,].
\end{aligned}$$

Figure 5.11a shows the network-induced delay of transmitting data from the observer to the predictor with the maximum network-induced delay $\delta_{\max} = 2$.

Now we assume the system matrices A, B_1, B_2, B_3 and C in (5.114) are unknown. To verify Algorithm 5.6, the convergence threshold ϵ is set to be $1 * 10^{-3}$ and the probing noises are selected as $3(\sin(k)\cos(0.8k)^2 + 3\sin(4k)^3)$, $5.5(\sin(2k)^3 \cos(9k) + 0.37\sin(1.1k)^2\cos(4k))$ and $0.9\sin(8k)^2$. The Luenberger type observer matrices are obtained below

$$G = \begin{bmatrix} 0 & 0 & 0 \\ 1 & 0 & 0 \\ 0 & 1 & 0 \end{bmatrix},\ L = \begin{bmatrix} -0.23 & -0.1661 & -0.1577 \\ 0.4407 & 0.3118 & 0.2559 \\ 0.1589 & 0.1535 & 0.2010 \end{bmatrix} \tag{5.115}$$

$$D_1 = \begin{bmatrix} 0.0770 \\ 0.0111 \\ -0.0724 \end{bmatrix},\ D_2 = \begin{bmatrix} -0.3116 \\ -0.0441 \\ 0.2253 \end{bmatrix},\ D_3 = \begin{bmatrix} 0.2925 \\ -0.1245 \\ 0.1271 \end{bmatrix} \tag{5.116}$$

$$T = \begin{bmatrix} -0.0624 & -0.6453 & 0.1739 \\ -0.0036 & -0.1031 & -0.0509 \\ 0.0436 & 0.4892 & 0.0283 \end{bmatrix}. \tag{5.117}$$

Then, by Algorithm 5.6, the convergence of the matrices $\hat{K}_i^j$ is achieved after 5 iterations shown in Fig. 5.11b. Figure 5.11c plots the residual signal, the output and the control inputs of the three players. From Fig. 5.11, the Nash equilibrium solution is found and the output of the networked three-player system is asymptotically stable. Thus, the simulation results verify the effectiveness of the proposed method.

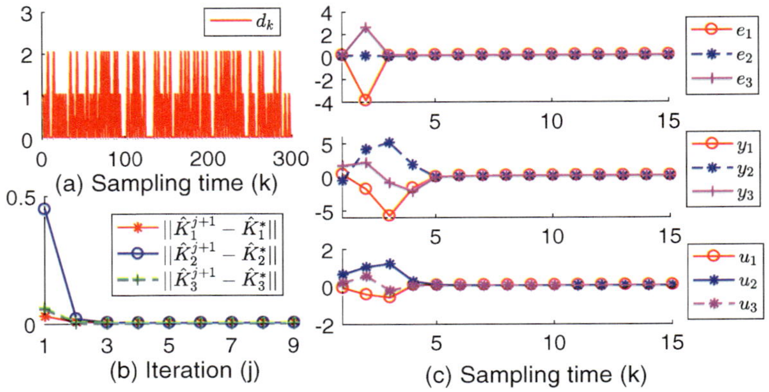

Fig. 5.11 **a** Network-induced delay **b** Convergence evolution of $\hat{K}_i^{j+1}$ matrices by Algorithm 5.6 and **c** Error of the observer, system output and optimal control input

5.3 Conclusion

In this section, a data-driven off-policy Q-learning method to solve the nonzero-sum game problem for linear DT multi-player systems subject to network-induced delay, unknown system model parameters and unmeasurable system state. By combining ADPRL with game theory and making appropriate mathematical manipulations, a novel off-policy game Q-learning algorithm is developed with the assistance of the virtual Smith predictor and state observer, such that the Nash equilibrium solution to the nonzero-sum game of networked multi-player systems is found using only available data. Numerical simulations verify the effectiveness of the proposed method.

References

Alipouri Y, Zhao S, Huang B (2020) Distributed data-driven observer for linear time invariant systems. Int J Adapt Control Signal Process 34(4):503–519

Al-Tamimi A, Lewis FL, Abu-Khalaf M (2007) Model-free Q-learning designs for linear discrete-time zero-sum games with application to H-infinity control. Automatica 43(3):473–481

Başar T, Olsder GJ (1998) Dynamic noncooperative game theory, 2nd edn. SIAM, PA

Beale S, Shafai B (1989) Robust control system design with a proportional integral observer. Int J Control 50(1):97–111

Chen G, Guo Z (2017) Distributed secondary and optimal active power sharing control for islanded microgrids with communication delays. IEEE Trans Smart Grid 10(2):2002–2014

Ding SX, Yin S, Wang Y, Yang Y, Ni B (2011) Data-driven design of observers and its applications. IFAC Proc Vol 44(1):11441–11446

Gao W, Jiang Z, Lewis FL, Wang Y (2018) Leader-to-formation stability of multiagent systems: an adaptive optimal control approach. IEEE Trans Autom Control 63(10):3581–3587

Jiang Y, Fan J, Chai T, Lewis FL, Li J (2017) Tracking control for linear discrete-time networked control systems with unknown dynamics and dropout. IEEE Trans Neural Netw Learn Syst 29(10):4607–4620

Jiang H, Zhang H, Zhang K, Cui X (2018) Data-driven adaptive dynamic programming schemes for non-zero-sum games of unknown discrete-time nonlinear systems. Neurocomputing 275:649–658

Jiang Y, Fan J, Gao W, Chai T, Lewis FL (2020) Cooperative adaptive optimal output regulation of nonlinear discrete-time multi-agent systems. Automatica 121:109149

Jiao Q, Modares H, Xu S, Lewis FL, Vamvoudakis KG (2016) Multi-agent zero-sum differential graphical games for disturbance rejection in distributed control. Automatica 69:24–34

Johnson M, Kamalapurkar R, Bhasin S, Dixon WE (2014) Approximate N-player nonzero-sum game solution for an uncertain continuous nonlinear system. IEEE Trans Neural Netw Learn Syst 26(8):1645–1658

Kiumarsi B, Lewis FL, Modares H, Karimpour A, Naghibi-Sistani MB (2014) Reinforcement Q-learning for optimal tracking control of linear discrete-time systems with unknown dynamics. Automatica 50(4):1167–1175

Kiumarsi B, Lewis FL, Naghibi-Sistani MB, Karimpour A (2015) Optimal tracking control of unknown discrete-time linear systems using input-output measured data. IEEE Trans Cybern 45(12):2770–2779

Kiumarsi B, Lewis FL, Jiang Z (2017) H_∞ control of linear discrete-time systems: off-policy reinforcement learning. Automatica 78:144–152

Li J, Modares H, Chai T, Lewis FL, Xie L (2017) Off-policy reinforcement learning for synchronization in multiagent graphical games. IEEE Trans Neural Netw Learn Syst 28(10):2434–2445

Li J, Chai T, Lewis FL, Ding Z, Jiang Y (2019) Off-policy interleaved Q-learning: optimal control for affine nonlinear discrete-time systems. IEEE Trans Neural Netw Learn Syst 30(5):1308–1320

Li J, Ding J, Chai T, Lewis FL (2020) Nonzero-sum game reinforcement learning for performance optimization in large-scale industrial processes. IEEE Trans Cybern 50(9):4132–4145

Liang H, Guo X, Pan Y, Huang T (2020) Event-triggered fuzzy bipartite tracking control for network systems based on distributed reduced-order observers. IEEE Trans Fuzzy Syst. https://doi.org/10.1109/TFUZZ.2020.2982618

Li J, Chai T, Lewis FL, Fan J, Ding Z, Ding J (2018a) Off-policy Q-learning: set-point design for optimizing dual-rate rougher flotation operational processes. IEEE Trans Ind Electron 65(5):4092–4102

Li Y, Li H, Ding X, Zhao G (2018b) Leader-follower consensus of multiagent systems with time delays over finite fields. IEEE Trans Cybern 49(8):3203–3208

Li Y, Li H, Sun W (2018c) Event-triggered control for robust set stabilization of logical control networks. Automatica 95:556–560

Linsenmayer S, Dimarogonas DV, Allgöwer F (2019) Periodic event-triggered control for networked control systems based on non-monotonic Lyapunov functions. Automatica 106:35–46

Liu W, Shi P (2019) Optimal linear filtering for networked control systems with time-correlated fading channels. Automatica 101:345–353

Liu S, Xie L, Quevedo DE (2016) Event-triggered quantized communication-based distributed convex optimization. IEEE Trans Control Netw Syst 5(1):167–178

Liu S, Liu PX, Wang X (2017) Stability analysis and compensation of network-induced delays in communication-based power system control: a survey. ISA Trans 66:143–153

Liu S, Liu PX, El Saddik A (2014a) Modeling and stability analysis of automatic generation control over cognitive radio networks in smart grids. IEEE Trans Syst Man Cybern Syst 45(2):223–234

Liu D, Li H, Wang D (2014b) Online synchronous approximate optimal learning algorithm for multi-player non-zero-sum games with unknown dynamics. IEEE Trans Syst Man Cybern Syst 44(8):1015–1027

Li J, Xiao Z, Li P, Ding Z (2020) Networked controller and observer design of discrete-time systems with inaccurate model parameters. ISA Trans 98:75–86

Luo B, Wu HN, Huang T (2014) Off-policy reinforcement learning for H_∞ control design. IEEE Trans Cybern 45(1):65–76

Lv Y, Ren X (2019) Approximate Nash solutions for multiplayer mixed-zero-sum game with reinforcement learning. IEEE Trans Syst Man Cybern Syst 49(12):2739–2750
Lv Y, Ren X, Na J (2020) Online Nash-optimization tracking control of multi-motor driven load system with simplified RL scheme. ISA Trans 98:251–262
Maass AI, Nešić D, Postoyan R, Dower PM (2019) Lp stability of networked control systems implemented on wirelesshart. Automatica 109:108514
Modares H, Lewis FL (2014) Linear quadratic tracking control of partially-unknown continuous-time systems using reinforcement learning. IEEE Trans Autom Control 59(11):3051–3056
Modares H, Lewis FL, Jiang ZP (2015) H_∞ tracking control of completely unknown continuous-time systems via off-policy reinforcement learning. IEEE Trans Neural Netw Learn Syst 26(10):2550–2562
Modares H, Nageshrao SP, Lopes GAD, Babuška R, Lewis FL (2016) Optimal model-free output synchronization of heterogeneous systems using off-policy reinforcement learning. Automatica 71:334–341
Na J, Lv Y, Zhang K, Zhao J (2020) Adaptive identifier-critic-based optimal tracking control for nonlinear systems with experimental validation. IEEE Trans Syst Man Cybern Syst
Odekunle A, Gao W, Davari M, Jiang ZP (2020) Reinforcement learning and non-zero-sum game output regulation for multi-player linear uncertain systems. Automatica 112:108672
Pham TN, Trinh H et al (2015) Load frequency control of power systems with electric vehicles and diverse transmission links using distributed functional observers. IEEE Trans Smart Grid 7(1):238–252
Poveda JI, Benosman M, Teel AR (2019) Hybrid online learning control in networked multiagent systems: a survey. Int J Adapt Control Signal Process 33(2):228–261
Qin J, Li M, Shi Y, Ma Q, Zheng WX (2019) Optimal synchronization control of multiagent systems with input saturation via off-policy reinforcement learning. IEEE Trans Neural Netw Learn Syst 30(1):85–96
Quevedo DE, Silva EI, Goodwin GC (2009) Subband coding for networked control systems. Int J Robust Nonlinear Control IFAC Affil J 19(16):1817–1836
Song R, Lewis FL, Wei Q (2016) Off-policy integral reinforcement learning method to solve nonlinear continuous-time multiplayer nonzero-sum games. IEEE Trans Neural Netw Learn Syst 28(3):704–713
Thomas P (2002) Fault detection and diagnosis in engineering systems-Janos J. Gertler; Marcel Dekker Inc., New York, 1998, ISBN 0-8247-9427-3. Control Eng Pract 9(10):1037–1038
Vamvoudakis KG (2015) Non-zero sum Nash Q-learning for unknown deterministic continuous-time linear systems. Automatica 61:274–281
Wang YL, Han QL (2018) Network-based modelling and dynamic output feedback control for unmanned marine vehicles in network environments. Automatica 91:43–53
Wang T, Gao H, Qiu J (2015) A combined adaptive neural network and nonlinear model predictive control for multirate networked industrial process control. IEEE Trans Neural Netw Learn Syst 27(2):416–425
Wang D, Ha M, Qiao J (2019) Self-learning optimal regulation for discrete-time nonlinear systems under event-driven formulation. IEEE Trans Autom Control 65(3):1272–1279
Wang Y, Han Q, Fei M, Peng C (2018a) Network-based T-S fuzzy dynamic positioning controller design for unmanned marine vehicles. IEEE Trans Cybern 48(9):2750–2763
Wang YL, Han QL, Fei MR, Peng C (2018b) Network-based T-S fuzzy dynamic positioning controller design for unmanned marine vehicles. IEEE Trans Cybern 48(9):2750–2763
Wang D, He H, Liu D (2017a) Adaptive critic nonlinear robust control: a survey. IEEE Trans Cybern 47(10):3429–3451
Wang B, Xu Y, Shen Z, Zou J, Li C, Liu H (2017b) Current control of grid-connected inverter with LCL filter based on extended-state observer estimations using single sensor and achieving improved robust observation dynamics. IEEE Trans Ind Electron 64(7):5428–5439
Wei Q, Liu D, Shi G (2014) A novel dual iterative Q-learning method for optimal battery management in smart residential environments. IEEE Trans Ind Electron 62(4):2509–2518

Xu H, Jagannathan S (2014) Neural network-based finite horizon stochastic optimal control design for nonlinear networked control systems. IEEE Trans Neural Netw Learn Syst 26(3):472–485
Yang R, Liu GP, Shi P, Thomas C, Basin MV (2013) Predictive output feedback control for networked control systems. IEEE Trans Ind Electron 61(1):512–520
Zhang W, Branicky MS, Phillips SM (2001) Stability of networked control systems. IEEE Control Syst Mag 21(1):84–99
Zhang H, Jiang H, Luo C, Xiao G (2016) Discrete-time nonzero-sum games for multiplayer using policy-iteration-based adaptive dynamic programming algorithms. IEEE Trans Cybern 47(10):3331–3340
Zhang H, Park JH, Yue D, Xie X (2018a) Finite-horizon optimal consensus control for unknown multiagent state-delay systems. IEEE Trans Cybern 50(2):402–413
Zhang H, Yue D, Dou C, Zhao W, Xie X (2018b) Data-driven distributed optimal consensus control for unknown multiagent systems with input-delay. IEEE Trans Cybern 49(6):2095–2105
Zhao D, Zhang Q, Wang D, Zhu Y (2015) Experience replay for optimal control of nonzero-sum game systems with unknown dynamics. IEEE Trans Cybern 46(3):854–865

Chapter 6
Interleaved Q-Learning

In this chapter, a novel off-policy interleaved Q-learning algorithm is presented for solving optimal control problem of affine nonlinear DT systems, using only the measured data along the system trajectories. Affine nonlinear feature of systems, unknown dynamics and off-policy learning approach pose tremendous challenges on approximating optimal controllers. To this end, the on-policy Q-learning method for optimal control of affine nonlinear DT systems is reviewed first, together with its proof of convergence. The bias of solution to Q-function based Bellman equation caused by adding probing noises to systems for satisfying persistent excitation is also analyzed when using the on-policy Q-learning approach. Then, a behavior control policy is introduced followed by an off-policy Q-learning algorithm. Meanwhile, the convergence of the proposed algorithm and no bias of solution to optimal control problem when adding probing noises to systems are investigated. Finally, three neural networks run by the interleaved Q-learning approach in the actor–critic framework. Thus, a novel off-policy interleaved Q-learning algorithm is derived and its convergence is proven. Illustrative examples are given to verify the effectiveness of the proposed method.

6.1 Optimal Control for Affine Nonlinear Systems

RL, one of machine learning tools, has become a powerful and practical tool for tackling optimal control problems. Increasingly large scales, high complexity of systems as well as growing requirements of cost, efficiency, energy, quality of products, etc. for practical industries, such as process industry, smart grid, smart residential energy systems, make data-driven control very promising for achieving optimum control processes (Chai et al. 2011, 2014; Wang et al. 2017a; Wei et al. 2014) under the help of larger and rich datasets. Q-learning, also known as action-dependent heuristic dynamic programming (ADHDP), is one of RL schemes, which combines adaptive

J. Li et al., *Reinforcement Learning*, Advances in Industrial Control,
https://doi.org/10.1007/978-3-031-28394-9_6

critics, RL technique with dynamic programming to solve optimal control problems (Wei et al. 2014; Kiumarsi et al. 2014, 2017; Al-Tamimi et al. 2007; Watkins and Dayan 1992; Tsitsiklis 1994; Doltsinis et al. 2014; Kim and Lewis 2010; Jiang et al. 2017; Lee et al. 2012). One of the strengths of Q-learning is that it is able to evaluate utility and update control policy without requiring models of the environment to be known a priori (Kiumarsi et al. 2014; Al-Tamimi et al. 2007).

It is well known that Q-learning has been studied for several decades aiming at MDPs (Watkins and Dayan 1992; Tsitsiklis 1994; Doltsinis et al. 2014), and the basic problem for which is to find a policy to minimize the expected cumulated costs (denoted by Q-function value) given state transition depending on only the present state-action pairs of the system, but not on its future and full past history. For the case of deterministic policy and deterministic state transition, increasing results using Q-learning to design an approximate optimal controller for the purpose of achieving optimum control performance have been reported, including for linear DT systems (Kiumarsi et al. 2014; Al-Tamimi et al. 2007; Kim and Lewis 2010; Jiang et al. 2017) and linear CT systems (Lee et al. 2012; Vamvoudakis 2017). Notice that the model-free optimal control for affine nonlinear systems using the Q-learning method has rarely been studied. This fact thus motivates this work for a better insight into how to design a Q-learning algorithm to learn optimal controllers only using data for affine nonlinear systems.

Moreover, one can find that the above-mentioned methods (Wei et al. 2014; Kiumarsi et al. 2014; Al-Tamimi et al. 2007; Kim and Lewis 2010; Jiang et al. 2017; Lee et al. 2012; Vamvoudakis 2017) are implemented by using on-policy Q-learning approaches. What kind of evaluating policy is called on-policy or off-policy? The essential difference between on-policy learning and off-policy learning lies on how to get data used for evaluating policy. If a target policy is evaluated using trajectories drawn from a behavior policy not the target policy, then this learning method is referred to as off-policy learning. Otherwise, it is known as on-policy learning (Kim and Lewis 2010; Jiang et al. 2017; Lee et al. 2012; Vamvoudakis 2017). Notice that off-policy learning algorithms have some distinct advantages over on-policy learning algorithms, as shown in Chap. 1.

Off-policy RL with the goal of finding the control policy for achieving optimal control of unknown dynamics has been attracted increasing attention in recent years. Included is for CT systems (Song et al. 2015; Luo et al. 2014; Jiang and Jiang 2012; Li et al. 2017b) and DT systems (Kiumarsi et al. 2017; Li et al. 2017a, 2018). Even though the property of nonlinearity poses the great challenge on off-policy-based RL for finding the optimal control policy without knowing the dynamics of systems, it is promising and practical since practical physical systems generally are nonlinear. Therefore, an off-policy interleaved Q-learning algorithm is presented to solve the optimal control of affine nonlinear DT systems in this chapter. Moreover, proving no bias of solution to the optimal control problem for the first time from the perspective of off-policy Q-learning for affine nonlinear DT systems, which is the extension of Kiumarsi et al. (2017) and Li et al. (2017a, 2018) where the off-policy RL for linear DT systems was concerned. Besides, an interleaved Q-learning approach for achieving approximate optimal control policy is developed by interleaving iteration of critic network and actor network, which is different from the traditional policy iteration and value iteration approaches.

6.1.1 Problem Statement

The optimal control problem of affine nonlinear DT systems is formulated and its standard solution by solving HJB equation is presented.

Consider the following affine nonlinear DT system:

$$x_{k+1} = f(x_k) + g(x_k)u_k \tag{6.1}$$

where $x_k \in \mathbb{R}^n$ and $u_k \in \mathbb{R}^m$ are the state and control input, respectively, $f(x_k) \in \mathbb{R}^n$ and $g(x_k) \in \mathbb{R}^{n\times m}$. Without loss of generality, suppose that (6.1) is drift free, i.e., $f(0) = 0$ and $g(0) = 0$; (6.1) can be stabilized on a prescribed compact set $\Omega \in \mathbb{R}^n$.

It is well known that it is the basic target for optimal control problem to find the control policy $u_k = u(x_k)$ which minimizes the infinite-horizon performance index expressed as

$$J(x_0) = \sum_{k=0}^{\infty} l(x_k, u_k) \tag{6.2}$$

where $l(x_k, u_k)$ is the utility function with $l(x_k, u_k) \geq 0$ for any x_k and u_k. In general, the utility function is chosen as a quadratic form $l(x_k, u_k) = x_k^T Q x_k + u_k^T R u_k$, where $Q \geq 0$ and $R > 0$ are respectively positive semi-definite matrix and positive definite matrix.

According to dynamic programming theory (Luo et al. 2016), the optimal value function should satisfy the DT HJB equation.

$$V^*(x_k) = \min_{u_k} \left(x_k^T Q x_k + u_k^T R u_k + V^*(x_{k+1}) \right). \tag{6.3}$$

From (6.3), solving the optimal control policy by minimizing the right-hand side of (6.3) yields the optimal value function $V^*(x_k)$. Based on the necessary condition for optimality, u_k^* can be obtained by taking the derivative of the right-hand side of (6.3) with respect to u_k. Thus, one has

$$u_k^* = -\frac{1}{2} R^{-1} g^T(x_k) \frac{\partial V^*(x_{k+1})}{\partial x_{k+1}}. \tag{6.4}$$

Substituting (6.4) into (6.3) yields DT HJB equation as

$$V^*(x_k) = x_k^T Qx_k + \frac{1}{4}\frac{\partial (V^*(x_{k+1}))^T}{\partial x_{k+1}} g(x_k)R^{-1}g^T(x_k)\frac{\partial V^*(x_{k+1})}{\partial x_{k+1}} + V^*(x_{k+1}). \tag{6.5}$$

Note that (6.5) is backward in time, and it is impossible to obtain x_{k+1} at the current time instant k. Especially for the affine nonlinear characteristics of (6.1), DT HJB equation (6.5) cannot be solved exactly. To overcome these challenging difficulties, various RL methods including heuristic dynamic programming (HDP), action-dependent HDP (Q-learning), dual heuristic dynamic programming (DHP), action-dependent DHP, globalized DHP have been reported for approximating the optimal solution of DT HJB equation instead of solving the analytical optimal solution (Prokhorov et al. 1995; Wang et al. 2017b). The followings introduce the Q-learning algorithm to approximately solve DT HJB equation (6.5).

6.1.2 On-Policy Q-Learning Formulation

This subsection focuses on three aspects:

(1) Review the on-policy Q-learning algorithm for finding the approximation value of the optimal control policy.
(2) Present a novel proof of convergence of on-policy Q-learning algorithm.
(3) Show the bias of solution to DT HJB equation (6.5) if probing noises are added into the systems for enriching data.

6.1.2.1 Derivation of Q-Learning Algorithm

Define the optimal action-dependent Q-function as

$$Q^*(x_k, u_k) = x_k^T Qx_k + u_k^T Ru_k + V^*(x_{k+1}). \tag{6.6}$$

Then, one has

$$V^*(x_k) = \min_{u_k} Q^*(x_k, u_k) = Q^*(x_k, u_k^*). \tag{6.7}$$

Combining with (6.3) yields the Q-function-based DT HJB equation

$$Q^*(x_k, u_k^*) = x_k^T Qx_k + (u_k^*)^T Ru_k^* + Q^*(x_{k+1}, u_{k+1}^*) \tag{6.8}$$

and the optimal control policy

$$u_k^* = -\frac{1}{2}R^{-1}g^T(x_k)\frac{\partial Q^*(x_{k+1}, u_{k+1}^*)}{\partial x_{k+1}}. \tag{6.9}$$

Referring to value iteration algorithms (Wei et al. 2014; Wang et al. 2017b), Algorithm 6.1 is given to learn the optimal control policy.

Algorithm 6.1 On-policy Q-learning

1: Initialize the optimal Q-function $Q^0(\cdot) = 0$, and set the iteration index $i = 0$;

2: Calculate the initial control u_k^0 by

$$u_k^0 = \arg\min_{u_k} \left(x_k^T Q x_k + u_k^T R u_k + Q^0(\cdot) \right); \tag{6.10}$$

3: Update the iterative Q-function:

$$Q^1(x_k, u_k^0) = x_k^T Q x_k + (u_k^0)^T R u_k^0 + Q^0(\cdot) = x_k^T Q x_k + (u_k^0)^T R u_k^0; \tag{6.11}$$

4: Update the sequence of action policies:

$$\begin{aligned} u_k^i &= \arg\min_{u_k} \left(x_k^T Q x_k + u_k^T R u_k + Q^i(x_{k+1}, u_{k+1}^{i-1}) \right) \\ &= \arg\min_{u_k} Q^{i+1}(x_k, u_k) \\ &= -\frac{1}{2} R^{-1} g^T(x_k) \frac{\partial Q^i(x_{k+1}, u_{k+1}^{i-1})}{\partial x_{k+1}} \end{aligned} \tag{6.12}$$

and a sequence of Q-functions:

$$Q^{i+1}(x_k, u_k^i) = x_k^T Q x_k + (u_k^i)^T R u_k^i + Q^i(x_{k+1}, u_{k+1}^{i-1}(x_{k+1})) \tag{6.13}$$

with $x_{k+1} = f(x_k) + g(x_k)u_k^i$;

5: If $\left\| Q^{i+1}(x_k, u_k^i) - Q^i(x_k, u_k^{i-1}) \right\| \leq \varepsilon \ (\varepsilon > 0)$, stop and obtain the approximate optimal control policy u_k^i; Otherwise, set $i = i + 1$ and go back to step 4.

6.1.2.2 Convergence Analysis of the On-Policy Q-Learning Algorithm

The following two lemmas are given to use for the proof of convergence of Algorithm 6.1.

Definition 6.1 (Prokhorov et al. 1995) A feedback control u_n defined on Ω_x is said to be admissible with respect to (6.2) if u_n is continuous on a compact set $\Omega_u \in \mathbb{R}^m$, $u(0) = 0$, u_n stabilizes system (6.1) on Ω_x, and $J(x_0)$ is finite $\forall x_0 \in \Omega_x$.

Lemma 6.1 *Suppose the sequence $\{Q^{i+1}\}$ to be defined as in (6.13). If system (6.1) is controllable and $Q^0(\cdot) = 0$, then the following conclusions hold.*

(a) Let μ^i be an arbitrary sequence of control policies, function W^{i+1} be defined as

$$W^{i+1}(x_k, \mu^i) = x_k^T Q x_k + (\mu^i)^T R\mu^i + W^i\left(f(x_k) + g(x_k)\mu^i, \mu^{i-1}\left(f(x_k) + g(x_k)\mu^i\right)\right) \tag{6.14}$$

and $W^0(\cdot) = 0$, then $Q^{i+1}(x_k, u_k^i) \le W^{i+1}(x_k, \mu^i)$ can be satisfied.

(b) There exists an upper bound $Y(x_k)$ such that $Q^{i+1}(x_k, u_k^i) \le W^{i+1}(x_k, \mu^i) \le Y(x_k)$, where W^{i+1} is obtained by letting μ^i be an admissible control policy.

(c) If (6.8) is solvable, then $Q^{i+1}(x_k, u_k^i) \le Q^*(x_k, u^*(x_k)) \le Y(x_k)$.

Proof (a) Notice that $Q^{i+1}(x_k, u_k^i)$ is the result of minimizing the right-hand side of (6.13) by using u_k^i obtained from (6.12), while $W^{i+1}(x_k, \mu^i)$ is achieved under an arbitrary control input referring to (6.14), then $Q^{i+1}(x_k, u_k^i) \le W^{i+1}(x_k, \mu^i)$ can be derived.

(b) Let$\mu^i = \eta(x_k)$to be an admissible control policy, and let $Q^0(\cdot) = W^0(\cdot) = 0$, one has the following difference:

$$\begin{aligned}
&W^{i+1}(x_k, \eta(x_k)) - W^i(x_k, \eta(x_k)) \\
&= W^i(x_{k+1}, \eta(x_{k+1})) - W^{i-1}(x_{k+1}, \eta(x_{k+1})) \\
&= W^{i-1}(x_{k+2}, \eta(x_{k+2})) - W^{i-2}(x_{k+2}, \eta(x_{k+2})) \\
&\vdots \\
&= W^2(x_{k+i-1}, \eta(x_{k+i-1})) - W^1(x_{k+i-1}, \eta(x_{k+i-1})) \\
&= W^1(x_{k+i}, \eta(x_{k+i})) - W^0(x_{k+i}, \eta(x_{k+i})) \\
&= W^1(x_{k+i}, \eta(x_{k+i})).
\end{aligned} \tag{6.15}$$

Rewriting (6.15) yields

$$\begin{aligned}
W^{i+1}(x_k, \eta(x_k)) &= W^1(x_{k+i}, \eta(x_{k+i})) - W^i(x_k, \eta(x_k)) \\
&= W^1(x_{k+i}, \eta(x_{k+i})) + W^1(x_{k+i-1}, \eta(x_{k+i-1})) \\
&+ W^{i-1}(x_k, \eta(x_k)) \\
&= W^1(x_{k+i}, \eta(x_{k+i})) + W^1(x_{k+i-1}, \eta(x_{k+i-1})) \\
&+ W^1(x_{k+i-2}, \eta(x_{k+i-2})) + W^{i-2}(x_k, \eta(x_k)) \\
&\vdots \\
&= W^1(x_{k+i}, \eta(x_{k+i})) + W^1(x_{k+i-1}, \eta(x_{k+i-1})) \\
&+ W^1(x_{k+i-2}, \eta(x_{k+i-2})) + \cdots + W^1(x_k, \eta(x_k)) \\
&= \sum_{n=0}^{i} x_{k+n}^T Q x_{k+n} + \eta^T(x_{k+n}) R \eta(x_{k+n}).
\end{aligned} \tag{6.16}$$

Since $\eta(x_k)$ is an admissible control policy, further, one has

$$W^{i+1}\left(x_k, \eta(x_k)\right) \leq Y(k) \tag{6.17}$$

where $Y(k) = \sum_{n=0}^{\infty} x_{k+n}^T Q x_{k+n} + \eta^T(x_{k+n}) R \eta(x_{k+n})$. Combining with (a), (b) holds, i.e., $Q^{i+1}(x_k, u_k^i) \leq W^{i+1}(x_k, \mu^i) \leq Y(x_k)$.

(c) If $\eta(x_k) = u^*(x_k)$, then $Q^{i+1}(x_k, u_k^i) \leq Q^*(x_k, u^*(x_k)) \leq Y(x_k)$ can be derived from (b). This completes the proof. □

Lemma 6.2 *Suppose the sequences u^i and Q^i to be defined as in (6.12) and (6.13). If $Q^0(\cdot) = 0$, then it follows $Q^i \leq Q^{i+1}$.*

Proof By (a) of Lemma 6.1, we have $Q^i \leq W^i$. Next we shall show $W^i \leq Q^{i+1}$ by using induction.

Since μ^i is an arbitrary sequence of control policies, then we let $\mu^i = u_k^{i+1}$. First, when $i = 0$, one has

$$Q^1(x_k, u_k^0) - W^0(x_k, u_k^0) = x_k^T Q x_k + (u_k^0)^T R u_k^0 \geq 0 \tag{6.18}$$

which means $Q^1 \geq W^0$.

Suppose that $W^{i-1}(x_k, u_k^{i-1}) \leq Q^i(x_k, u_k^{i-1})$ holds, then one has

$$\begin{aligned}
&Q^{i+1}(x_k, u_k^i) - W^i(x_k, u_k^i) \\
&\quad = Q^i\left(f(x_k) + g(x_k)u_k^i, u^{i-1}\left(f(x_k) + g(x_k)u_k^i\right)\right) \\
&\quad - W^{i-1}\left(f(x_k) + g(x_k)u_k^i, u^{i-1}\left(f(x_k) + g(x_k)u_k^i\right)\right) \\
&\quad = Q^i\left(x_{k+1}, u^{i-1}(x_{k+1})\right) - W^{i-1}\left(x_{k+1}, u^{i-1}(x_{k+1})\right) \geq 0.
\end{aligned} \tag{6.19}$$

By induction, it can conclude $W^i \leq Q^{i+1}$. Since $Q^i \leq W^i$, then $Q^i \leq Q^{i+1}$ holds. This completes the proof. □

Theorem 6.1 *For the iterative control policy u^i and the iterative Q-function Q^i respectively defined as in (6.12) and (6.13), if $Q^0(\cdot) = 0$, then Q^i converges to the optimal value Q^* and u^i converges to the optimal control policy u^* as $i \to \infty$, i.e., $\lim_{i\to\infty} Q^i = Q^*$ and $\lim_{i\to\infty} u^i = u^*$.*

Proof From Lemmas 6.1 and 6.2, one can conclude that the iterative Q-function Q^i converges, which leads to u^i converging as well. We are now in a position to prove that they, respectively, converge to the optimal value Q^* and the optimal control policy u^* as $i \to \infty$.

By (6.7), one has

$$Q^*(x_k, u_k^*) \leq Q^{i+1}(x_k, u_k^i). \tag{6.20}$$

Combining (6.20) with the conclusion (c) of Lemma 6.1 yields

$$\lim_{i\to\infty} Q^i(x_k, u_k^{i-1}) = Q^*(x_k, u_k^*). \tag{6.21}$$

Thus, one has $\lim_{i\to\infty} u^i = u^*$ by referring to (6.9) and (6.12). This completes the proof. □

Remark 6.1 Solving $Q^{i+1}(x_k, u_k^i)$ in terms of (6.13) when implementing Algorithm 6.1 generally needs to add probing noise for satisfying PE condition like (Wei et al. 2014; Kiumarsi et al. 2014; Al-Tamimi et al. 2007). Kiumarsi et al. (2017) has shown that incorrect solutions resulting in incorrect optimal control policy would be caused by probing noise if using on-policy HDP method for optimal control of linear DT systems. This conclusion will be proven to still hold by the sequel for the case of affine nonlinear systems using on-policy Q-learning algorithm.

6.1.2.3 Bias of Solution Analysis for On-Policy Q-Learning Algorithm

Lemma 6.3 *Suppose that probing noise e_k is added to the control policy u_k^i in Algorithm 6.1. Let $\tilde{Q}^{i+1}$ be the solution to (6.13) with $\tilde{u}_k^i = u_k^i + e_k$, $e_k \neq 0$, then $\tilde{Q}^{i+1}$ is not the solution to (6.13) with $e_k = 0$.*

Proof Let (6.13) be Bellman equation without probing noise, i.e., $e_k = 0$. If probing noise is added into the system (6.1), i.e., $\tilde{u}_k^i = u_k^i + e_k$ ($e_k \neq 0$) acts as control input to generate data using for performance evaluation, then (6.1) and (6.13), respectively, become the forms as follows:

$$\tilde{x}_{k+1} = f(x_k) + g(x_k)u_k^i + g(x_k)e_k \tag{6.22}$$

and

$$\tilde{Q}^{i+1}(x_k, u_k^i) = x_k^T Q x_k + (u_k^i)^T R u_k^i + \tilde{Q}^i(\tilde{x}_{k+1}, u_{k+1}^i(\tilde{x}_{k+1})). \tag{6.23}$$

By considering (6.1) in (6.23), one has

$$\begin{aligned}\tilde{Q}^{i+1}(x_k, u_k^i) &= x_k^T Q x_k + (u_k^i)^T R u_k^i \\ &\quad + \tilde{Q}^i\left(x_{k+1} + g(x_k)e_k, u_{k+1}^i(x_{k+1} + g(x_k)e_k)\right).\end{aligned} \tag{6.24}$$

Contrasting (6.13) with (6.24) shows that $\tilde{Q}^{i+1}$ is not the same as Q^{i+1}, which might lead to incorrect the control update since

$$u_k^{i+1} = -\frac{1}{2}R^{-1}g^T(x_k)\frac{\partial \tilde{Q}^{i+1}\left(x_{k+1} + g(x_k)e_k, u_{k+1}^i(x_{k+1} + g(x_k)e_k)\right)}{\partial x_{k+1}}. \tag{6.25}$$

This completes the proof. □

6.2 Off-Policy Q-Learning Technique

This section devotes to proposing an off-policy Q-learning algorithm and proving the convergence of the proposed off-policy Q-learning algorithm, as well as analyzing no bias of solution even though probing noise is added into the systems for reaching PE condition.

6.2.1 *Off-Policy and Q-Learning*

On-policy and off-policy are two kinds of RL methods. On-policy methods evaluate or improve the same policy as the one that is applied to the systems for generating data. While, in the off-policy methods, there exist two types of unrelated control policies, one is called behavior policy used to generate date for implementing learning, and the other is target or estimation policy, which is evaluated and improved to approximate the optimal control policy (Kim and Lewis 2010; Jiang et al. 2017; Lee et al. 2012; Vamvoudakis 2017; Kiumarsi et al. 2017; Song et al. 2015; Li et al. 2017a, b, 2018; Luo et al. 2016).

Q-learning can be implemented by on-policy (Wei et al. 2014; Kiumarsi et al. 2014; Al-Tamimi et al. 2007) or off-policy approach (Kiumarsi et al. 2017; Li et al. 2017a, 2018) depending on updating Q-function value by using data from a behavior policy or the target policy. What is showed Q-learning in Algorithm 6.1 is actually an on-policy method because it updates its Q-values using the trajectories drawn from the evaluated action.

6.2.2 *Derivation of Off-Policy Q-Learning Algorithm*

Introducing an auxiliary variable u_k^i into system (6.1) yields

$$x_{k+1} = f(x_k) + g(x_k)u_k^i + g(x_k)(u_k - u_k^i) \tag{6.26}$$

where u_k is called the behavior policy and u_k^{i-1} is viewed as the target policy needed to be evaluated and improved. It is well known that (6.12) and (6.13) are, respectively, equivalent to

$$u_k^i = -\frac{1}{2}R^{-1}g^T(x_k)\frac{\partial Q^i\left(f(x_k) + g(x_k)u_k^{i-1}, u_{k+1}^{i-1}\right)}{\partial\left(fx_k) + g(x_k)u_k^{i-1}\right)} \tag{6.27}$$

and

$$Q^{i+1}(x_k, u_k^i) = x_k^T Q x_k + (u_k^i)^T R u_k^i \\ + Q^i \left(f(x_k) + g(x_k)u_k^i, u_{k+1}^{i-1}\left(f(x_k) + g(x_k)u_k^i\right)\right). \tag{6.28}$$

Along the trajectory of (6.26)–(6.28) can be respectively rewritten as

$$u_k^i = -\frac{1}{2}R^{-1}g^T(x_k)\frac{\partial Q^i\left\{x_{k+1}^{i-1}, u^{i-1}\left(x_{k+1}^{i-1}\right)\right\}}{\partial\left(x_{k+1}^{i-1}\right)} \tag{6.29}$$

and

$$Q^{i+1}(x_k, u_k^i) - Q^i\left\{x_{k+1} - g(x_k)(u_k - u_k^i), u^{i-1}\left(x_{k+1} - g(x_k)\left(u_k - u_k^i\right)\right)\right\} \\ = x_k^T Q x_k + (u_k^i)^T R u_k^i \tag{6.30}$$

where $x_{k+1}^{i-1} = x_{k+1} - g(x_k)\left(u_k - u_k^{i-1}\right)$. The following Algorithm 6.2 is to show how to implement off-policy learning for approximating the optimal control policy.

Algorithm 6.2 Off-policy Q-learning

1: Initialize the optimal Q-function $Q^i(\cdot) = 0$, and set the iteration index $i = 0$;
2: Calculate the initial control u_k^0 by

$$u_k^0 = \arg\min_{u_k}\left(x_k^T Q x_k + u_k^T R u_k + Q^0\left(x_{k+1}^0, u_{k+1}^{-1}\left(x_{k+1}^0\right)\right)\right); \tag{6.31}$$

3: Update the iterative Q-function:

$$Q^1(x_k, u_k^0) = x_k^T Q x_k + (u_k^0)^T R u_k^0; \tag{6.32}$$

4: Update the sequence of action policies by (6.30) and the sequence of the iterative Q-functions by (6.29);
5: If $\left\| Q^{i+1}(x_k, u_k^i) - Q^i(x_k, u_k^{i-1}) \right\| \le \varepsilon\ (\varepsilon > 0)$, stop and obtain the approximate optimal control policy u_k^i; Otherwise, set $i = i + 1$ and go to step 4.

Theorem 6.2 *$(Q^{i+1},\ u^i)$ is the solution of (6.12) and (6.13) if and only if it is the solution of (6.29) and (6.30).*

Proof One can find that if $(Q^{i+1},\ u^i)$ is the solution of (6.12) and (6.13), then it also make (6.27) and (6.28) hold for $\forall x_k \in \Omega_x$ (Ω_x is a compact set). For the state x_k generated by (6.26), substituting $x_{k+1} - g(x_k)(u_k - u_k^i) = f(x_k) + g(x_k)u_k^i$ into (6.27) and (6.28) yields (6.29) and (6.30), so the solution $(Q^{i+1},\ u^i)$ of (6.12) and (6.13) can satisfy (6.29) and (6.30) as well. Next, we shall prove that the solution of (6.29) and (6.30) is also the solution of (6.12) and (6.13). Substituting (6.26) into (6.29) and (6.30) yields (6.27) and (6.28), further gets (6.13) and (6.12). This completes the proof. □

Remark 6.2 Note that the solutions of Algorithms 6.1 and 6.2 are equivalent as shown in Theorem 6.2. Moreover, the convergence of Algorithm 6.1 has been proved in Theorem 6.1, therefore, if (Q^{i+1}, u^i) can be solved correctly from Algorithm 6.2, then $\lim_{i\to\infty} Q^i = Q^*$ and $\lim_{i\to\infty} u^i = u^*$ can be concluded.

Remark 6.3 The Q-learning in Algorithm 6.2 is definitely an off-policy approach, since the target control policy is updated but not to be applied to the real systems during learning due to the introduction of an arbitrary stabilizing behavior policy u_k used to generate data and enrich data exploration, which is a remarkable feature possessed by the off-policy learning as opposed to the on-policy learning.

6.2.3 *No Bias of Off-Policy Q-Learning Algorithm*

In Kiumarsi et al. (2017), it was shown that adding probing noise does not result in biased solution for optimal control of linear DT systems using off-policy RL learning. Here, we extend that result to affine nonlinear DT systems for finding the optimal control policy by using off-policy Q-learning.

Theorem 6.3 *Suppose that the probing noise e_k is added to the behavior policy in Algorithm 6.2. Let $(\bar{Q}^{i+1}, \bar{u}^i)$ be the solution to (6.29) and (6.30) with $\bar{u}_k = u_k + e_k$, $e_k \neq 0$, then $(\bar{Q}^{i+1}, \bar{u}^i)$ is also the solution to (6.29) and (6.30) with $e_k = 0$.*

Proof A probing noise is added into the behavior control policy, that is, $\bar{u}_k = u_k + e_k$. By Algorithm 6.2, $\bar{u}_k^0 = u_k^0$ and $\bar{Q}^1(x_k, \bar{u}_k^0) = Q^1(x_k, u_k^0)$ hold. We assume $\bar{u}_k^{i-1} = u_k^{i-1}$ and $\bar{Q}^i(x_k, \bar{u}_k^{i-1}) = Q^i(x_k, u_k^{i-1})$ hold, and substituting $\bar{u}_k$ into (6.26) yields

$$\begin{aligned} \bar{x}_{k+1} &= f(x_k) + g(x_k)(u_k + e_k) \\ \bar{x}_{k+1}^{i-1} = \bar{x}_{k+1} - g(x_k)(\bar{u}_k - \bar{u}_k^{i-1}) &= f(x_k) + g(x_k)u_k^{i-1} = x_{k+1}^{i-1}. \end{aligned} \tag{6.33}$$

By (6.29), one has

$$\bar{u}_k^i = -\frac{1}{2}R^{-1}g^T(x_k)\frac{\partial \bar{Q}^i\left\{\bar{x}_{k+1}^{i-1}, \bar{u}^{i-1}\left(\bar{x}_{k+1}^{i-1}\right)\right\}}{\partial\left(\bar{x}_{k+1}^i\right)}. \tag{6.34}$$

Due to $\bar{u}_k^{i-1} = u_k^{i-1}$ and $\bar{Q}^i(x_k, \bar{u}_k^{i-1}) = Q^i(x_k, u_k^{i-1})$, (6.34) becomes

$$\bar{u}_k^i = -\frac{1}{2}R^{-1}g^T(x_k)\frac{\partial Q^i\left\{x_{k+1}^{i-1}, u^{i-1}\left(x_{k+1}^{i-1}\right)\right\}}{\partial\left(x_{k+1}^{i-1}\right)}. \tag{6.35}$$

By comparing (6.29) and (6.35), one can conclude that $\bar{u}_k^i = u_k^i$ resulting in $\bar{x}_{k+1}^i = x_{k+1}^i$ by referring to (6.33). Note that

$$\bar{Q}^{i+1}(x_k, \bar{u}_k^i) = x_k^T Q x_k + (\bar{u}_k^i)^T R\bar{u}_k^i + \bar{Q}^i\left(\bar{x}_{k+1}^i, \bar{u}^{i-1}(\bar{x}_{k+1}^i)\right) \tag{6.36}$$

substituting $\bar{u}_k^i = u_k^i$ and $\bar{x}_{k+1}^i = x_{k+1}^i$ into (6.36), one has

$$\begin{aligned}\bar{Q}^{i+1}(x_k, u_k^i) &= x_k^T Q x_k + (u_k^i)^T R u_k^i \\ &\quad + \bar{Q}^i \left\{x_{k+1} - g(x_k)(u_k - u_k^i), u^i \left(x_{k+1} - g(x_k)(u_k - u_k^i)\right)\right\}. \end{aligned} \tag{6.37}$$

By comparing (6.30) with (6.37), one can conclude that $\bar{Q}^{i+1} = Q^{i+1}$. By mathematical induction, $(\bar{Q}^{i+1}, \bar{u}^i) = (Q^{i+1}, u^i)$ even though $e_k \neq 0$.

Therefore, adding the probing noise during implementing the proposed off-policy Q-learning Algorithm 6.2 cannot produce bias of solution. This completes the proof. □

Remark 6.4 In contrast to the off-policy Q-learning method (Luo et al. 2016), the developed off-policy Q-learning Algorithm 6.2 in this chapter has a different learning strategy shown in (6.29) and (6.30). More importantly, even though probing noise is added to the system for satisfying PE condition, no bias of solution can be guaranteed and is proved for the first time from the perspective of Q-learning, whereas off-policy RL for linear DT systems was taken into account in Kiumarsi et al. (2017), Li et al. (2017a, 2018).

6.3 Neural Network-Based Off-Policy Interleaved Q-Learning

In this section, three neural networks are used to approximate $Q^i(x_k, u_k^{i-1})$, u_k^i and the affine nonlinear system (6.1) by function value approximation approach. Algorithm 6.2 is implemented based on interleaved-learning critic and actor structure by NNs. Therefore, this is a data-driven approximate optimal control strategy without the knowledge of system model.

6.3.1 Model Neural Network

Note that updating Q-function and control policy in Algorithm 6.2 requires $g(x_k)$ to be known a priori, but it is difficult to know $g(x_k)$ in real applications. Actually $g(x_k) = \frac{\partial x_{k+1}}{\partial u_k}$, so the following three-layer NN (Mu et al. 2017; Liu and Wei 2013) is used to approximate the dynamics of system (6.1) for estimating $g(x_k)$ by using $\frac{\partial \hat{x}_{k+1}}{\partial u_k}$:

$$\hat{x}_{k+1} = \omega_x^T \sigma \left(v_x^T \begin{bmatrix} x_k \\ u_k \end{bmatrix}\right) \tag{6.38}$$

where ω_x and v_x are respectively the weights of the hidden layer to the output layer and the weights of the input layer to the hidden layer. $\sigma\left(v_x^T\begin{bmatrix}x_k\\u_k\end{bmatrix}\right)\in\mathbb{R}^l$ is the activation function vector, $[\sigma(z)]_l=(e^z-e^{-z})/(e^z+e^{-z})$ and l is the number of neurons in the hidden layer. To train the NN (6.38), the gradient descent algorithm is used to update the weight ω_x.

$$\omega_x(k+1)=\omega_x(k)-\eta_c\frac{\partial E_{xk}}{\partial\omega_x} \tag{6.39}$$

where e_{xk} and E_{xk} are respectively the approximate error and the squared error of model network, and they are defined as

$$E_{xk}=\frac{1}{2}e_{xk}^Te_{xk},\ e_{xk}=\hat{x}_{k+1}-x_{k+1}. \tag{6.40}$$

Thus, one has

$$\frac{\partial E_{xk}}{\partial\omega_x}=\frac{\partial E_{xk}}{\partial e_{xk}}\frac{\partial e_{xk}}{\partial\hat{x}_{k+1}}\frac{\partial\hat{x}_{k+1}}{\partial\omega_x}=\sigma\left(v_x^T\begin{bmatrix}x_k\\u_k\end{bmatrix}\right)e_{xk}^T. \tag{6.41}$$

6.3.2 Actor Neural Network

We employ the actor NN to approximate actor u_k^i given by

$$\hat{u}_k^i=(\hat{\omega}_{ak}^i)^T\sigma(Z_a(k)) \tag{6.42}$$

where $Z_a(k)=v_a^Tx_k$, $\hat{\omega}_{ak}^i$ and v_a respectively are the weights of the hidden layer to the output layer and the weights of the input layer to the hidden layer. Training $\hat{\omega}_{ak}^i$ is implemented by using gradient descent algorithm.

$$\hat{\omega}_{ak}^{i+1}=\hat{\omega}_{ak}^i-\eta_a\frac{\partial E_{ak}^i}{\partial\hat{\omega}_{ak}^i} \tag{6.43}$$

where e_{ak}^i and E_{ak}^i respectively are the approximate error and the squared error of actor network, and they are defined as

$$E_{ak}^i=\frac{1}{2}(e_{ak}^i)^Te_{ak}^i,\ e_{ak}^i=\hat{u}_k^i-u_k^i \tag{6.44}$$

and

$$\begin{aligned}\frac{\partial E_{ak}^i}{\partial \hat{\omega}_{ak}^i} &= \frac{\partial E_{ak}^i}{e_{ak}^i}\frac{e_{ak}^i}{\partial \hat{u}_k^i}\frac{\partial \hat{u}_k^i}{\partial \hat{\omega}_{ak}^i} = e_{ak}^i \sigma(Z_a(k)) \\ u_k^i &= -\frac{1}{2}R^{-1}\left(\frac{\partial \hat{x}_{k+1}}{\partial u_k}\right)^T \frac{\partial \hat{Q}^i\{x_{k+1}^{i-1}, u^{i-1}(x_{k+1}^{i-1})\}}{\partial (x_{k+1}^{i-1})}.\end{aligned} \tag{6.45}$$

6.3.3 Critic Neural Network

A critic neural network is used to allow approximate the iterative Q-function. The critic NN is given by the form of the three-layer neural network

$$\hat{Q}^{i+1}(x_k, \hat{u}_k^i) = (\hat{\omega}_{c,k}^{i+1})^T \sigma(\hat{Z}_c(k)) \tag{6.46}$$

where $\hat{Z}_c^i(k) = v_c^T \begin{bmatrix} x_k \\ \hat{u}_k^i \end{bmatrix}$, ω_{ci} and v_c respectively are the weights of the hidden layer to the output layer and the weights of the input layer to the hidden layer. (6.46) can be used to approximate $\hat{Q}^i\left(x_{k+1}, \hat{u}_{k+1}^{i-1}\right)$ as

$$\hat{Q}^i\{x_{k+1}^i, \hat{u}^{i-1}(x_{k+1}^i)\} = (\hat{\omega}_{c,k}^i)^T \sigma\left(v_c^T \begin{bmatrix} x_{k+1}^i \\ \hat{u}^i(x_{k+1}^i) \end{bmatrix}\right) \tag{6.47}$$

where

$$\begin{aligned} x_{k+1}^i = x_{k+1} - \frac{\partial \hat{x}_{k+1}}{\partial u_k}(u_k - \hat{u}_k^i), \ \frac{\partial \hat{x}_{k+1}}{\partial u_k} &= \omega_x I_1 (\bar{I} - \sigma^2\left(v_x^T \begin{bmatrix} x_k \\ u_k \end{bmatrix}\right) \\ I_1 = \begin{bmatrix} 0_{n\times n} & 0_{n\times m} \\ 0_{m\times n} & I_{m\times m}\end{bmatrix}, \ \bar{I} &= \begin{bmatrix} 1 \\ \vdots \\ 1 \end{bmatrix}_{(n+m)\times 1}. \end{aligned} \tag{6.48}$$

Then, the squared error and approximate error of the critic network are respectively defined as E_{ck}^i and e_{ck}^i.

$$E_{ck}^i = \frac{1}{2}(e_{c,k}^i)^T e_{ck}^i \tag{6.49}$$

$$\begin{aligned} e_{ck}^i &= \hat{Q}^i(x_k, \hat{u}_k^{i-1}) - Q^i(x_k, \hat{u}_k^{i-1}) \\ &= \hat{Q}^i(x_k, \hat{u}_k^{i-1}) - \left(x_k^T Q x_k + (\hat{u}_k^{i-1})^T R \hat{u}_k^{i-1} + \hat{Q}^{i-1}\{x_{k+1}^{i-1}, \hat{u}^{i-2}(x_{k+1}^{i-1})\}\right). \end{aligned} \tag{6.50}$$

The gradient descent algorithm is used to update the weight for the critic network, which is given as follows:

$$\hat{\omega}_{ck}^{i+1} = \hat{\omega}_{ck}^{i} - \eta_c \frac{\partial E_{ck}^{i}}{\partial \hat{\omega}_{ck}^{i}} \tag{6.51}$$

where

$$\frac{\partial E_{c,k}^{i}}{\partial \hat{\omega}_{c,k}^{i}} = \frac{\partial E_{c,k}^{i}}{\partial e_{c,k}^{i}} \frac{\partial e_{c,k}^{i}}{\partial \hat{Q}^{i}(x_k, \hat{u}_k^{i})} \frac{\partial \hat{Q}^{i}(x_k, \hat{u}_k^{i-1})}{\partial \hat{\omega}_{c,k}^{i}} = e_{c,k}^{i} \sigma(\hat{Z}_c^{i-1}(k)). \tag{6.52}$$

6.3.4 *Interleaved Q-Learning*

The following presents an off-policy interleaved Q-learning Algorithm 6.3 based on interleaved-iteration critic and actor networks.

Algorithm 6.3 Off-policy interleaved Q-learning

1: Data collection: Collect system data x_k and store them in the sample sets by using the behavior control policy u_k;
2: Initialization: Initialize weight vector ω_x, v_x for the affine nonlinear neural network;
3: Train the model network:

  3.1: Train weight in terms of (6.39) using the measured data until $e_{xk} \le \varepsilon_x$ ($\varepsilon_x > 0$);
  3.2: Get the trained weight ω_x. Let $k = 0$;

4: Interleaved iteration:

  4.1: Let the initial iterative Q-function $Q^0(\cdot) = 0$ and further calculate the initial control u_k^0 by (6.31). Let the iteration index $i = 0$;
  4.2: Train critic network: Calculate $\hat{Q}^{i+1}(x_k, \hat{u}_k^i)$, $\hat{Q}^i\{x_{k+1}^i, \hat{u}^i(x_{k+1}^i)\}$ and e_{ck}^i using (6.46), (6.47) and (6.50) to update the critic weight $\hat{\omega}_{ck}^{i+1}$ once using (6.51);
  4.3: Train actor network: Calculate $\hat{u}_k^i$, u_k^i and e_{ak}^i using (6.42) and (6.45) to update the actor weight $\hat{\omega}_{ak}^{i+1}$ once using (6.43);
  4.4: Check $\left\| \hat{Q}^i(x_k, \hat{u}_k^{i-1}) - \hat{Q}^{i+1}(x_k, \hat{u}_k^i) \right\| \le \varepsilon_Q$ ($\varepsilon_Q > 0$), if it is not satisfied, go to step 4.2, otherwise get $\hat{u}_k^i$;

5: Set $k = k + 1$, and go back to step 4.

Remark 6.5 One can easily find that no information on dynamics of affine nonlinear systems is required when learning the optimal control policy by constructing three neural networks in Algorithm 6.3.

Remark 6.6 In Algorithms 6.1 and 6.2 of this chapter, the neural network weight $\hat{\omega}_{ck}^{i+1}$ is kept training with time k until it converges for each iterative index $i + 1$ and then the actor weight $\hat{\omega}_{ak}^{i}$ is trained by the same approach, which is the traditional value iteration RL method. While, in interleaved Q-learning Algorithm 6.3, for each time k, critic network and actor network are interleaved iterated with iterative index

i until convergence, and they finally converge with increasing time k. Actually, the proposed interleaved Q-learning is a kind of variant of generalized value iteration, is more easily implemented for the practical applications.

Remark 6.7 Notice that, in Algorithm 6.3, the critic neural network for Q-function value is updated off-line by using an entire set of data under the PE condition, instead of on-line updating it.

Theorem 6.4 *Let the optimal performance index function and the optimal control policy be expressed by*

$$Q^*(x_k, u_k^*) = (\omega_{ck}^*)^T \sigma(Z_c(k)) \tag{6.53}$$

and

$$u_k^* = (\omega_{ak}^*)^T \sigma(Z_a(k)) \tag{6.54}$$

respectively, where $Z_c(k) = v_c^T \begin{bmatrix} x_k \\ u_k^* \end{bmatrix}$. *Let the actor and critic networks be regulated by (6.43) and (6.51), respectively. Let* $\bar{\omega}_{ck}^i = \hat{\omega}_{ck}^i - \omega_{ck}^*$, $\bar{\omega}_{ak}^i = \hat{\omega}_{ak}^i - \omega_{ak}^*$, *if there exist* $W_c > 0$ *and* $W_a > 0$ *satisfying*

$$\begin{aligned} \eta_c &< \frac{1}{\left\|\sigma(\hat{Z}_c^i(k))\right\|^2}, \ \eta_a < \frac{1}{\|\sigma(Z_a(k))\|^2} \\ \left\|e_{ck}^i\right\|^2 &> \frac{W_c\left\|\sigma(\hat{Z}_c^i(k))\right\|^2}{\eta_q}, \ \left\|e_{ak}^i\right\|^2 > \frac{W_a\|\sigma(Z_a(k))\|^2}{\eta_u} \end{aligned} \tag{6.55}$$

then the errors $\bar{\omega}_{ck}^i$ *and* $\bar{\omega}_{ak}^i$ *both converge to zero, as* $i \to \infty$, *where* $\eta_q = 2 - \eta_c\left\|\sigma(\hat{Z}_c(k+1))\right\|^2 - \left\|\sigma(\hat{Z}_c(k+1))\right\|^2$, $\eta_u = 2 - \eta_a\|\sigma(Z_a(k))\|^2 - \|\sigma(Z_a(k))\|^2$.

Proof By (6.43) and (6.51), one has

$$\bar{\omega}_{ck}^{i+1} = \bar{\omega}_{ck}^i - \eta_c e_{ck}^i \sigma(\hat{Z}_c^{i-1}(k)) \tag{6.56}$$

and

$$\bar{\omega}_{ak}^{i+1} = \bar{\omega}_{ak}^i - \eta_a e_{ak}^i \sigma(Z_a(k)). \tag{6.57}$$

Choose a Lyapunov function candidate as

$$V(\bar{\omega}_{c,k}^i, \bar{\omega}_{a,k}^i) = \operatorname{tr}\left((\bar{\omega}_{c,k}^i)^T \bar{\omega}_{c,k}^i + (\bar{\omega}_{a,k}^i)^T \bar{\omega}_{a,k}^i\right). \tag{6.58}$$

Let $V_1(i) = (\bar{\omega}_{ck}^i)^T \bar{\omega}_{ck}^i$ and $V_2(i) = (\bar{\omega}_{a,k}^i)^T \bar{\omega}_{a,k}^i$, so the following holds:

$$
\begin{aligned}
\Delta V_1(i) &= \mathrm{tr}\left((\bar{\omega}_{ck}^{i+1})^T \bar{\omega}_{ck}^{i+1} - (\bar{\omega}_{ck}^{i})^T \bar{\omega}_{ck}^{i}\right) \\
&= \eta_c^2 \left(e_{ck}^i \sigma(\hat{Z}_c(k+1))\right)^T e_{ck}^i \sigma\left(\hat{Z}_c(k+1)\right) - 2\eta_c \left(e_{ck}^i \sigma\left(\hat{Z}_c^{i-1}(k)\right)^T \bar{\omega}_{ck}^i\right) \\
&= \eta_c^2 \left\| e_{ck}^i \right\|^2 \left\| \sigma(\hat{Z}_c^{i-1}(k)) \right\|^2 \\
&\quad - 2\eta_c \left(e_{ck}^i \sigma\left(\hat{Z}_c^{i-1}(k)\right)^T (\hat{\omega}_{ck}^i - \omega_{ck}^i + \omega_{ck}^i - \omega_{ck}^*)\right)
\end{aligned} \tag{6.59}
$$

and

$$
\begin{aligned}
\Delta V_2(i) &= \mathrm{tr}\left((\bar{\omega}_{ak}^{i+1})^T \bar{\omega}_{ak}^{i+1} - (\bar{\omega}_{ak}^{i})^T \bar{\omega}_{ak}^{i}\right) \\
&= \eta_a^2 \left(e_{ak}^i \sigma(Z_a(k))\right)^T e_{ak}^i \sigma\left(Z_a(k)\right) - 2\eta_a \left(e_{ak}^i \sigma\left(Z_a(x_k)\right)^T \bar{\omega}_{ak}^i\right) \\
&= \eta_a^2 \left\| e_{ak}^i \right\|^2 \left\| \sigma(Z_a(k)) \right\|^2 - 2\eta_a e_{ak}^i \sigma(Z_a(k))^T \left(\hat{\omega}_{ak}^i - \omega_{ak}^i + \omega_{ak}^i - \omega_{ak}^*\right).
\end{aligned} \tag{6.60}
$$

We assume

$$
Q^i\left\{x_{k+1}^i, \hat{u}^{i-1}(x_{k+1}^i)\right\} = (\omega_{ck}^i)^T \sigma\left(v_c^T \begin{bmatrix} x_{k+1}^i \\ \hat{u}^{i-1}(x_{k+1}^i) \end{bmatrix}\right), \; u_k^i = (\omega_{ak}^i)^T Z_a(k)
$$

then we have

$$
\begin{aligned}
e_{ck}^i &= \hat{Q}^i(x_k, \hat{u}_k^{i-1}) - Q^i(x_k, \hat{u}_k^{i-1}) = \sigma\left(\hat{Z}_c^{i-1}(k)\right)^T (\hat{\omega}_{ck}^i - \omega_{ck}^i) \\
e_{ak}^i &= \hat{u}_k^i - u_k^i = \sigma\left(Z_a(k)\right)^T (\hat{\omega}_{ak}^i - \omega_{ak}^i).
\end{aligned}
$$

By the analysis in Remark 6.2, one can know that $\lim_{i\to\infty}(\omega_{ck}^i - \omega_{ck}^*) = 0$ and $\lim_{i\to\infty}(\omega_{ak}^i - \omega_{ak}^*) = 0$. So there must exist $W_c > 0$ and $W_a > 0$, such that $\left\|\omega_{ck}^i - \omega_{ck}^*\right\|^2 \le W_c$, $\left\|\omega_{ak}^i - \omega_{ak}^*\right\|^2 \le W_a$ hold. Thus, one has

$$
\begin{aligned}
\Delta V_1(i) &= -\eta_c \left\| e_{ck}^i \right\|^2 \left(2 - \eta_c \left\| \sigma(\hat{Z}_c^{i-1}(k)) \right\|^2\right) + 2\left(e_{ck}^i \sigma^T(\hat{Z}_c^{i-1}(k))(\omega_{ck}^* - \omega_{ck}^i)\right) \\
&\le -\eta_c \left\| e_{ck}^i \right\|^2 \left(2 - \eta_c \left\| \sigma(\hat{Z}_c^{i-1}(k)) \right\|^2\right) \\
&\quad + \eta_c \left(\left\| e_{ck}^i \right\|^2 + \left\| \sigma^T(\hat{Z}_c^{i-1}(k))(\omega_{ck}^* - \omega_{ck}^i) \right\|^2 \right) \\
&= \eta_c \left(\left\| e_{ck}^i \right\|^2 \eta_q - W_c \left\| \sigma(\hat{Z}_c^{i-1}(k)) \right\|^2 \right)
\end{aligned} \tag{6.61}
$$

and

$$\begin{aligned}\Delta V_2(i) &= -\eta_a \left\|e_{ak}^i\right\|^2 \left(2 - \eta_a \|\sigma(Z_a(k))\|^2\right) + 2\left(e_{ak}^i \sigma^T(Z_a(k))(\omega_{ak}^* - \omega_{ak}^i)\right) \\ &\leq -\eta_a \left\|e_{ak}^i\right\|^2 \left(2 - \eta_a \|\sigma(Z_a(k))\|^2\right) \\ &+ \eta_a \left(\left\|e_{ak}^i\right\|^2 + \left\|\sigma^T(Z_a(k))(\omega_{ak}^* - \omega_{ak}^i)\right\|^2\right) \\ &\leq -\eta_a \left(\left\|e_{ak}^i\right\|^2 \eta_u - W_a \|\sigma(Z_a(k))\|^2\right).\end{aligned} \tag{6.62}$$

If (6.55) holds, then $\Delta V(i) = \Delta V_1(i) + \Delta V_2(i) < 0$. Hence, $\lim_{i\to\infty} \bar{\omega}_{ck}^i = 0$ and $\lim_{i\to\infty} \bar{\omega}_{ak}^i = 0$. This completes the proof. □

Remark 6.8 Since the analytical solution (Q^{i+1}, u^i) is quite hard to achieve, NN approximation is necessary for presenting a numerical solution of them. But it has to point out that the reconstruction errors inherently exist due to the facts of $Q^*(x_k, u_k^*) = (\omega_{ck}^*)^T \sigma(Z_c(k)) + \varepsilon_1(x_k)$ and $u_k^* = (\omega_{ak}^*)^T \sigma(Z_a(k)) + \varepsilon_2(x_k)$, where $\varepsilon_1(x_k)$ and $\varepsilon_2(x_k)$ are bounded reconstruction errors. This means that $\hat{\omega}_{ck}^i - \omega_{ck}^*$ and $\hat{\omega}_{ak}^i - \omega_{ak}^*$ are both bounded, whose details can be seen in Yang and Jagannathan (2011). Hence, we claim that an approximate optimal solution of the HJB equation (6.8) is actually obtained instead of the exact optimal one.

6.3.5 *Optimal Control for Linear Systems*

For linear DT system given as

$$x_{k+1} = Ax_k + Bu_k. \tag{6.63}$$

Actually, the optimal Q-function is a quadratic form

$$Q^*(x_k, u_k) = \left(\begin{bmatrix} x_k \\ u_k \end{bmatrix}^T \otimes \begin{bmatrix} x_k \\ u_k \end{bmatrix}^T\right) \text{vec}(H) \tag{6.64}$$

since $J(x_k) = (x_k^T \otimes x_k^T)\text{vec}(P)$ if $u_k = -Kx_k$, where $H \geq 0$, $P \geq 0$ and

$$H = \begin{bmatrix} H_{xx} & H_{xu} \\ (H_{xu})^T & H_{uu} \end{bmatrix} = \begin{bmatrix} A^T PA + Q & A^T PB \\ B^T PA & B^T PB + R \end{bmatrix} \tag{6.65}$$

$$P = \begin{bmatrix} I \\ -K \end{bmatrix}^T H \begin{bmatrix} I \\ -K \end{bmatrix}. \tag{6.66}$$

The DT HJB Eq. (6.8) is reduced as

$$(z_k^T \otimes z_k^T)\text{vec}(H) = x_k^T Q x_k + u_k^T R u_k + (z_{k+1}^T \otimes z_{k+1}^T)\text{vec}(H) \tag{6.67}$$

where $z_k = \left[x_k^T \; u_k^T\right]^T$. Equation (6.9) correspondingly becomes

$$u_k^* = -H_{uu}^{-1}(H_{xu})^T x_k. \tag{6.68}$$

Thus, (6.12) and (6.13) are correspondingly rewritten as

$$(z_k^T \otimes z_k^T)\text{vec}(H^{i+1}) = x_k^T Q x_k + (u_k^i)^T R u_k^i + (z_{k+1}^T \otimes z_{k+1}^T)\text{vec}(H^i) \tag{6.69}$$

and

$$u_k^i = -(H_{uu}^i)^{-1}(H_{xu}^i)^T x_k. \tag{6.70}$$

To implement the off-policy Q-learning algorithm for linear system (6.63), (6.26) is correspondingly changed into

$$x_{k+1} = A_c x_k + B(u_k - u_k^i) \tag{6.71}$$

where $A_c = A - BK^i$ and $u_k^j = -K^i x_k$. Notice that (6.69) is equivalent to the following form:

$$\begin{aligned}\begin{bmatrix} I \\ -K^i \end{bmatrix}^T H^{i+1} \begin{bmatrix} I \\ -K^i \end{bmatrix} &= Q + (K^i)^T R K^i \\ &\quad + (A - BK^i)^T \begin{bmatrix} I \\ -K^i \end{bmatrix}^T H^i \begin{bmatrix} I \\ -K^i \end{bmatrix} (A - BK^i).\end{aligned} \tag{6.72}$$

Thus, (6.29) is correspondingly changed into

$$\begin{aligned}Q^{i+1}(x_k, u_k^i) &- x_k^T A_c{}^T \begin{bmatrix} I \\ -K^j \end{bmatrix}^T H^i \begin{bmatrix} I \\ -K^j \end{bmatrix} A_c x_k \\ &= x_k^T \begin{bmatrix} I \\ -K^i \end{bmatrix}^T H^{i+1} \begin{bmatrix} I \\ -K^i \end{bmatrix} x_k - (x_{k+1} - B(u_k - u_k^i))^T \\ &\times \begin{bmatrix} I \\ -K^i \end{bmatrix}^T H^i \begin{bmatrix} I \\ -K^i \end{bmatrix} (x_{k+1} - B(u_k - u_k^i)) \\ &= x_k^T (Q + (K^i)^T R K^i) x_k.\end{aligned} \tag{6.73}$$

Since P^{i+1} and H^{i+1} have the relationship shown in (6.65) and (6.66), then the following off-policy Q-function based Bellman equation holds.

$$
\begin{aligned}
&x_k^T \begin{bmatrix} I \\ -K^i \end{bmatrix}^T H^{i+1} \begin{bmatrix} I \\ -K^i \end{bmatrix} x_k - x_{k+1}^T \begin{bmatrix} I \\ -K^i \end{bmatrix}^T H^i \begin{bmatrix} I \\ -K^i \end{bmatrix} x_{k+1} \\
&+ 2x_k^T H_{xu}^i (u_k + K^i x_k) + u_k^T (H_{uu}^i - R)(u_k + K^i x_k) \\
&- (K^i x_k)^T (H_{uu}^i - R)(u_k + K^i x_k) \\
&= x_k^T (Q + (K^i)^T R K^i) x_k. \qquad (6.74)
\end{aligned}
$$

Properly manipulating (6.74) yields the following form:

$$
\theta^i(k)\text{vec}(H^{i+1}) = \rho_k^i \qquad (6.75)
$$

where

$$
\begin{aligned}
\rho_k^i &= x_k^T Q x_k + u_k^T R u_k + \left(x_{k+1}^T \begin{bmatrix} I \\ -K^i \end{bmatrix}^T \right) \otimes \left(x_{k+1}^T \begin{bmatrix} I \\ -K^i \end{bmatrix}^T \right) \text{vec}(H^i) \\
&\quad - 2 \left(x_k^T \otimes (u_k + K^i x_k)^T \right) \text{vec}(H_{xu}^i) - (u_k^T \otimes u_k^T)\text{vec}(H_{uu}^i) \\
\theta^i(k) &= \left(x_k^T \begin{bmatrix} I \\ -K^i \end{bmatrix}^T \right) \otimes \left(x_k^T \begin{bmatrix} I \\ -K^i \end{bmatrix}^T \right).
\end{aligned}
$$

When finding H^{i+1}, thus K^{j+1} can be calculated as

$$
K^{i+1} = (H_{uu}^{i+1})^{-1} (H_{xu}^{i+1})^T. \qquad (6.76)
$$

Algorithm 6.4 is reduced as follows for the case of linear system.

Algorithm 6.4 Off-policy interleaved Q-learning for linear systems

1: Data collection: Collect system data x_k and store them in the sample sets $\theta^i(k)$ and ρ^i by using the behavior control policy u_k;
2: Initialization: Choose the initial stabilizing gains K^0, and let the initial iterative matrix H^0. Set $i = 0, k = 0$;
3: Implementing Q-learning: Calculate H^{i+1} in (6.75) using the collected data in Step 1, and then K^{i+1} can be updated in terms of (6.76);
4: If $\| K^{i+1} - K^i \| \le l$ (l is some small positive number), then stop the iteration, and let $k = k + 1$, go back to Step 3, and thus the optimal control policy has been obtained. Otherwise, let $i = i + 1$ and go back to Step 3.

Remark 6.9 A distinctive feature existed in Algorithm 6.4 for the special case of linear systems is to approximate the optimal control policy gain without knowing system matrices A and B, even no need of identifying system model using neural networks or something similar.

6.4 Illustrative Examples

In this section, the proposed off-policy interleaved Q-learning algorithm is applied to three examples to show its effectiveness. Simulations are operated to show the no bias of solutions when adding probing noise to systems if we use this developed off-policy Q-learning algorithm. Moreover, simulations show the implementation and control performance of the proposed algorithm.

Example 6.1 Consider the following open-loop unstable system:

$$x_{k+1} = \begin{bmatrix} -1 & 2 \\ 2.2 & 1.7 \end{bmatrix} x_k + \begin{bmatrix} 2 \\ 1.6 \end{bmatrix} u_k. \tag{6.77}$$

Choose $Q = 6$ and $R = 1$. First, the optimal solution P^* was calculated by using command "dare" in MATLAB. Thus, the optimal Q-function matrix H^* and the optimal controller gain K^* can be respectively obtained in terms of (6.65) and (6.68).

$$\begin{aligned} H^* &= \begin{bmatrix} 96.4653 & -95.5203 & -96.9205 \\ -95.5203 & 289.2950 & 281.7826 \\ -96.9205 & 281.7826 & 281.3409 \end{bmatrix} \\ K^* &= \begin{bmatrix} 0.3445 & -1.0016 \end{bmatrix}. \end{aligned} \tag{6.78}$$

Using three different probing noises, the unbiasedness of the off-policy Q-learning algorithm is verified compared with the on-policy Q-learning algorithm. The probing noise is, respectively, considered as follows:

1.

$$e_k = 1.1(0.5\sin^2(2.0k)\cos(10.1k) + 0.9\sin^2(1.102k)\cos(4.001k); \tag{6.79}$$

2.

$$e_k = 2.97(0.5\sin^2(2.0k)\cos(10.1k) + 0.9\sin^2(1.102k)\cos(4.001k); \tag{6.80}$$

3.

$$e_k = 3.2(0.5\sin^2(2.0k)\cos(10.1k) + 0.9\sin^2(1.102k)\cos(4.001k). \tag{6.81}$$

Table 6.1 respectively lists the convergence results of the iterative controller gain and means and variances of their differences from the theoretical optimal controller gains by using the on-policy Q-learning algorithm and the developed off-policy Q-learning algorithm under the above-mentioned three cases. In Table 6.1, "N" denotes unavailable. For probing noise 1, the PE condition is not satisfied when implementing the on-policy Q-learning Algorithm 6.1, thus this algorithm cannot work. For probing

Table 6.1 Comparisons between on-policy and off-policy learning

	On-policy Q-learning			Off-policy Q-learning		
Probing noise	Controller gain	Mean	Variation	Controller gain	Mean	Variation
1	N	N	N	$K^{10} = [0.3441\ \ -1.0017]$	$6.7075e^{-4}$	$1.7353e^{-5}$
2	$K^1 = [0.4839\ \ -1.0025]$	N	N	$K^{10} = [0.3445\ \ -1.0016]$	$3.2443e^{-4}$	$1.7353e^{-5}$
3	$K^3 = [0.1669\ \ -0.7327]$	N	N	$K^{10} = [0.3445\ \ -1.0016]$	$3.2849e^{-4}$	$1.7353e^{-5}$

noise 2 and 3, the PE condition is satisfied only at the first iteration and the third iteration, respectively. The learned controller gain shown in Table 6.1 cannot stabilize system (6.77) (see Fig. 6.1a with using K^3 of probing noise 3). It shows that the learned controller gains are incorrect. However, for all three probing noises, the controller gains can converge to the theoretical optimal values when implementing off-policy interleaved Q-learning after 10 iterations (see Fig. 6.2), which shows that adding probing noise cannot produce bias on learning solution of LQT problem unlike on-policy Q-learning. Figure 6.1b, c shows the state trajectories of the system and cost variation under the learned optimal control policy, respectively.

Example 6.2 Now the developed off-policy interleaved Q-learning algorithm is verified in the following inverted pendulum system (Beard 1995):

$$\begin{bmatrix} x_{1(k+1)} \\ x_{2(k+1)} \end{bmatrix} = \begin{bmatrix} x_{1k} + \Delta t x_{2k} \\ \frac{g}{\iota}\Delta t \sin(x_{1k}) + (1 - \kappa\iota\Delta t)x_{2k} \end{bmatrix} + \begin{bmatrix} 0 \\ \frac{\Delta t}{m\iota^2} \end{bmatrix} u_k \tag{6.82}$$

where the sampling interval $\Delta t = 0.1$ s, $m = 0.5$ kg and $\iota = 1.0545$ m are the mass and length of the pendulum bar, respectively. Let $\kappa = 8.5415$ and $g = 3.1002\,\mathrm{m/s^2}$ be the frictional factor and the gravitational acceleration, respectively. Let the initial state be $x_0 = [0.3\ \ -0.3]^T$, the structures of the inverted pendulum network, the critic and action networks be 3-6-1, 3-8-1 and 2-2-1, respectively. Choose $Q = \mathrm{diag}(1, 1)$ and $R = 0.1$.

Let the learning rates of the inverted pendulum network, the critic and action networks respectively be 0.1, 0.3 and 0.1. Let the training errors be 0.02 for these three neural networks. Figure 6.3a shows the results of regulating neural network weights. Implementing the off-policy interleaved Q-learning Algorithm 6.3 yields the training or iteration results of the critic and actor networks as shown in Fig. 6.3b, c. Thus, the approximate optimal control policy is learned, Fig. 6.4a presents the approximation of the optimal Q-function $Q^*(x_k, u^*(x_k))$. In the real operation of the inverted pendulum system, external disturbance and measurement errors are not completely

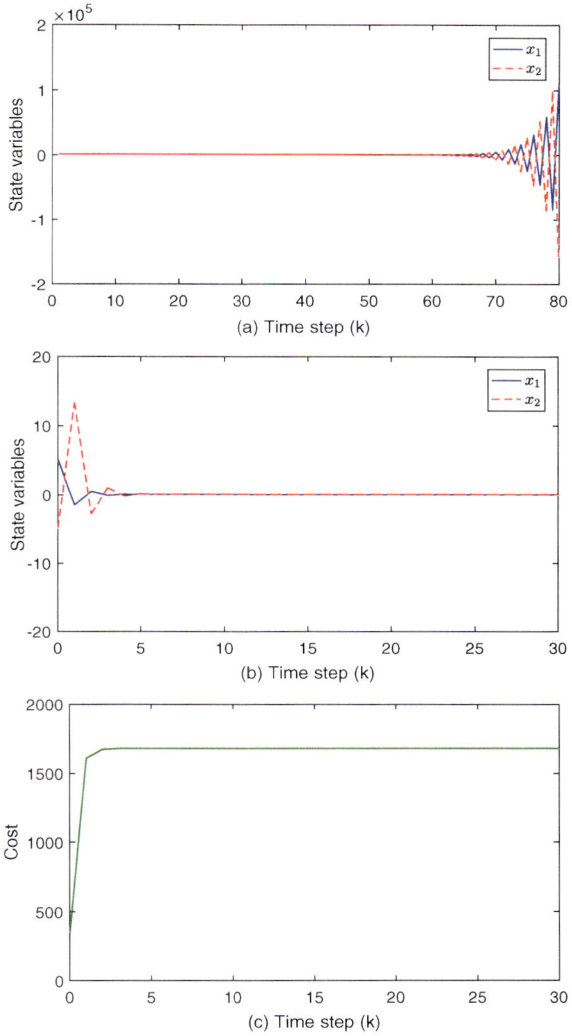

Fig. 6.1 Curves of state trajectories **a** on-policy Q-learning, **b** off-policy Q-learning and **c** the cost using off-policy Q-learning

avoided, so they are combined and assumed as $0.2e^{-0.0001k}\sin([2k\ 0]^T)$ and put it into (6.82). Figure 6.4b, c are given to show the system states under the approximate optimal control policy and the trajectory of the approximate optimal control policy, respectively. The performance $J^*(x_0)$ along with the system trajectories under the learned optimal control policy is plotted in Fig. 6.4d.

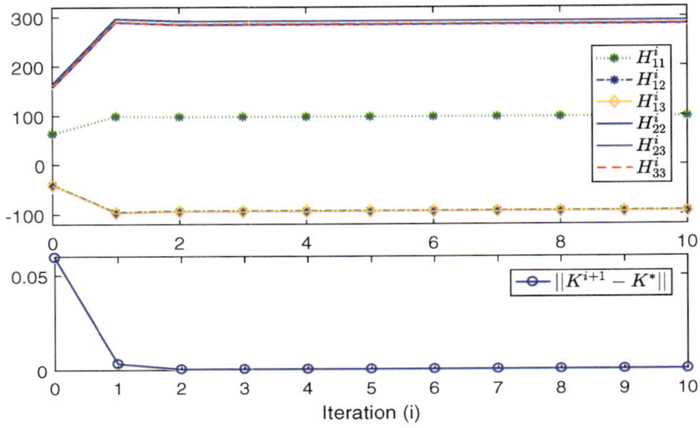

Fig. 6.2 Convergence of matrices H^i and K^i

Example 6.3 Consider the following three-dimensional DT affine nonlinear system (Mu et al. 2017):

$$\begin{bmatrix} x_{1(k+1)} \\ x_{2(k+1)} \\ x_{2(k+1)} \end{bmatrix} = \begin{bmatrix} x_{1k}x_{2k} \\ x_{1k}^2 - 0.5\sin(x_{2k}) \\ x_{3k} \end{bmatrix} + \begin{bmatrix} 0 \\ 1 \\ -1 \end{bmatrix} u_k. \tag{6.83}$$

Let $x_0 = [-0.5\ 0.5\ 1]^T$, $Q = \mathrm{diag}(1, 1, 1)$ and $R = 0.1$. The model network, the critic network and the action network are built with the structures 4-3, 4-8-1 and 3-1, respectively. And the learning rates of these three networks are all set to be 0.1. For the critic network, the entries of input layer to hidden layer weight matrix are randomly generated in $[-0.15,\ 0.05]$ and then kept unchanged.

Under the probing noise $e_k = \mathrm{rand}(1, 1) * 0.2$, the performance is tested on 5 trials under the same scenario (the same initial neural network weights ω_x, $\hat{\omega}_{ak}^i$, $\hat{\omega}_{ck}^i$ and v_c). The results of 5 trials by using off-policy interleaved Q-learning Algorithm 6.3 are listed in Table 6.2. Approximate optimal control policy is quite hard to be found by using on-policy learning as not only neural network approximation but also adding probing noise might produce biased iterative Q-function solutions, as shown in these 5 trials wherein four testings are failed and the not good performance is obtained in one successful testing compared with the off-policy Q-learning method. Whereas adding probing noise would not take any effect on precise solution of iterative Q-function and adequate exploration can be satisfied by using arbitrary behavior control policy if the off-policy Q-learning algorithm is employed. Additionally, the iterative target control policy with probing noise has to act on the real system to learn the optimal control policy when running on-policy learning, which inevitably produces negative impact on performance of systems. Figure 6.5a, b give the curves of state trajectories and the approximate optimal control laws that make the accumulated cost respectively reach 5.6789 and 195.1209 by using off-policy interleaved Q-learning and on-policy interleaved Q-learning.

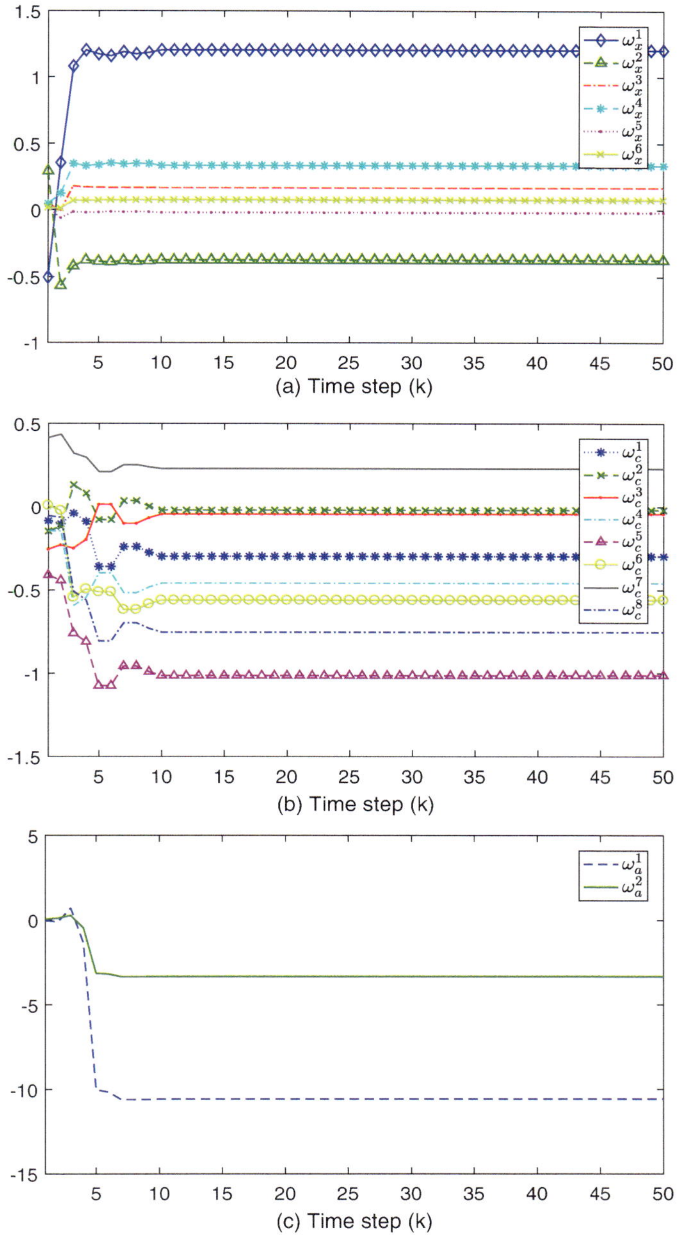

Fig. 6.3 Updating process of weights of **a** inverted pendulum NN, **b** critic NN and **c** actor NN

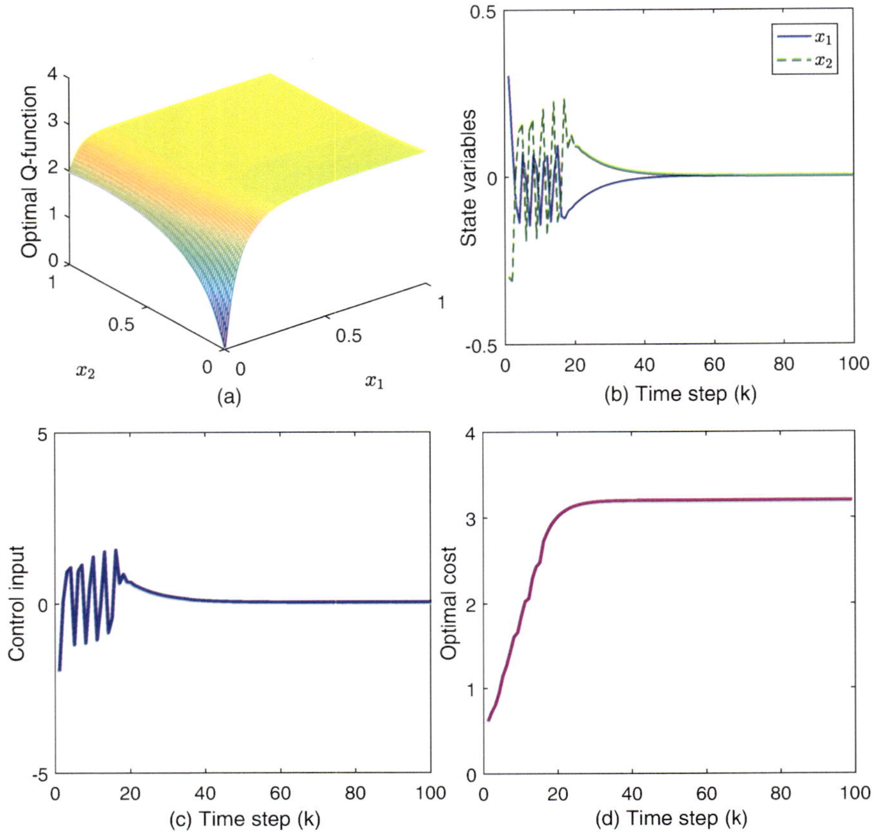

Fig. 6.4 Simulation results under the approximate optimal control policy

Table 6.2 Performance comparisons of on-policy versus off-policy learning

Off-policy interleaved Q-learning		
	Approximate optimal cost $J(x_0)$	Operation time (s)
1	3.4618	1.692
2	7.8311	1.662
3	5.6789	1.293
4	5.7520	1.248
5	6.1557	1.505
Average	5.7757	1.48
Standard deviation	1.5599	0.2046
On-policy interleaved Q-learning		
1	195.1209	1.018

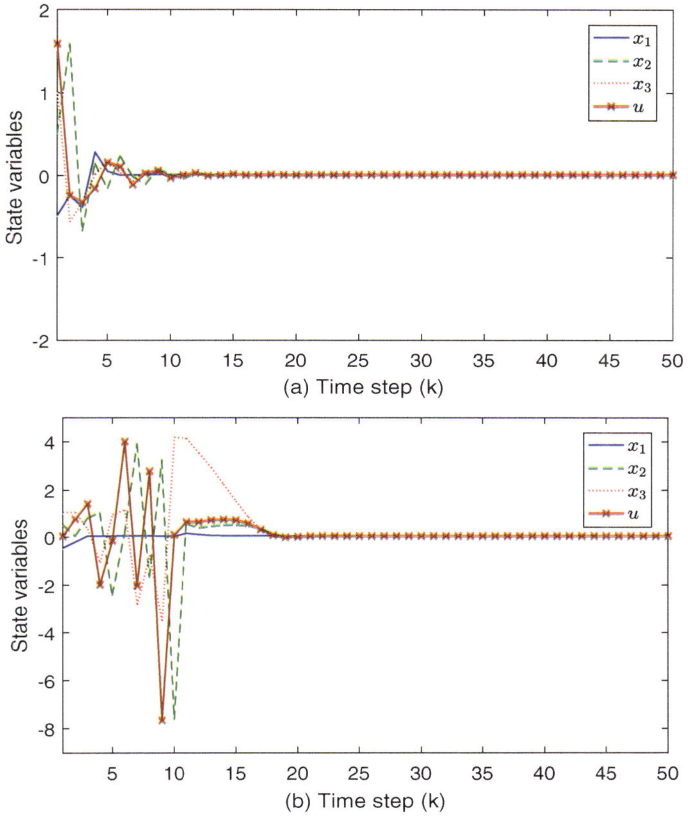

Fig. 6.5 The curves of state and control input **a** using off-policy interleaved Q-learning and **b** using on-policy interleaved Q-learning

6.5 Conclusion

This section focuses on presenting a novel off-policy interleaved Q-learning method for approximating the optimal control policy to achieve the optimum of affine nonlinear DT systems without knowing the dynamics of models. Based on the existing on-policy Q-learning methods for solving the optimal control problem, an off-policy Q-learning algorithm is developed and further the critic and actor structure based off-policy interleaved Q-learning algorithm is proposed. The rigorously theoretical proofs on the less sensitivity of solution of optimality problem to probing noise and the convergence of the proposed off-policy interleaved Q-learning are presented. Simulation examples have demonstrated the effectiveness of the proposed method.

References

Al-Tamimi A, Lewis FL, Abu-Khalaf M (2007) Model-free Q-learning designs for linear discrete-time zero-sum games with application to H-infinity control. Automatica 43(3):473–481

Beard RW (1995) Improving the closed-loop performance of nonlinear systems. PhD thesis, Citeseer

Chai T, Ding J, Wu F (2011) Hybrid intelligent control for optimal operation of shaft furnace roasting process. Control Eng Pract 19(3):264–275

Chai T, Qin SJ, Wang H (2014) Optimal operational control for complex industrial processes. Annu Rev Control 38(1):81–92

Doltsinis S, Ferreira P, Lohse N (2014) An MDP model-based reinforcement learning approach for production station ramp-up optimization: Q-learning analysis. IEEE Trans Syst Man Cybern Syst 44(9):1125–1138

Jiang Y, Jiang ZP (2012) Computational adaptive optimal control for continuous-time linear systems with completely unknown dynamics. Automatica 48(10):2699–2704

Jiang Y, Fan J, Chai T, Lewis FL, Li J (2017) Tracking control for linear discrete-time networked control systems with unknown dynamics and dropout. IEEE Trans Neural Netw Learn Syst 29(10):4607–4620

Kim JH, Lewis FL (2010) Model-free H_∞ control design for unknown linear discrete-time systems via Q-learning with LMI, vol 46. Elsevier

Kiumarsi B, Lewis FL, Jiang Z (2017) H_∞ control of linear discrete-time systems: off-policy reinforcement learning. Automatica 78:144–152

Kiumarsi B, Lewis FL, Modares H, Karimpour A, Naghibi-Sistani MB (2014) Reinforcement Q-learning for optimal tracking control of linear discrete-time systems with unknown dynamics. Automatica 50(4):1167–1175

Lee JY, Park JB, Choi YH (2012) Integral Q-learning and explorized policy iteration for adaptive optimal control of continuous-time linear systems. Automatica 48(11):2850–2859

Li J, Kiumarsi B, Chai T, Lewis FL, Fan J (2017a) Off-policy reinforcement learning: optimal operational control for two-time-scale industrial processes. IEEE Trans Cybern 47(12):4547–4558

Li J, Modares H, Chai T, Lewis FL, Xie L (2017b) Off-policy reinforcement learning for synchronization in multiagent graphical games. IEEE Trans Neural Netw Learn Syst 28(10):2434–2445

Li J, Chai T, Lewis FL, Fan J, Ding Z, Ding J (2018) Off-policy Q-learning: set-point design for optimizing dual-rate rougher flotation operational processes. IEEE Trans Ind Electron 65(5):4092–4102

Liu D, Wei Q (2013) Policy iteration adaptive dynamic programming algorithm for discrete-time nonlinear systems. IEEE Trans Neural Netw Learn Syst 25(3):621–634

Luo B, Wu HN, Huang T (2014) Off-policy reinforcement learning for H_∞ control design. IEEE Trans Cybern 45(1):65–76

Luo B, Liu D, Huang T, Wang D (2016) Model-free optimal tracking control via critic-only Q-learning. IEEE Trans Neural Netw Learn Syst 27(10):2134–2144

Mu C, Wang D, He H (2017) Novel iterative neural dynamic programming for data-based approximate optimal control design. Automatica 81:240–252

Prokhorov DV, Santiago RA, Wunsch DC II (1995) Adaptive critic designs: a case study for neurocontrol. Neural Netw 8(9):1367–1372

Song R, Lewis FL, Wei Q, Zhang H (2015) Off-policy actor-critic structure for optimal control of unknown systems with disturbances. IEEE Trans Cybern 46(5):1041–1050

Tsitsiklis JN (1994) Asynchronous stochastic approximation and Q-learning. Mach Learn 16(3):185–202

Vamvoudakis KG (2017) Q-learning for continuous-time linear systems: a model-free infinite horizon optimal control approach. Syst Control Lett 100:14–20

Wang D, He H, Zhong X, Liu D (2017a) Event-driven nonlinear discounted optimal regulation involving a power system application. IEEE Trans Ind Electron 64(10):8177–8186

Wang D, Liu D, Mu C, Zhang Y (2017b) Neural network learning and robust stabilization of nonlinear systems with dynamic uncertainties. IEEE Trans Neural Netw Learn Syst 29(4):1342–1351

Watkins CJ, Dayan P (1992) Q-learning. Mach Learn 8(3–4):279–292

Wei Q, Liu D, Shi G (2014) A novel dual iterative Q-learning method for optimal battery management in smart residential environments. IEEE Trans Ind Electron 62(4):2509–2518

Yang Q, Jagannathan S (2011) Reinforcement learning controller design for affine nonlinear discrete-time systems using online approximators. IEEE Trans Syst Man Cybern Part B (Cybern) 42(2):377–390

Chapter 7
Off-Policy Game Reinforcement Learning

This chapter deals with the optimal synchronization control problem for CT multi-agent systems based on graphical games and the cooperative optimal control problem for DT multi-player systems based on nonzero-sum games. First, we developed an off-policy RL algorithm to solve optimal synchronization of multi-agent systems. In contrast to traditional control protocols, which require complete knowledge of agent dynamics, the presented algorithm is a model-free approach, in that it solves the optimal synchronization problem without knowing any knowledge of the agent dynamics. It is shown that the optimal distributed policies found by the proposed algorithm satisfy the global Nash equilibrium and synchronize all agents to the leader. Then, an off-policy game Q-learning algorithm is proposed for solving linear DT multi-player game problems based on dynamic programming and Q-learning methods, such that the Nash equilibrium is reached under the learned control policies. The proposed algorithm does not require the system model parameters to be known a priori and fully utilizes measurable data to learn the Nash equilibrium solution. Moreover, there is no bias of Nash equilibrium solution when implementing the proposed algorithm even though probing noises are added for ensuring the PE condition. Simulation results illustrate the effectiveness of the proposed method.

7.1 Graphical Game for Optimal Synchronization

MASs have attracted compelling attention in the last two decades because of their potential applications in a variety of disciplines including engineering, social science and natural science (Xu and Liu 2011; Yu et al. 2014; Olfati-Saber 2006). The distributed synchronization problem of MASs has been considered with the goal of developing control protocols based on local information of each agent and its neighbors to make all agents reach an agreement on certain quantities of interest

J. Li et al., *Reinforcement Learning*, Advances in Industrial Control,
https://doi.org/10.1007/978-3-031-28394-9_7

or track a reference trajectory (Olfati-Saber 2006; Movric and Lewis 2014; Cheng and Savkin 2011; Lewis et al. 2013). Most existing results on synchronization of multi-agent systems did not impose optimality, and therefore, are generally far from optimal.

Optimal distributed control considered in the literature (Lewis et al. 2013; Dunbar and Murray 2006; Vamvoudakis et al. 2012; Semsar-Kazerooni and Khorasani 2009; Dong 2010; Wang et al. 2014; Liu and Jia 2012) is desirable design, because it minimizes a predefined performance function that captures energy consumption, efficiency and accuracy of the solution. However, these existing methods require either complete knowledge of the system dynamics (Lewis et al. 2013; Dunbar and Murray 2006; Vamvoudakis et al. 2012; Semsar-Kazerooni and Khorasani 2009; Dong 2010) or at least partial knowledge of the system dynamics (Wang et al. 2014). Adaptive leader-follower synchronization is presented in Liu and Jia (2012) for MASs using the model reference adaptive control approach to avoid the requirement of knowing the system dynamics. However, adaptive approaches are generally far from optimal and this method still requires partial knowledge of the system dynamics.

With the rapid development and extensive applications of digital sensor technology, extensive data-carrying system information can be collected. It is desired to use these data to develop model-free data-based optimal control protocols (Luo et al. 2014; Hou and Wang 2013; Bian et al. 2014). The robust adaptive dynamic programming (RADP) method is adopted in Bian et al. (2014) to design decentralized optimal controllers for large-scale systems without requiring the knowledge of system dynamics. Note that synchronization and distributed optimal controller design are out of the concerns of Bian et al. (2014). The RL technique is a promising technique that can be used to learn optimal solutions to control problems in real time using only measured data along the system trajectories. Q-learning (Kaya and Alhajj 2005) as a behavior-dependent heuristic dynamic programming has been developed to learn optimal synchronization protocols for MASs. However, these results are limited to Markov process and discrete-time systems. Thus, we concentrate on the optimal synchronization of multi-agent CT systems with completely unknown dynamics, while the data-driven optimal control problem of MASs is challenging because of the coupling of the agent dynamics as a result of data exchange between them. That is, an off-policy RL algorithm is presented to learn optimal control protocols for CT MASs using only measured data in this section. The off-policy RL and multi-agent graphical games are brought together, where the dynamics and performance objective of each agent are affected by it and its neighbors in a graphical topology.

The organization structure of this section is as follows. Section 7.1.1 introduces graph theory concepts and some definitions that are used throughout the chapter. Section 7.1.2 defines the optimal synchronization problem and investigates global Nash equilibrium and stability of the optimal solution. Section 7.1.3 develops an off-policy RL algorithm to learn optimal controllers using data generated from agents. Section 7.1.4 presents the simulation results.

7.1.1 Preliminaries

We first introduce the graph theory (Movric and Lewis 2014; Vamvoudakis et al. 2012). The synchronization problem of MASs is then defined.

7.1.1.1 Communication Graph

Consider a graph denoted by $\mathcal{G} = (\mathcal{V}, \mathcal{E})$ with a set of vertices $\mathcal{V} = \{v_1, v_2, \ldots, v_N\}$ and a set of edges or arcs $\mathcal{E} \subseteq \mathcal{V} \times \mathcal{V}$. $E = [e_{ij}]$ is called the connectivity matrix with $e_{ij} > 0$ if $(v_j, v_i) \in \mathcal{E}$, and otherwise $e_{ij} = 0$. $(v_j, v_i) \in \mathcal{E}$ indicates that there exists an edge starting from the vertex v_j to v_i in a directed graph. A graph is called simple if there are no repeated edges or self-loops $(v_i, v_i) \in \mathcal{E}$ for $\forall i$. Denote the set of neighbors of node v_i as $N_i = \{v_j : (v_j, v_i) \in \mathcal{E}\}$. $\mathcal{D} = \text{diag}(d_1 \ldots d_N)$ is called in-degree matrix with the weighted degree $d_i = \sum_j e_{ij}$ of node v_i (i.e., the ith row sum of E). Define the graph Laplacian matrix as $L = \mathcal{D} - E$, in which the summing elements of every row is equal to zero.

A directed path is a sequence of edges (v_{i_1}, v_{i_2}), (v_{i_2}, v_{i_3}), $\ldots$, $(v_{i_{j-1}}, v_{i_j})$ with $(v_{i_{l-1}}, v_{i_l}) \in \mathcal{E}$ for $l \in \{2, \ldots, j\}$, which is a path starting from the node v_{i_1} and ending at v_{i_j}. A directed graph is strongly connected if there is a directed path for every two distinct vertices. The graph is said to be strongly connected if there is a directed path from v_i to v_j for any distinct nodes $v_i, v_j \in \mathcal{V}$ (Movric and Lewis 2014; Vamvoudakis et al. 2012).

7.1.1.2 Synchronization of Multi-agent Systems

Consider the N systems or agents with identical dynamics

$$\dot{x}_i = Ax_i + Bu_i, \tag{7.1}$$

where $x_i = x_i(t) \in \mathbb{R}^n$ denotes the state vector, $u_i = u_i(t) \in \mathbb{R}^p$ $(i = 1, 2 \ldots, N)$ denotes the control input. A and B are matrices of appropriate dimensions. The dynamics of the command generator or leader with state x_0 is given by

$$\dot{x}_0 = Ax_0. \tag{7.2}$$

Assumption 7.1 The pair (A, B) is controllable. The pair $(A, \sqrt{Q_i})$ is observable.

The local neighborhood tracking error δ_i of agent i is defined as

$$\delta_i = \sum_{j \in N_i} e_{ij}(x_i - x_j) + g_i(x_i - x_0), \tag{7.3}$$

where $g_i \geq 0$ is the pinning gain for agent i with $g_i \neq 0$ if agent i has direct access to the leader and $g_i = 0$ otherwise.

Assumption 7.2 The graph is strongly connected and the leader is pinned to at least one node.

Let $\xi = x - \underline{x}_0$ be the global synchronization error (Vamvoudakis et al. 2012), where $\underline{x}_0 = \underline{1} \otimes x_0$, $\underline{1} = [1,\ 1,\ \ldots,\ 1]^T \in\in \mathbb{R}^N$. The global error vector of the MASs with the command generator is given from (7.3) by

$$\delta = ((L + G) \otimes I_n)\xi, \tag{7.4}$$

where G is a diagonal matrix with diagonal entries equal to the pinning gains g_i. Using (7.1), (7.2) and (7.3) yields the local neighborhood tracking error dynamics as

$$\dot{\delta}_i = A\delta_i + (d_i + g_i)Bu_i - \sum_{j \in N_i} e_{ij} Bu_j \tag{7.5}$$

which is further expressed by the compact form

$$\dot{\delta} = (I_N \otimes A)\delta + ((L + G) \otimes B)u. \tag{7.6}$$

This is the dynamics of the overall neighborhood errors, where $\delta = \left[\delta_1^T\ \delta_2^T\ \ldots\ \delta_N^T\right]^T$ and $u = \left[u_1^T\ u_2^T\ \ldots\ u_N^T\right]^T$.

Synchronization Problem: Design local control protocols u_i in (7.1) to synchronize the state of all agents in $\mathcal{G}$ to the trajectory of the leader, i.e., $\lim_{t\to\infty} x_i(t) = x_0(t)$ for $\forall i$, $i = 1, 2, \ldots, N$ or $\lim_{t\to\infty} \xi(t) = \lim_{t\to\infty}(x(t) - \underline{x}_0) = 0$.

Remark 7.1 Our objective is to make $\lim_{t\to\infty} \xi(t) = 0$. In the subsequent development, we show how to make $\lim_{t\to\infty} \delta(t) = 0$. According to (7.4), one has $\|\xi(t)\| \leq \frac{1}{\sigma_{\min}(L+G)} \|\delta(t)\|$, where $\sigma_{\min}(L + G)$ denotes the smallest singular value of the matrix $L + G$. By Assumption 7.2, $\sigma_{\min}(L + G) > 0$, so that $\lim_{t\to\infty} \delta(t) = 0 \Rightarrow \lim_{t\to\infty} \xi(t) = 0$.

Therefore, to solve the synchronization problem, one can design a control protocol for each agent i to guarantee asymptotic stability of the local neighborhood tracking error dynamics (7.5). In the subsequent Sects. 7.1.2 and 7.1.3, it is shown how to design local control protocols to stabilize the error dynamics (7.5) in an optimal manner by minimizing a predefined performance function for each agent.

7.1.2 Multi-agent Graphical Games

Optimal synchronization of MASs on graphs is discussed in the framework of multi-agent graphical games. This section will show how to find optimal protocols for every agent and also shows that the optimal response makes all agents synchronize to the leader and reach a global Nash equilibrium.

7.1.2.1 Graphical Game for Dynamical Multi-agent Systems

Define a local quadratic performance index for each agent as

$$J_i(\delta_i(t_0), u_i, u_{-i}) = \int_{t_0}^{\infty} \left(\delta_i^T Q_i \delta_i + u_i^T R_i u_i\right) dt, \tag{7.7}$$

where Q_i and R_i are positive semi-definite and positive definite matrices, respectively.

Minimizing (7.7) subject to (7.5) is a graphical game, since both the dynamics and the performance function for each agent depends on the agent and its neighbors (Kaya and Alhajj 2005). In graphical games, the focus is on the global Nash equilibrium. The definition of global Nash equilibrium is given as follows.

Definition 7.1 (Başar and Olsder 1998) A global Nash equilibrium solution for an N-player game is given by a N-tuple of policies $\{u_1^*, u_2^*, \ldots, u_N^*\}$ if it satisfies

$$J_i^* \triangleq J_i(\delta_i(t_0), u_i^*, u_{\mathcal{G}-i}^*) \leq J_i(\delta_i(t_0), u_i, u_{\mathcal{G}-i}^*) \tag{7.8}$$

for all $i \in N$ and $\forall u_i, u_{\mathcal{G}-i}$ $(u_{\mathcal{G}-i} = \{u_j : j \in \mathcal{V}, j \neq i\})$. The N-tuple of game values $\{J_1^*, J_2^*, \ldots, J_N^*\}$ is said to be a Nash equilibrium outcome of the N-player game.

From (7.5), one can find that the performance index (7.7) depends on agent i and its neighbors. Thus, global Nash equilibrium (7.8) can be written as

$$J_i^* \triangleq J_i(\delta_i(t_0), u_i^*, u_{-i}^*) \leq J_i(\delta_i(t_0), u_i, u_{-i}^*)$$

since

$$\begin{aligned} J_i(\delta_i(t_0), u_i^*, u_{\mathcal{G}-i}^*) &= J_i(\delta_i(t_0), u_i^*, u_{-i}^*) \\ J_i(\delta_i(t_0), u_i, u_{\mathcal{G}-i}^*) &= J_i(\delta_i(t_0), u_i, u_{-i}^*), \end{aligned}$$

where $u_{-i} = \{u_j : j \in N_i\}$.

7.1.2.2 Coupled HJB Equations for Solving Graphical Games

Interpreting the control input u_i as a policy dependent on the local neighborhood tracking error $\delta_i(t)$, the value function corresponding to the performance index (7.7) is introduced as

$$V_i(\delta_i(t)) = \int_{t}^{\infty} \left(\delta_i^T Q_i \delta_i + u_i^T R_i u_i\right) d\tau. \tag{7.9}$$

By taking the derivative of $V_i(\delta_i(t))$ with respect to time t along with the trajectory of the local neighborhood tracking error $\delta_i(t)$, the Bellman equation is given in terms of the Hamiltonian function as

$$H_i(\delta_i, \nabla V_i, u_i, u_{-i}) = \delta_i^T Q_i \delta_i + u_i^T R_i u_i + \nabla V_i^T \left(A\delta_i + (d_i + g_i)Bu_i - \sum_{j \in N_i} e_{ij} B u_j \right) = 0. \tag{7.10}$$

The optimal response of agent i to fixed policies u_{-i} can be derived by minimizing Hamiltonian function with respect to u_i as follows:

$$u_i^*(t) = \arg\min_{u_i} \left[H_i(\delta_i, \nabla V_i^*, u_i, u_{-i}) \right] \tag{7.11}$$

which yields

$$u_i^*(t) = -\frac{1}{2}(d_i + g_i) R_i^{-1} B_i^T \nabla V_i^*, \tag{7.12}$$

where $\nabla_i = \partial V_i / \partial \delta_i$ stands for a gradient operator.

Let all neighbors of agent i select control polices given by (7.12) and substitute (7.12) into (7.10), then one gets the following coupled cooperative game HJB equations.

$$\begin{aligned} H_i(\delta_i, \nabla V_i^*, u_i^*, u_{-i}^*) &= \delta_i^T Q_i \delta_i + \frac{1}{4}(d_i + g_i)^2 (\nabla V_i^*)^T B R_i^{-1} B^T \nabla V_i^* \\ &+ (\nabla V_i^*)^T \Bigg(A\delta_i - \frac{1}{2}(d_i + g_i)^2 B R_i^{-1} B^T \nabla V_i^* \\ &+ \frac{1}{2} \sum_{j \in N_i} e_{ij}(d_i + g_i) B R_j^{-1} B^T \nabla V_j^* \Bigg) = 0. \end{aligned} \tag{7.13}$$

We now show that under a certain assumption, these coupled HJB equations can be simplified and resemble the coupled AREs that appear in standard linear quadratic multi-player nonzero-sum games (Başar and Olsder 1998; Starr and Ho 1969).

Assumption 7.3 The cost function is quadratic and is given by $V_i = \delta_i^T P_i \delta_i$ where P_i is a positive definite matrix.

Using Assumption 7.3, (7.13) can be written as the form shown in Lemma 7.1.

Lemma 7.1 *Under Assumption 7.3, coupled cooperative game HJB equations (7.13) are equivalent to the coupled AREs*

$$\begin{aligned} &2\delta_i^T P_i^T \left(A\delta_i + \sum_{j \in N_i} e_{ij}(d_j + g_j) B R_j^{-1} B^T P_j \delta_j \right) \\ &+ \delta_i^T Q_i \delta_i - (d_i + g_i)^2 \delta_i^T P_i^T B R_i^T B^T P_i \delta_i = 0 \end{aligned} \tag{7.14}$$

and optimal response (7.12) becomes

$$u_i^*(t) = -(d_i + g_i)R_i^{-1}B^T P_i\delta_i. \tag{7.15}$$

Proof For the quadratic cost function $V_i = \delta_i^T P_i \delta_i$, one has $\nabla V_i = 2P_i\delta_i$. Substituting this into (7.12) leads to (7.15). On the other hand, substituting the optimal response (7.15) into the coupled cooperative game HJB equations (7.13) gives (7.14). The proof is completed. □

Equation (7.14) is similar to the coupled AREs in Başar and Olsder (1998); Starr and Ho (1969) for standard nonzero-sum game problems.

Remark 7.2 As (7.15) shows, if the cost function V_i satisfies Assumption 7.3, then the derived optimal controller only depends on the local neighborhood error δ_i of agent i. Based on (7.3), it is concluded that the optimal response (7.15) is in fact distributed under Assumption 7.3.

Remark 7.3 It is noted that none of the upcoming analyses or proofs require Assumption 7.3. However, if solutions can be found for (7.14), then these are also solutions of (7.13). In standard multi-player nonzero-sum games, there is only one state dynamic equation and it is known that the values are quadratic in the state (Başar and Olsder 1998; Starr and Ho 1969). However, in graphical games, each agent has its own dynamics. It has not been shown that the values are quadratic in the local states. That is, in general, Assumption 7.3 may not hold.

7.1.2.3 Stability and Global Nash Equilibrium for the Proposed Solution

To achieve global Nash equilibrium, one needs to calculate the optimal response for every agent i by solving N coupled partial differential HJB equations (7.13) for the N player game problem. The following Theorem 7.1 shows that if all agents select their own optimal response and the communication topology graph is strongly connected, then system (7.5) is asymptotically stable for all i $(i = 1, 2, \ldots, N)$. Therefore, all agents synchronize. Meanwhile, all of N agents are in global Nash equilibrium.

Theorem 7.1 *Make Assumption 7.2. Let V_i be smooth solutions to HJB equations (7.13) and design the control policies u_i^* as (7.12). Then:*

a. *Systems (7.5) are asymptotically stable, and therefore, by Assumption 7.2, all agents are synchronized to the leader;*
b. *$\left[u_1^*, u_2^*, \ldots, u_N^*\right]$ are global Nash equilibrium policies and the corresponding Nash equilibrium outcomes are $J_i^*(\delta_i(0)) = V_i$ $(i = 1, 2, \ldots, N)$.*

Proof a. Let V_i be Lyapunov function candidates. Then, taking derivative of V_i with respect to time t along with the trajectory of the local neighborhood tracking error $\delta_i(t)$, one has

$$\frac{dV_i}{dt} = \nabla V_i^T \dot{\delta} = \nabla V_i^T \left(A\delta_i + (d_i + g_i)Bu_i - \sum_{j\in N_i} e_{ij} Bu_j \right). \tag{7.16}$$

Using (7.10), this becomes

$$\frac{dV_i}{dt} + (\delta_i^T Q_i \delta_i + u_i^T R_i u_i) = H_i(\delta_i, \nabla V_i, u_i, u_{-i}). \tag{7.17}$$

On the other hand, based on (7.10) and (7.13), and using the same development as of Lewis et al. (2012), one can get

$$\begin{aligned} H_i(\delta_i, \nabla V_i, u_i, u_{-i}) = {} & (u_i - u_i^*)^T R_i (u_i - u_i^*) \\ & + \frac{2}{(d_i + g_i)} \sum_{j\in N_i} e_{ij} (u_i^*)^T R_i (u_j - u_j^*). \end{aligned} \tag{7.18}$$

Therefore

$$\begin{aligned} \frac{dV_i}{dt} + (\delta_i^T Q_i \delta_i + u_i^T R_i u_i) = {} & (u_i - u_i^*)^T R_i (u_i - u_i^*) \\ & + \frac{2}{(d_i + g_i)} \sum_{j\in N_i} e_{ij} (u_i^*)^T R_i (u_j - u_j^*). \end{aligned} \tag{7.19}$$

Selecting $u_i = u_i^*$ and $u_j = u_j^*$ gives

$$\frac{dV_i}{dt} + (\delta_i^T Q_i \delta_i + u_i^T R_i u_i) = 0. \tag{7.20}$$

Since the matrices $Q_i \geq 0$ and $R_i > 0$, then $\frac{dV_i}{dt} < 0$ hold for all agents. Therefore, systems (7.5) are asymptotically stable and so all agents synchronize to the leader.

b. Since (a) holds for the selected control polices, then $\delta_i(t) \to 0$ when $t \to \infty$. For Lyapunov functions $V_i(\delta_i)$ $(i = 1, 2, \ldots, N)$, satisfying $V_i(0) = 0$, we have $V_i(\delta_\infty) = 0$. Thus, performance index (7.7) can be written as

$$J_i(\delta_i(0), u_i, u_{-i}) = \int_0^\infty \left(\delta_i^T Q_i \delta_i + u_i^T R_i u_i\right) dt + V_i(\delta_i(0)) + \int_0^\infty \dot{V} dt \tag{7.21}$$

or

$$\begin{aligned} J_i(\delta_i(0), u_i, u_{-i}) = {} & \int_0^\infty \left(\delta_i^T Q_i \delta_i + u_i^T R_i u_i\right) dt + V_i(\delta_i(0)) \\ & + \int_0^\infty \left(\nabla V_i^T \left(A\delta_i + (d_i + g_i)Bu_i - \sum_{j\in N_i} e_{ij} Bu_j \right) \right) dt. \end{aligned} \tag{7.22}$$

If V_i satisfy (7.13) and u_i^*, u_{-i}^*, given by (7.12), are optimal control polices, then by completing the square, one has

$$J_i(\delta_i(0), u_i, u_{-i}) = V_i(\delta_i(0)) + \int_0^\infty \Big(u_i^T R_i u_i + \nabla V_i^T (d_i + g_i) B u_i$$
$$-\nabla V_i^T \sum_{j \in N_i} e_{ij} B u_j + (u_i^*)^T R_i u_i^* dt + \nabla V_i^T \sum_{j \in N_i} e_{ij} B u_j^* \Big) dt$$
$$= V_i(\delta_i(0)) + \int_0^\infty \left((u_i - u_i^*)^T R_i (u_i - u_i^*) - \nabla V_i^T \sum_{j \in N_i} e_{ij} B(u_j^* - u_j) \right) dt. \tag{7.23}$$

If $u_i = u_i^*$ and $u_j = u_j^*$, then $J_i^*(\delta_i(0), u_i^*, u_{-i}^*) = V_i(\delta_i(0))$.
Define

$$J_i(\delta_i(0), u_i, u_{-i}^*) = V_i(\delta_i(0)) + \int_0^\infty (u_i - u_i^*)^T R_i (u_i - u_i^*) dt$$

then, it is clear that $J_i^*(\delta_i(0), u_i^*, u_{-i}^*) < J_i(\delta_i(0), u_i, u_{-i}^*)$ holds for all i ($i = 1, 2, \ldots, N$). Therefore, global Nash equilibrium is reached and the proof is completed. □

From coupled cooperative game HJB equations (7.13), one can see that designing optimal control polices for agents (7.1) requires solving two issues. One issue is that coupled cooperative game HJB equations are N nonlinear partial differential equations (PDE), which makes them hard or even impossible to solve analytically. The other is that the system matrices A and B need to be completely known to find the solutions. An off-policy RL algorithm is designed in the next section to overcome these difficulties.

7.1.3 *Off-Policy Reinforcement Learning Algorithm*

In Vamvoudakis et al. (2012), the graphical game was solved but full knowledge of all agent dynamics is needed. The off-policy RL allows the solution of optimality problems without knowing any knowledge of the agent dynamics. This section presents an off-policy learning algorithm for the synchronization of MAS that does not require any knowledge of the dynamics.

To this end, off-policy Bellman equations are first derived, and then an actor–critic NN structure is used to evaluate the value function and find an improved control policy for each agent. Then, an iterative off-policy RL algorithm is given to learn approximate optimal control policies that make the MASs reach global Nash equilibrium and meanwhile guarantee the synchronization of all agents to the command generator. In off-policy RL, a behavior policy is applied to the system to generate the data for learning, and a different policy, called the target policy, is evaluated and updated using measured data.

7.1.3.1 Derivation of Off-Policy Bellman Equations

A model-based RL algorithm is first given which is used to compare with the results given by the off-policy RL algorithm. Then, off-policy Bellman equations are derived and it is shown that they have the same solution as the coupled cooperative game HJB equations (7.13).

Definition 7.2 (Vamvoudakis et al. 2012) For given system (7.5), a control input u_i is called to be admissible with respect to cost (7.7) if u_i is continuous, $u_i(0) = 0$, u_i stabilizes system (7.5) and V_i is finite.

Algorithm 7.1 is presented to learn the optimal control policies by using knowledge of system models.

Algorithm 7.1 Model-based on-policy reinforcement learning

1: Initialize the agents with admissible control policies $u_i^{(0)}$ for $\forall i$ and set $s = 0$;
2: Evaluate policies by solving V_i:

$$H_i(\delta_i, \nabla V_i^{(s+1)}, u_i^{(s)}, u_{-i}^{(s)}) = \delta_i^T Q_i \delta_i + \left(u_i^{(s)}\right)^T R_i u_i^{(s)} \\ + \left(\nabla V_i^{(s+1)}\right)^T \left(A\delta_i + (d_i + g_i)Bu_i^{(s)} - \sum_{j \in N_i} e_{ij} Bu_j^{(s)}\right) = 0 \tag{7.24}$$

where s denotes iteration index;
3: Find improved control policies using u_i:

$$u_i^{(s)} = -\frac{1}{2}(d_i + g_i)R_i^{-1} B^T \nabla V_i^{(s)}; \tag{7.25}$$

4: Stop when $\left\| V_i^s - V_i^{s+1} \right\| \leq \varepsilon$ with a small constant ε.

Remark 7.4 The result (Vamvoudakis et al. 2012) has shown that under a weak coupling assumption, Algorithm 7.1 converges to the solution of coupled cooperative game HJB equations (7.13) if all agents update their control policies in terms of (7.25) at each iteration. This conclusion holds under the condition that the initial control policies are admissible.

Algorithm 7.1 provides a method to learn control policies that achieve the global Nash equilibrium and synchronization. However, Algorithm 7.1 requires the knowledge of agent dynamics during the implementation of the iterative process. To obviate this requirement and present a model-free approach, off-policy Bellman equations are next presented. An off-policy RL algorithm is then provided to learn solutions of coupled cooperative game HJB equations (7.13) to obtain distributed approximate optimal control policies.

Introducing auxiliary variables $u_i^{(s)}$ and $u_{-i}^{(s)}$ for each agent dynamics (7.5), one has

$$\begin{aligned}\dot{\delta}_i &= A\delta_i + (d_i + g_i)Bu_i^{(s)} - \sum_{j\in N_i} e_{ij}Bu_j^{(s)} \\ &\quad + (d_i + g_i)B(u_i - u_i^{(s)}) + \sum_{j\in N_i} e_{ij}B(u_j - u_j^{(s)}).\end{aligned} \tag{7.26}$$

In (7.26), u_i are interpreted as behavior policies actually applied to the system. By contrast, $u_i^{(s)}$ are the target policies learned in Algorithm 7.1.

Differentiating $V_i^{(s+1)}$ with respect to agent (7.26) yields

$$\begin{aligned}\frac{dV_i^{(s+1)}(\delta_i)}{dt} &= \nabla V_i^{(s+1)}\dot{\delta}_i = \left(\nabla V_i^{(s+1)}\right)^T \left(A\delta_i + (d_i + g_i)Bu_i^{(s)} - \sum_{j\in N_i} e_{ij}Bu_j^{(s)}\right) \\ &\quad + \left(\nabla V_i^{(s+1)}\right)^T \left((d_i + g_i)B(u_i - u_i^{(s)}) - \sum_{j\in N_i} e_{ij}B(u_j - u_j^{(s)})\right).\end{aligned} \tag{7.27}$$

Using (7.24) and (7.25) in (7.27) gives

$$\begin{aligned}\frac{dV_i^{(s+1)}(\delta_i)}{dt} &= -\delta_i^T Q_i\delta_i - (u_i^{(s)})^T R_i u_i^{(s)} - 2(u_i^{(s+1)})^T R_i(u_i - u_i^{(s)}) \\ &\quad + 2(d_i + g_i)^{-1}\sum_{j\in N_i} e_{ij}(u_i^{(s+1)})^T R_i(u_j - u_j^{(s)}).\end{aligned} \tag{7.28}$$

The integral reinforcement learning idea (Vrabie and Lewis 2009) is now used to develop Bellman equations for evaluating the value of target policies. Integrating both sides of (7.28) on the interval $[t, t']$ yields the following off-policy Bellman equations:

$$
\begin{aligned}
&V_i^{(s+1)}(\delta_i) - V_i^{(s+1)}(\delta_i^{'}) \\
&= \int_t^{t^{'}} \left((\delta_i(\tau))^T Q_i \delta_i(\tau) + \left(u_i^{(s)}(\delta_i(\tau))\right)^T R_i u_i^{(s)}(\delta_i(\tau)) \right) d\tau \\
&- 2 \int_t^{t^{'}} \left(\left(u_i^{(s+1)}(\delta_i(\tau))\right)^T R_i \left(u_i^{(s)}(\delta_i(\tau)) - u_i(\delta_i(\tau))\right) \right) d\tau \\
&+ 2(d_i + g_i)^{-1} \int_t^{t^{'}} \sum_{j \in N_i} e_{ij} \\
&\times \left(\left(u_i^{(s+1)}(\delta_i(\tau))\right)^T R_i \left(u_j^{(s)}(\delta_j(\tau)) - u_j(\delta_j(\tau))\right) \right) d\tau, \qquad (7.29)
\end{aligned}
$$

where $\delta_i = \delta_i(t)$, $\delta_i^{'} = \delta_i(t^{'})$.

Theorem 7.2 shows that the value function and improved policy solutions found by solving the off-policy Bellman equations (7.29) are identical to the value function solution found by (7.24) and the improved policy found by (7.25), simultaneously.

Let $\vartheta_i(\delta_i)$ be the set of all admissible control policies for agent i.

Theorem 7.2 *Let $V_i^{(s+1)}(\delta_i)$ satisfy $V_i^{(s+1)}(\delta_i) \geq 0$, $V_i^{(s+1)}(0) = 0$ and $u_i^{(s+1)}(\delta_i) \in \vartheta_i(\delta_i)$. Then, the solution $\left(V_i^{(s+1)}(\delta_i), u_i^{(s+1)}(\delta_i)\right)$ to (7.29) is identical to the solution $V_i^{(s+1)}(\delta_i)$ to (7.24) and $u_i^{(s+1)}(\delta_i)$ to (7.25), at the same time.*

Proof a. Sufficiency: One can conclude that if $V_i^{s+1}(0) = 0$ and $u_i^{s+1}(\delta_i)$ satisfies (7.24) and (7.25), then they are the solution to (7.29) by taking the derivative of (7.29).

b. Necessity: The necessity proof can be completed if the uniqueness of solution of (7.29) is shown. It is now shown by contradiction that the solution to (7.29) is unique. Suppose that (7.29) has another solution $(W_i(\delta_i), v_i(\delta_i))$, where $W_i(\delta_i) \geq 0$, $W_i(0) = 0$ and $v_i(\delta_i) \in \vartheta_i(\delta_i)$.

For any function $p(t)$, one has

$$
\begin{aligned}
&\lim_{\Delta t \to 0} \frac{1}{\Delta t} \left(\int_t^{t+\Delta t} p(\tau) d\tau \right) \\
&= \lim_{\Delta t \to 0} \frac{1}{\Delta t} \left(\int_0^{t+\Delta t} p(\tau) d\tau - \int_0^t p(\tau) d\tau \right) = \frac{d}{dt} \int_0^t p(\tau) d\tau = p(t).
\end{aligned}
\tag{7.30}
$$

Taking the derivative of $V_i(\delta_i)$ and using (7.30) yield

$$
\begin{aligned}
\frac{dV_i^{(s+1)}(\delta_i)}{dt} &= \lim_{\Delta t \to 0} \frac{1}{\Delta t} \left(V_i^{(s+1)}(\delta_i(t + \Delta t)) - V_i^{(s+1)}(\delta_i(t)) \right) \\
&= 2 \lim_{\Delta t \to 0} \frac{1}{\Delta t} \int_t^{t+\Delta t} \left(u_i^{(s+1)}(\delta_i(\tau))\right)^T R_i \left(u_i^{(s)}(\delta_i(\tau)) - u_i(\delta_i(\tau))\right) d\tau
\end{aligned}
$$

$$
\begin{aligned}
&- 2(d_i + g_i)^{-1} \lim_{\Delta t \to 0} \frac{1}{\Delta t} \int_t^{t+\Delta t} \sum_{j \in N_i} e_{ij} \left(u_i^{(s+1)}(\delta_i(\tau)) \right)^T R_i \\
&\times \left(u_j^{(s)}(\delta_j(\tau)) - u_j(\delta_j(\tau)) \right) d\tau - \lim_{\Delta t \to 0} \frac{1}{\Delta t} \left(\int_t^{t+\Delta t} \delta_i^T(\tau) Q_i \delta_i(\tau) \right. \\
&\left. + (u_i^{(s)})^T(\delta_i(\tau)) R_i^{-1} u_i^{(s)}(\delta_i(\tau)) \right) d\tau \qquad (7.31)
\end{aligned}
$$

which is equivalent to (7.29). Since $(W_i(\delta_i), v_i(\delta_i))$ is also assumed a solution to (7.29), we have

$$
\begin{aligned}
\frac{dW_i(\delta_i)}{dt} &= -\delta_i^T Q_i \delta_i - \left(u_i^{(s)}\right)^T R_i^{-1} u_i^{(s)} - 2\left(v_i^{(s+1)}\right)^T R_i (u_i - u_i^{(s)}) \\
&\quad + 2(d_i + g_i)^{-1} \sum_{j \in N_i} e_{ij} \left(v_i^{(s+1)}\right)^T R_i (u_j - u_j^{(s)}). \qquad (7.32)
\end{aligned}
$$

Subtracting (7.32) from (7.31) yields

$$
\begin{aligned}
\frac{d(V_i^{(s+1)}(\delta_i) - W_i(\delta_i))}{dt} &= 2\left(u_i^{(s+1)} - v_i^{(s+1)}\right)^T R_i (u_i^{(s)} - u_i) - 2(d_i + g_i)^{-1} \\
&\quad \times \sum_{j \in N_i} e_{ij} \left(u_i^{(s+1)} - v_i^{(s+1)}\right)^T R_i (u_j^{(s)} - u_j) \qquad (7.33)
\end{aligned}
$$

which means that the above equation holds for any u_i and u_{-i}. Letting $u_i = u_i^{(s)}$ and $u_j = u_j^{(s)}$ $(j \in N_i)$ yields

$$
\frac{d(V_i^{(s+1)}(\delta_i) - W_i(\delta_i))}{dt} = 0. \qquad (7.34)
$$

Thus, we have $V_i^{(s+1)}(\delta_i) = W_i(\delta_i) + c$, where c is a real constant. Due to $V_i^{(s+1)}(0) = 0$ and $W_i(0) = 0$, then $c = 0$. Thus, we have $V_i^{(s+1)}(\delta_i) = W_i(\delta_i)$ for $\forall \delta_i$. From (7.34), the following form holds for $\forall u_i$ and u_j, $j \in N_i$:

$$
\left(u_i^{(s+1)} - v_i^{(s+1)}\right)^T R_i \left((u_i^{(s)} - u_i) - 2(d_i + g_i)^{-1} \sum_{j \in N_i} e_{ij} (u_j^{(s)} - u_j) \right) = 0. \qquad (7.35)
$$

For $\forall u_i \in \vartheta_i(\delta_i)$ and $u_j \in \vartheta_j(\delta_j)$, we have $u_i^{(s+1)} = v_i^{(s+1)}$. Therefore, we have a conclusion that contradicts to the assumption. The proof is completed. □

Based on Theorem 7.2, the following model-free off-policy RL Algorithm 7.2 is presented.

Algorithm 7.2 Off-policy RL for multi-agent games

1: Begin with admissible initial control policies $u_i^{(0)}$ for $\forall i$, and set $s = 0$;
2: Solve $V_i^{(s+1)}$ and $u_i^{(s+1)}$ from off-policy Bellman equation (7.29), where s denotes the iteration index;
3: Stop when $\| V_i^{(s)} - V_i^{(s+1)} \| \leq \varepsilon$.

Remark 7.5 How to find the admissible initial control policies for systems with completely unknown system dynamics has been clarified in Modares and Lewis (2014). If systems (7.5) can be a priori known to be stable, then the initial policies can be chosen as $u_i^{(0)} = 0$, which can guarantee the admissibility of the initial policy without requiring any knowledge of the dynamics of systems (7.5). Otherwise, suppose that systems (7.5) have nominal models A_N and B_N satisfying $A = A_N + \Delta A$ and $B = B_N + \Delta B$, where ΔA and ΔB are unknown part of systems (7.5). In this case, robust control methods such as H_∞ control with the nominal models A_N and B_N can be used to yield an admissible initial policy.

Remark 7.6 Since Theorem 7.2 shows that the solution of Algorithm 7.1 with (7.24) and (7.25) is equivalent to the solution of off-policy Bellman equations (7.29), the convergence of Algorithm 7.2 can be guaranteed since (Vamvoudakis et al. 2012) and (Saridis and Lee 1979) have proven that the RL Algorithm 7.1 converges. Compared with (7.24) and (7.25), (7.29) does not need knowledge of agent dynamics (A, B).

Remark 7.7 Off-policy Algorithm 7.2 can learn the solution to the HJB equations (7.13) without requiring any knowledge of the agent dynamics because the Bellman equations (7.29) do not contain the system matrices A and B of the agents. This is because the information about the agent dynamics is embedded in the local neighborhood error δ_i and control input u_i, as well as u_{-i}. The most prominent advantage of learning the optimal solution by using (7.29) is that the resulting optimal control protocol does not suffer from model inaccuracy or simplifications made in identifying system models.

7.1.3.2 Using Actor–Critic Structure for Off-Policy Learning in Graphical Games

To solve $V_i^{(s+1)}$ and $u_i^{(s+1)}$ in (7.29), multiple actor–critic NNs based approach is developed. According to Weierstrass high-order approximation theorem (Abu-Khalaf and Lewis 2005), $V_i^{(s+1)}$ and $u_i^{(s+1)}$ can be approximately expressed as linear combination of independent basis functions. Thus, the following N critic NN and actor NN are constructed to approximate the optimal cost function V_i^* and the optimal control law u_i^*:

$$\hat{V}_i^{(s)}(\delta_i) = \phi_i(\delta_i^T) W_{vi}^{(s)} \tag{7.36}$$

$$\hat{u}_{il}^{(s)}(\delta_i) = \psi_{il}(\delta_i^T) W_{u_{il}}^{(s)}, \tag{7.37}$$

where $\phi_i(\delta_i) \in R^{h_v}$ is the activation function vector with h_v neurons in the critic NN hidden layer of agent i. $\psi_{il}(\delta_i) \in R^{h_{lu_i}}$ $(l = 1, 2, \ldots, p)$ is the activation function vectors with h_{lu_i} neurons in the lth sub-actor NN hidden layer for agent i. For $\forall s = 0, 1, \ldots$, $W_{vi}^{(s)}$ and $W_{u_{il}}^{(s)}$ are weight vectors of critic and actor NNs, respectively. Expression (7.37) can be rewritten as the compact form

$$\begin{aligned}\hat{u}_i^{(s)}(\delta_i) &= \left[\hat{u}_{i1}^{(s)}(\delta_i)\ \hat{u}_{i2}^{(s)}(\delta_i)\ \ldots\ \hat{u}_{ip}^{(s)}(\delta_i)\right]^T \\ &= \left[\psi_{i1}(\delta_i^T)W_{u_{i1}}^{(s)}\ \psi_{i2}(\delta_i^T)W_{u_{i2}}^{(s)}\ \ldots\ \psi_{ip}(\delta_i^T)W_{u_{ip}}^{(s)}\right]^T. \end{aligned} \tag{7.38}$$

If we make Assumption 7.3 that the value function is quadratic in the form of $V_i = \delta_i^T P_i \delta_i$ for linear systems (7.5) with multiple control inputs, then the activation function vectors of critic NNs and actor NNs are in fact $\phi_i(\delta_i)^T = \delta_i^T \otimes \delta_i^T$ and $\psi_{il}(\delta_i)^T = \delta_i^T$, respectively.

To estimate the solution $\left(V_i^{(s+1)}(\delta_i), u_i^{(s+1)}(\delta_i)\right)$, the weighted residual method is used here Luo et al. (2014); Vrabie and Lewis (2009). To this end, $V_i^{(s+1)}(\delta_i)$ and $u_i^{(s+1)}(\delta_i)$ are respectively substituted by $\hat{V}_i^{(s+1)}(\delta_i)$ and $\widehat{u}_i^{(s+1)}(\delta_i)$ to yield the following residual error:

$$\begin{aligned}\sigma_i^{(s)}(\delta_i(t), u(t), \delta_i^{'}(t)) &= \left(\phi_i(\delta_i(t)) - \phi_i(\delta_i^{'}(t))\right)^T W_{vi}^{s+1} \\ &+ 2\sum_{l_1=1}^{p}\sum_{l_2=1}^{p} r_{l_1,l_2}\int_t^{t'}\left(\psi_{il_1}\delta_i^T(\tau)W_{u_{il_1}}^{(s)} - u_{il_1}\delta_i(\tau)\right)(\psi_{il_2}(\delta_i(\tau)))^T W_{u_{il_2}}^{(s+1)}d\tau \\ &- \int_t^{t'}\delta_i^T(\tau)Q_i\delta_i(\tau)d\tau - \sum_{l_1=1}^{p}\sum_{l_2=1}^{p} r_{l_1,l_2}\int_t^{t'}\left(W_{u_{il_1}}^{(s)}\right)^T\psi_{il_1}(\delta_i(\tau)) \\ &\times (\psi_{il_2}(\delta_i(\tau)))^T W_{u_{il_2}}^{(s)}d\tau - 2(d_i+g_i)^{-1}\sum_{l_1=1}^{p}\sum_{l_2=1}^{p} r_{l_1,l_2}\int_t^{t'}\sum_{i\in N_i} e_{ij} \\ &\times \psi_{il_1}(\delta_i(\tau))^T W_{u_{il_1}}^{(s)}\left(\psi_{jl_2}(\delta_i(\tau))^T W_{u_{jl_2}}^{(s)} - u_{jl_2}(\delta_i(\tau))\right)d\tau. \end{aligned} \tag{7.39}$$

The above expression can be rewritten as

$$\sigma_i^{(s)}(\delta_i(t), u(t), \delta_i^{'}(t)) = \bar{\rho}_i^{(s)}(\delta_i(t), u(t), \delta_i^{'}(t))W_i^{(s+1)} - \pi_i^{(s)}(\delta_i(t)), \tag{7.40}$$

where

$$\begin{aligned} W_i^{(s+1)} &\triangleq \left[\left(W_{vi}^{(s+1)}\right)^T\ \left(W_{u_{i1}}^{(s+1)}\right)^T\ \ldots\ \left(W_{u_{ip}}^{(s+1)}\right)^T\right]^T \\ \pi_i^{(s)}(\delta_i) &\triangleq \rho_Q(\delta_i) + \sum_{l_1=1}^{p}\sum_{l_2=1}^{p} r_{l_1,l_2}\left(W_{u_{il_1}}^{(s)}\right)^T \rho_\psi^{l_1,l_2}(\delta_i)W_{u_{il_2}}^{(s)} \end{aligned}$$

$$\bar{\rho}_i^{(s)}(\delta_i(t), u(t), \delta_i^{'}(t)) \triangleq \Big[\rho_{\Delta\varphi}^T(\delta_i(t), \delta_i^{'}(t))\ 2\theta^{(s)1}(\delta_i(t), \delta_j(t), u(t)) \ldots$$
$$2\theta^{(s)p}(\delta_i(t), \delta_j(t), u(t))\Big]$$
$$\theta^{(s)l_2}(\delta_i(t), \delta_j(t), u(t)) \triangleq \sum_{l_1}^{p} r_{l_1,l_2}\left(\left(W_{u_{il_1}}^{(s)}\right)^T \rho_{\psi}^{l_1,l_2}(\delta_i(t)) - \rho_{u\psi}^{l_1,l_2}(\delta_i(t), u_i(t))\right)$$
$$- 2(d_i + g_i)^{-1}\sum_{l_2}^{p}\sum_{j\in N_i} e_{ij} r_{l,l_2}\Big(W_{u_{jl_2}}^{(s)}\rho_{\psi}^{l_1,l_2}(\delta_i(t), \bar{\delta}_j(t))$$
$$-\rho_{u,\psi}^{l,l_2}(\delta_i(t), u_j(t))\Big)\ (l = 1, 2, \ldots, p)$$

with

$$\rho_{\Delta\phi}(\delta_i(t), \delta_i^{'}(t)) \triangleq \left[\phi(\delta_i(t)) - \phi(\delta_i^{'}(t))\right]$$
$$\rho_Q(\delta_i) \triangleq \int_t^{t'} \delta_i^T(\tau) Q_i \delta_i(\tau) d\tau$$
$$\rho_{\psi}^{l_1,l_2}(\delta_i(t)) \triangleq \int_t^{t'} \psi_{i,l_1}(\delta_i(\tau))\big(\psi_{i,l_2}(\delta_i(\tau))\big)^T d\tau$$
$$\rho_{u,\psi}^{l_1 l_2}(\delta_i(t), u_i(t)) \triangleq \int_t^{t'} u_{i,l_1}(\tau)\left(\psi_{i,l_2}(\delta_i(\tau))\right)^T d\tau$$
$$\rho_{\psi}^{l_1 l_2}(\delta_i(t), \delta_j(t)) \triangleq \int_t^{t'} \psi_{j,l_2}(\delta_j(\tau))(\psi_{i,l_1}(\delta_i(\tau)))^T d\tau$$
$$\rho_{u,\psi}^{l_1 l_2}(\delta_i(t), u_j(t)) \triangleq \int_t^{t'} u_{j,l_2}(\tau)(\psi_{i,l_1}(\delta_i(\tau)))^T d\tau,$$

where $l_1, l_2 = 1, 2, \ldots, p$, $m = h_v + \sum_{i=1}^{p} h_{lu_i}$ is the length of the vector $\bar{\rho}_i^{(s)}$.

To compute the weight vector $W_i^{(s+1)}$, the residual error $\sigma_i^{(s)}(\delta_i, u, \delta_i^{'})$ is projected onto $d\sigma_i^{(s)}(\delta_i, u, \delta_i^{'})/dW_i^{(s+1)}$ and the projective value is set to be zero on domain $\mathbb{D}_i$, i.e.,

$$\langle d\sigma_i^{(s)}(\delta_i, u, \delta_i^{'})/dW_i^{(s+1)}, \sigma_i^{(s)}(\delta_i, u, \delta_i^{'})\rangle_{\mathbb{D}_i} = 0. \tag{7.41}$$

Using (7.40) and the definition of derivative, (7.41) becomes

$$W_i^{(s+1)} = \langle\bar{\rho}_i^{(k)}(\delta_i(t), u(t), \delta_i^{'}(t)), \bar{\rho}_i^{(k)}(\delta_i(t), u(t), \delta_i^{'}(t))\rangle_{\mathbb{D}_i}^{-1}$$
$$\times \langle\bar{\rho}_i^{(k)}(\delta_i(t), u(t), \delta_i^{'}(t)), \pi_i^{(k)}(\delta_i(t))\rangle_{\mathbb{D}_i}. \tag{7.42}$$

Since

$$\langle\bar{\rho}_i^{(k)}(\delta_i(t), u(t), \delta_i^{'}(t)), \bar{\rho}_i^{(k)}(\delta_i(t), u(t), \delta_i^{'}(t))\rangle_{\mathbb{D}_i} = \frac{I_{\mathbb{D}_i}}{M_i}\left(Z_i^{(s)}\right)^T Z_i^{(s)} \tag{7.43}$$

and

$$\langle\bar{\rho}_i^{(k)}(\delta_i(t), u(t), \delta_i^{'}), \pi_i^{(k)}(\delta_i(t))\rangle_{\mathbb{D}_i} = \frac{I_{\mathbb{D}_i}}{M_i}\left(Z_i^{(s)}\right)^T \eta_i^{(s)}. \tag{7.44}$$

Thus, we have

$$W_i^{(s+1)} = \left[(Z_i^{(s)})^T Z_i^{(s)}\right]^{-1} \left(Z_i^{(s)}\right)^T \eta_i^{(s)}, \tag{7.45}$$

where

$$\begin{aligned}
Z_i^{(s)} &= \left[\left(\bar{\rho}_i^{(s)}(\delta_{i,1}, u_1, \delta_{i,1}^{'})\right)^T \dots \left(\bar{\rho}_i^{(s)}(\delta_{i,M}, u_{M_i}, \delta_{i,M}^{'})\right)^T\right]^T \\
\eta_i^{(s)} &= \left[\pi^{(s)}(\delta_{i,1}) \dots \pi^{(s)}(\delta_{i,M})\right]^T \\
I_{\mathbb{D}_i} &\triangleq \int_{\mathbb{D}_i} d(\delta_i(t), u(t), \delta_i^{'}(t)) \\
\delta_{M_i} &= \left\{(\delta_{i,k}, u_k, \delta_{i,k}^{'}) | (\delta_{i,k}, u_k, \delta_{i,k}^{'}) \in \mathbb{D}_i, k = 1, 2, \dots, M_i\right\}
\end{aligned}$$

and M_i is the size of sample set δ_{M_i}.

The data in sample set δ_{M_i} are collected from the real application system on domain $\mathbb{D}_i$. Exploratory prescribed behavior policies can be used to assure that the data set is rich. To compute weight vector $W_i^{(s+1)}$, the trapezoidal rule is used to approximate the definite integrals that appeared in (7.45). Thus, we have

$$\begin{aligned}
\rho_{\Delta\varphi}(\delta_{i,k}, \delta_{i,k}^{'}) &\triangleq \varphi_i(\delta_{i,k}) - \varphi_i(\delta_{i,k}^{'}) \\
\rho_Q(\delta_{i,k}) &\triangleq \tfrac{\Delta t}{2}\left(\delta_{i,k}^T Q_i + (Q_{i,k}^{'})^T Q_i \delta_{i,k}^{i}\right) \\
\rho_{\psi}^{l_1,l_2}(\delta_i(t)) &\triangleq \left(\psi_{i,l_1}(\delta_{i,k})\left(\psi_{i,l_2}(\delta_{i,k})\right)^T + \psi_{i,l_1}(\delta_{i,k}^{'})\left(\psi_{i,l_2}(\delta_{i,k}^{'})\right)^T\right) \\
\rho_{u,\psi}^{l_1,l_2}(\delta_{ik}, u_{ik}) &\triangleq \tfrac{\Delta t}{2}\left(u_{i,l_1,k}(\psi_{i,l_2}(\delta_{ik}))^T + u_{i,l_1,k}(\psi_{i,l_2}(\delta_{i,k}^{'}))^T\right) \\
\rho_{\psi}^{l_1,l_2}(\delta_{i,k}, \delta_{j,k}) &\triangleq \tfrac{\Delta t}{2}\left(\psi_{j,l_2}(\delta_{j,k})(\psi_{i,l_1}(\delta_{i,k}))^T + \psi_{j,l_2}(\delta_{j,k}^{'})(\psi_{i,l_1}(\delta_{i,k}^{'}))^T\right) \\
\rho_{u_k,\psi}^{l_1,l_2}(\delta_{ik}, u_{jk}) &\triangleq \tfrac{\Delta t}{2}\left(u_{j,l_2,k}(\psi_{i,l_1}(\delta_{i,k}))^T + u_{j,l_2,k}(\psi_{i,l_1}(\delta_{i,k}^{'}))^T\right).
\end{aligned}$$

Remark 7.8 Solving (7.45) for NN weights $W_i^{(s+1)}$ requires $Z_i^{(s)} \in R^{M_i \times m_i}$ to be full column rank. Therefore, the size M_i of the sample set δ_{M_i} should be larger than $m_i = hv + \sum_{l=1}^{p} h_{lu_i}$, i.e., $M_i \geq m_i$. This requires satisfaction of a proper persistence of excitation condition to assure that $\mathrm{Rank}(Z_i^{(s)}) = m_i$ is satisfied (Luo et al. 2014; Vamvoudakis et al. 2012).

Using the approximate control policy (7.37), closed-loop system (7.5) is presented by

$$\dot{\delta}_i = A\delta_i + (d_i + g_i)B_i\hat{u}_i^{(s)} - \sum_{j\in N_i} e_{ij}B_j\hat{u}_j^{(s)}. \tag{7.46}$$

Selecting a Lyapunov function candidate as the approximate value function $\hat{V}_i^{(s)}$, the next sufficient condition is presented to guarantee the stability of (7.46).

Theorem 7.3 *The closed-loop system (7.46) is asymptotically stable with approximate control policies $\hat{u}_i^{(s)}$ and $\hat{u}_{-i}^{(s)}$, if the following condition holds:*

$$\delta_i^T Q_i \delta_i + (u_i^{(s)})^T R_i u_i^{(s)} > \left[\nabla V_i^{(s+1)}\right]^T \left((d_i + g_i) B \epsilon_{u_i}^{(s)} - \sum_{j \in N_i} e_{ij} B \epsilon_{u_j}^{(s)}\right), \quad (7.47)$$

where $\epsilon_{u_i}^{(s)} = \hat{u}_i^{(s)}(\delta_i) - u_i^{(s)}(\delta_i)$ $(i = 1, 2, \ldots, N)$ *denotes the error between the approximate value of* $u_i^{(s)}(\delta_i)$ *and its real value.*

Proof Differentiating $V_i^{(s+1)}$ along with the dynamics of agent i (7.46) yields

$$\begin{aligned} \frac{dV_i^{(s+1)}(\delta_i)}{dt} &= \left[\nabla V_i^{(s+1)}\right]^T \left(A\delta_i + (d_i + g_i) B \hat{u}_i^{(s)} - \sum_{j \in N_i} e_{ij} B \hat{u}_j^{(s)}\right) \\ &= \left[\nabla V_i^{(s+1)}\right]^T \left(A\delta_i + (d_i + g_i)(u_i^{(s)} + \epsilon_{u_i}^{(s)}) - \sum_{j \in N_i} e_{ij} B(u_j^{(s)} + \epsilon_{u_j}^{(s)})\right). \end{aligned} \quad (7.48)$$

Using (7.24), this becomes

$$\begin{aligned} \frac{dV_i^{(s+1)}}{dt} &= \left[\nabla V_i^{(s+1)}\right]^T \left((d_i + g_i) B \epsilon_{u_i}^{(s)} - \sum_{j \in N_i} e_{ij} B \epsilon_{u_j}^{(s)}\right) \\ &\quad - \delta_i^T Q_i \delta_i - (u_i^{(s)})^T R_i u_i^{(s)}. \end{aligned} \quad (7.49)$$

If (7.47) holds, then $\frac{dV_i^{(s+1)}}{dt} < 0$, which indicates asymptotic stability of closed-loop system (7.46). The proof is completed. □

Remark 7.9 Referring to the definition of the performance index and the derived optimal control policy, one can appropriately choose activation function sets and their size h_{lu_i} $(l = 1, 2, \ldots, p)$ for actor NNs, such that $\epsilon_{u_i}^{(s)}$ can be made arbitrarily small (Luo et al. 2014, 2015; Vrabie and Lewis 2009) based on the uniform approximation property of NNs. Under Assumption 7.3, one clearly knows that the activation function vector of actor NNs is $\psi_{il}(\delta_i)^T = \delta_i^T$ $(i = 1, 2, \ldots, N;\ l = 1, 2, \ldots, p)$ due to the linear optimal control policy u_i dependent on δ_i. Thus, $\epsilon_{u_i}^{(s)}$ can be made arbitrarily small based on the uniform approximation property of NNs. In this sense, the condition in Theorem 7.3 can be satisfied, so the asymptotic stability of (7.46) is guaranteed.

7.1.3.3 Reinforcement Learning Algorithm Using Off-Policy Approach

Summarizing the results given in Sects. 7.1.3.1 and 7.1.3.2 yields the following off-policy RL algorithm for obtaining the optimal control policy.

Algorithm 7.3 Data-based off-policy RL for multi-agent game

1: Collect real system data $(\delta_{i,k}, u_{ik}, u_{jk}, \delta'_{i,k})$ from agent i for the sample set $\mathbb{D}$ ($\mathbb{D} = \cup_{i=1}^{n} \mathbb{D}_i$) using different control input u, where $i = 1, 2, \ldots, N$ and $j \in N_i$;
2: Set the initial critic NN weight vector $W_{iv}^{(0)}$, and choose the initial actor NN weight vector $W_{uil}^{(0)}$ $(l = 1, 2, \ldots, p)$, such that $\hat{u}_i^{(0)}$ is admissible. Let $s = 0$;
3: Compute $Z^{(i)}$ and $\eta^{(i)}$ to update $W_i^{(s+1)}$ in terms of (7.45);
4: Let $s = s + 1$. If $\left| W_i^{(s)} - W_i^{(s-1)} \right| \leq \ell$ (ℓ is a small positive number), then stop the iteration and employ $W_i^{(s)}$ to (7.37) in order to derive the final control policy. Otherwise, go back to Step 3 and continue.

Remark 7.10 Algorithm 7.3 consists of two stages. In the first stage, the algorithm collects data obtained by applying prescribed behavior control policies to the agents. This data includes the local neighborhood tracking error and the control inputs of the agent and its neighbors. In the second stage, the weights of critic and actor NNs are updated in real time to learn the value corresponding to the target policy under evaluation and find an improved target policy. The behavior policy for each agent can be an exploratory policy to assure that the collected data are rich enough and the target policy for each agent is learned using these collected data without actually being applied to the agents.

Remark 7.11 Note that the final control policy learned by Algorithm 7.3 is static, while the optimal control policy is dynamical in terms of (7.12) for the case of dynamical neighbor graphs. That is why the results of this paper cannot be extended to the case of dynamical neighbor graphs.

Theorem 7.4 *The approximate control policies $\hat{u}_i^{(s)}$ derived by Algorithm 7.3 converge to the global Nash equilibrium solution u_i^* as $s \to \infty$, i.e., $\lim_{s\to\infty} \hat{u}_i^{(s)} = u_i^*$ $(i = 1, 2, \ldots, N)$.*

Proof It is clear that $\hat{u}_i^{(s)}$ is used to approximate the target policy $u_i^{(s)}$ in off-policy Bellman equations (7.29). As stated in Remark 7.9, if the activation function sets and their size h_{lu_i} $(l = 1, 2, \ldots, p)$ for actor NNs are chosen appropriately, $\epsilon_{u_i}^{(s)} = \hat{u}_i^{(s)} - u_i^{(s)}$ can be made arbitrarily small (Luo et al. 2014, 2015; Vrabie and Lewis 2009). This means that for any $\varepsilon_1 > 0$, it follows

$$\left| \hat{u}_i^{(s)} - u_i^{(s)} \right| \leq \varepsilon_1. \tag{7.50}$$

Theorem 7.1 has shown that the equivalence between the solutions $(V_i^{(s)}, u_i^{(s)})$ derived by off-policy Bellman equations (7.29) and $(V_i^{(s)}, u_i^{(s)})$ learned by Algorithm 7.1 given by (7.24) and (7.25). Actually, (Vamvoudakis et al. 2012; Saridis and Lee 1979) showed that $V_i^{(s)}$ in (7.24) and $u_i^{(s)}$ with the form of (7.25) can respectively converge to the solutions of coupled cooperative game HJB equations (7.13) and global Nash equilibrium solution u_i^*, which means that for any $\varepsilon_2 > 0$

$$\left|u_i^{(s)} - u_i^*\right| \leq \varepsilon_2 \text{ (if } s \to \infty). \tag{7.51}$$

Combining (7.50) with (7.51), therefore, yields for any $\varepsilon > 0$

$$\left|\hat{u}_i^{(s)} - u_i^*\right| \leq \left|\hat{u}_i^{(s)} - u_i^{(s)}\right| + \left|u_i^{(s)} - u_i^*\right| \leq \varepsilon_1 + \varepsilon_2 = \varepsilon \tag{7.52}$$

if $s \to \infty$. Therefore, $\lim_{s\to\infty} \hat{u}_i^{(s)} = u_i^*$. The proof is completed. □

Remark 7.12 The most significant advantage of Algorithm 7.3 is that the knowledge of the agent dynamics is not required for learning approximate optimal control protocols. No systems identification is needed, and therefore, the negative impacts brought by identifying inaccurate models are eliminated without compromising the accuracy of optimal control protocols. This is in contrast to the existing model-based optimal control for MAS (Movric and Lewis 2014; Dunbar and Murray 2006; Vamvoudakis et al. 2012; Dong 2010; Wang et al. 2014; Liu and Jia 2012).

Remark 7.13 For the usage of model-free off-policy RL presented in Algorithm 7.3, it may be challenging for higher dimensional systems because of the requirement of heavy computational resources and slow convergence. However, Recursive Least Squares (RLS) or Least Mean Squares (LMS) can be used to replace the Batch Least Squares (7.45) to save data storage space for speeding up the convergence.

7.1.4 Simulation Examples

This subsection verifies the effectiveness of the proposed off-policy RL algorithm for optimal control of MASs.

Example 7.1 Consider the five-node communication graph shown in Fig. 7.1. The leader is pinned to node 3, thus $g_3 = 1$ and $g_i = 0\,(i \neq 3)$. The weight of each edge is set to be 1. Therefore, the graph Laplacian matrix corresponding to Fig. 7.1 becomes

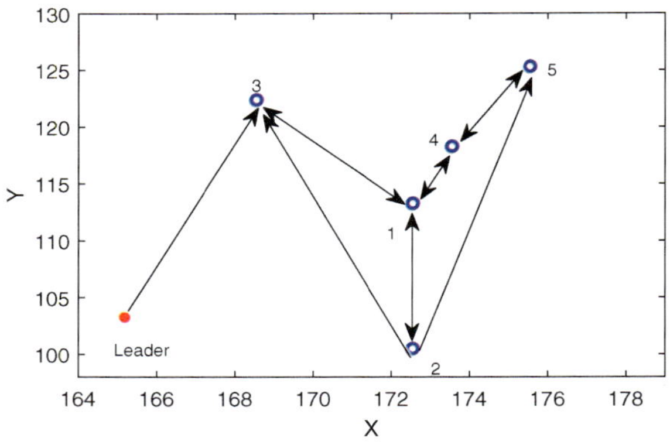

Fig. 7.1 Communication graph

$$L = \begin{bmatrix} 3 & -1 & -1 & -1 & 0 \\ -1 & 1 & 0 & 0 & 0 \\ -1 & -1 & 2 & 0 & 0 \\ -1 & 0 & 0 & 2 & -1 \\ 0 & -1 & 0 & -1 & 2 \end{bmatrix}. \tag{7.53}$$

The weighting matrices in the performance function (7.7) are specified as

$$Q_1 = Q_2 = Q_3 = Q_4 = Q_5 = \begin{bmatrix} 1\,0\,0\,0 \\ 0\,1\,0\,0 \\ 0\,0\,1\,0 \\ 0\,0\,0\,1 \end{bmatrix}$$
$$R_1 = 1,\ R_2 = 0.1,\ R_3 = 0.2,\ R_4 = 0.01,\ R_5 = 0.02.$$

Each agent is assumed a fourth-order system, and therefore, the local neighborhood error of each agent i can be expressed as $\delta_i = [\delta_{i1}\ \delta_{i2}\ \delta_{i3}\ \delta_{i4}]^T$. Under Assumption 7.3, the activation function vectors for the critic NNs and the actor NNs are chosen as

$$\phi_i(\delta_i) = [\delta_{i1}^2\ \delta_{i1}\delta_{i2}\ \delta_{i1}\delta_{i3}\ \delta_{i1}\delta_{i4}\ \delta_{i2}^2\ \delta_{i2}\delta_{i3}\ \delta_{i2}\delta_{i4}\ \delta_{i3}^2\ \delta_{i3}\delta_{i4}\ \delta_{i4}^2]^{\mathrm{T}}$$
$$\psi_{i1}(\delta_i) = [\delta_{i1}\ \delta_{i2}\ \delta_{i3}\ \delta_{i4}]^T\ (i = 1, 2, 3, 4, 5).$$

Consider the dynamics of agent i $(i = 1, 2, 3, 4, 5)$ as

$$\dot{x}_i = \begin{bmatrix} -1 & 1 & 0 & 0 \\ -0.5 & 0.5 & 0 & 0 \\ 0 & 0 & -2 & 0 \\ 0 & 0 & 0 & -3 \end{bmatrix} x_i + \begin{bmatrix} 0 \\ 1 \\ 1 \\ 1 \end{bmatrix} u_i \tag{7.54}$$

and the leader

$$\dot{x}_0 = \begin{bmatrix} -1 & 1 & 0 & 0 \\ -0.5 & 0.5 & 0 & 0 \\ 0 & 0 & -2 & 0 \\ 0 & 0 & 0 & -3 \end{bmatrix} x_0. \tag{7.55}$$

Note that the leader is marginally stable with poles at $s = -0.5$, $s = 0$, $s = -2$ and $s = -3$. The initial critic NN weights are set to be zero, and the initial admissible controllers are respectively chosen as

Fig. 7.2 Three representative critic NN weights

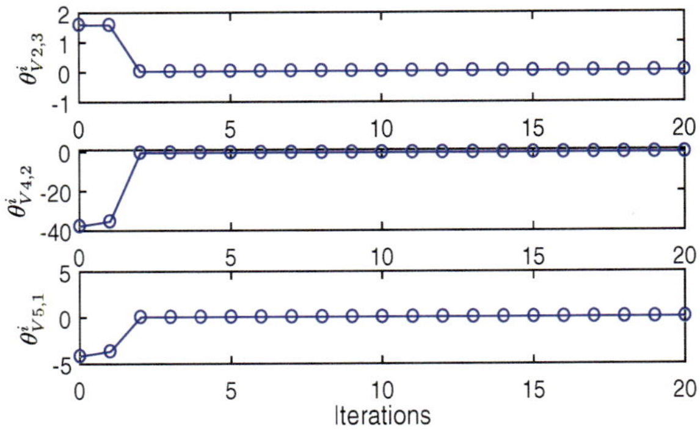

Fig. 7.3 Three representative actor NN weights

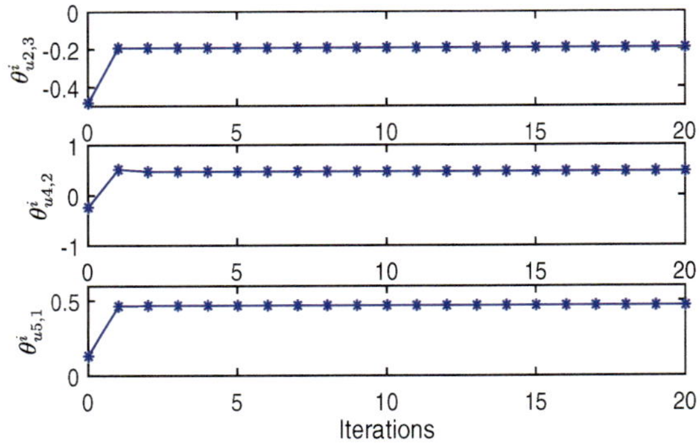

$$u_1^0 = \left[-0.1203\ -0.4374\ 0.3639\ 0.1591\right] \delta_1$$

$$u_2^0 = \left[-0.1512\ 0.0918\ 0.1976\ 0.2376\right] \delta_2$$

$$u_3^0 = \left[-0.3068\ -0.3585\ 0.3603\ 0.5889\right] \delta_3$$

$$u_4^0 = \left[-0.2141\ -0.2674\ 0.2739\ 0.1529\right] \delta_4$$

$$u_5^0 = \left[-0.2802\ 0.0577\ -0.7025\ 0.1208\right] \delta_5.$$

Setting the value of the convergence criterion $\ell = 10^{-4}$, Algorithm 7.3 is implemented. Figures 7.2 and 7.3 respectively show the convergence of the critic NN weights and the actor NN weights. Applying the optimal control policy (7.37) to agents, the first two state trajectories and the last two state trajectories of each agent are respectively given in Figs. 7.4 and 7.5, which show that all three agents are synchronized with the leader.

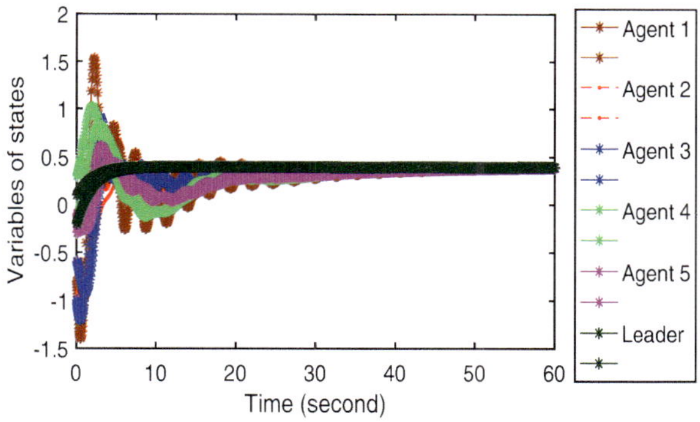

Fig. 7.4 Synchronization of the first two states of all agents with the leader

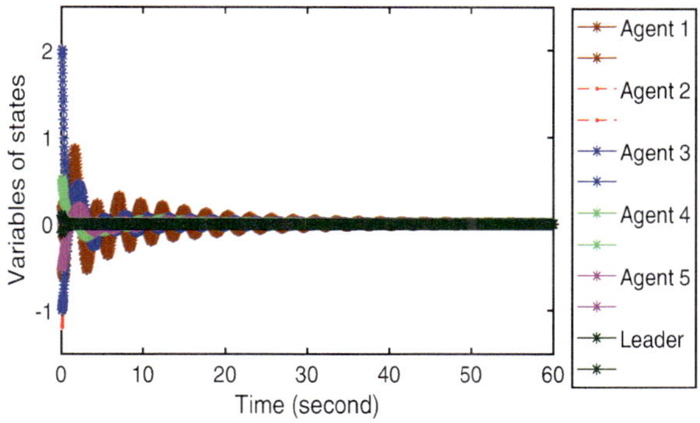

Fig. 7.5 Synchronization of the last two states of all agents with the leader

7.2 Nonzero-Sum Game for Cooperative Optimization

With the increase of system size, system complexity and the application of multiple subsystems, the use of multi-player systems or multi-controller systems is increasing, which brings control problems that should also draw our attention, such as the cooperative optimal control problem based on nonzero-sum games. In nonzero-sum or zero-sum multi-player games, each player makes efforts to optimize its own performance or reward by learning feedback from the environment and improving its behavior. In Al-Tamimi et al. (2007b), the application of model-free Q-learning zero-sum game to H_∞ problem has been studied. Vamvoudakis et al. (2017) systematically summarized game theory-based RL methods to solve two-player games including DT and CT systems. Notice that the off-policy RL algorithm has been proposed in Vamvoudakis et al. (2017) for linear DT multi-player systems, then whether the off-policy Q-learning method can be used to study the optimal control problem of the completely unknown linear DT multi-player games or not? And if it can work, then how to design the off-policy Q-learning algorithm for achieving the Nash equilibrium of linear DT multi-player games using only measured data is the key point in this section.

Therefore, considering the advantages of off-policy RL methods over on-policy RL methods, we present an off-policy game Q-learning algorithm linear DT multi-player systems by integrating game theory, Q-learning techniques and off-policy RL methods. This algorithm can achieve the Nash equilibrium of multi-player systems based on nonzero-sum games without requiring the system dynamic model parameters using measurable data only. Besides, no bias and bias of Nash equilibrium solution when adding probing noises into multi-player systems are rigorously proved, which are the extension of where the systems with single player or agent are concerned.

This section is organized as follows. In Sect. 7.2.1, the nonzero-sum games problem of linear DT multi-player systems is formulated. Section 7.2.2 devotes to solving the nonzero-sum games. In Sect. 7.2.3, an on-policy game Q-learning algorithm is proposed and the bias of the solution is analyzed. In Sect. 7.2.4, an off-policy game Q-learning algorithm is proposed, and the rigorous proof of the unbiased solution is presented. The effectiveness of the proposed algorithm is verified by numerical simulations, and the comparisons between the off-policy game Q-learning algorithm and the on-policy game Q-learning algorithm are carried out in Sect. 7.2.5.

7.2.1 Problem Statement

In this subsection, the optimal control problem of linear DT multi-player systems is formulated. Moreover, the value function and the Q-function defined in terms of the cost function of each player are proved to be linear quadratic forms.

Consider the following linear DT multi-player system:

$$x_{k+1} = Ax_k + \sum_{i=1}^{n} B_{ik} u_{ik}, \tag{7.56}$$

where $x_k = x(k) \in \mathbb{R}^p$ is the system state, $u_{ik} = u_i(k) \in \mathbb{R}^{m_i}$ $(i = 1, 2, \ldots, n)$ are the control inputs. $A \in \mathbb{R}^{p\times p}$, $B_i \in \mathbb{R}^{p\times m_i}$ and k is the sampling time instant. The full state of the system (7.56) can be accessed by each of the agents or players i. The target of each player is to minimize its own performance index by its efforts, regardless of the performance of other players. The performance index J_i of each player i $(i = 1, 2, \ldots, n)$ is defined as the accumulative sum of utility functions from time instant 0 to infinity as given below (Li et al. 2017)

$$J_i(x_0, u_1, \ldots, u_n) = \sum_{k=0}^{\infty} \left(x_k^T Q_i x_k + \sum_{q=1}^{n} u_{qk}^T R_q u_{qk} \right), \tag{7.57}$$

where Q_i and R_q are respectively positive semi-definite matrices and positive definite matrices. x_0 represents the initial state of system (7.56) at time instant 0.

Minimizing (7.57) subject to (7.56) is indeed a standard multi-player nonzero-sum games problem, and all players will finally reach the Nash equilibrium. The objective of this article is to find the stabilizable control policies $u_{1k}, u_{2k}, \ldots, u_{nk}$ by using RL combined with game theory, such that the performance index of each player shown in (7.57) is minimized.

The definition of admissible control policies is given below, which is useful for Assumption 7.4 and Lemma 7.2.

Definition 7.3 (Al-Tamimi et al. 2007b; Vamvoudakis and Lewis 2011) Control policies $u_1(x_k), u_2(x_k), \ldots, u_n(x_k)$ are called admissible with respect to (7.57) on $\Omega \in \mathbb{R}^p$, if $u_1(x_k), u_2(x_k), \ldots, u_n(x_k)$ are continuous on Ω, $u_1(0) = 0$, $u_2(0) = 0$, $\ldots, u_n(0) = 0$, $u_1(x_k), u_2(x_k), \ldots, u_n(x_k)$ stabilize (7.56) on Ω and (7.57) is finite $\forall x_0 \in \Omega$.

Assumption 7.4 (Vamvoudakis and Lewis 2011) The n-player system (7.56) is controllable and there exists at least one set of admissible control policies.

According to performance indicator (7.57), suppose there exists a set of admissible control policies $u_1(x), u_2(x), \ldots, u_n(x)$, one can respectively define the following optimal value function and the optimal Q-function for each player i $(i = 1, 2, \ldots, n)$ as Li et al. (2017):

$$V_i^*(x_k) = \min_{u_i} \sum_{l=k}^{\infty} \left(x_l^T Q_i x_l + \sum_{q=1}^{n} u_{ql}^T R_q u_{ql} \right) \tag{7.58}$$

and

$$Q_i^*(x_k, u_{ik}, u_{-ik}) = x_k^T Q_i x_k + \sum_{q=1}^{n} u_{qk}^T R_q u_{qk} + V_i^*(x_{k+1}) \tag{7.59}$$

$u_{-ik} = \left[u_{1k}^T, \ldots, u_{i-1,k}^T, u_{i+1,k}^T, \ldots, u_{nk}^T\right]^T$. Thus, the following relation holds:

$$V_i^*(x_k) = Q_i^*(x_k, u_{ik}^*, u_{-ik}) \ (i = 1, 2, \ldots, n). \tag{7.60}$$

Lemma 7.2 *Suppose that the control policies $u_{ik} = -K_i x_k$ and they are admissible, the value function $V_i(x_k)$ and the Q-function $Q_i(x_k, u_{ik}, u_{-ik})$ of each player i can be respectively expressed as the following quadratic forms:*

$$V_i(x_k) = x_k^T P_i x_k \tag{7.61}$$

and

$$Q_i(x_k, u_{ik}, u_{-ik}) = z_k^T H_i z_k, \tag{7.62}$$

where P_i and H_i are positive definite matrices, and

$$z_k = \begin{bmatrix} x_k^T & u_1^T & u_2^T & \dots & u_n^T \end{bmatrix}^T. \tag{7.63}$$

Proof

$$\begin{aligned} V_i(x_k) &= \sum_{l=k}^{\infty}(x_l^T Q_i x_l + \sum_{q=1}^{n} u_{ql}^T R_q u_{ql}) \\ &= \sum_{l=k}^{\infty}\left[x_l^T Q_i x_l + \sum_{q=1}^{n}(-K_q x_l)^T R_q(-K_q x_l)\right] \\ &= \sum_{l=0}^{\infty} x_{l+k}^T \left[Q_i + \sum_{q=1}^{n}(K_q)^T R_q (K_q)\right] x_{l+k}, \end{aligned} \tag{7.64}$$

where $x_{l+k} = (A - \sum_{i=1}^{n} B_i K_i)^l x_k = G^l x_k$. Further, one has

$$V_i(x_k) = \sum_{l=0}^{\infty} x_k^T (G^l)^T \left[Q_i + \sum_{q=1}^{n}(K_q)^T R_q (K_q)\right](G^l)x_k \tag{7.65}$$

then, one has

$$V_i(x_k) = x_k^T P_i x_k, \tag{7.66}$$

where

$$P_i = \sum_{l=0}^{\infty}(G^l)^T \left[Q_i + \sum_{q=1}^{n}(K_q)^T R_q (K_q)\right](G^l).$$

Then, one has

$$\begin{aligned} Q_i(x_k, u_{ik}, u_{-ik}) &= x_k^T Q_i x_k + \sum_{q=1}^{n} u_{qk}^T R_q u_{qk} + V_i^*(x_{k+1}) \\ &= x_k^T Q_i x_k + \sum_{q=1}^{n} u_{qk}^T R_q u_{qk} + \left(Ax_k + \sum_{i=1}^{n} B_i u_i\right)^T P_i \left(Ax_k + \sum_{i=1}^{n} B_i u_i\right) \\ &= z_k^T H_i z_k, \end{aligned} \tag{7.67}$$

where

$$H_i = \begin{bmatrix} H_{i,xx} & H_{i,xu_1} & H_{i,xu_2} & \cdots & H_{i,xu_n} \\ H_{i,xu_1}^T & H_{i,u_1u_1} & H_{i,u_1u_2} & \cdots & H_{i,u_1u_n} \\ H_{i,xu_2}^T & H_{i,u_1u_2}^T & H_{i,u_2u_2} & \cdots & H_{i,u_2u_n} \\ \vdots & \vdots & \vdots & \ddots & \vdots \\ H_{i,xu_n}^T & H_{i,u_1u_n}^T & H_{i,u_2u_n}^T & \cdots & H_{i,u_nu_n} \end{bmatrix}$$
$$= \begin{bmatrix} A^T P_i A + Q_i & A^T P_i B_1 & \cdots & A^T P_i B_n \\ (A^T P_i B_1)^T & B_1^T P_i B_1 + R_1 & \cdots & B_1^T P_i B_n \\ (A^T P_i B_2)^T & (B_1^T P_i B_2)^T & \cdots & B_2^T P_i B_n \\ \vdots & \vdots & \ddots & \vdots \\ (A^T P_i B_n)^T & (B_1^T P_i B_n)^T & \cdots & B_n^T P_i B_n + R_n \end{bmatrix}. \tag{7.68}$$

By (7.66) and (7.67), one can get

$$P_i = M^T H_i M, \tag{7.69}$$

where

$$M = \begin{bmatrix} I & -K_1^T & -K_2^T & \cdots & -K_n^T \end{bmatrix}^T.$$

This completes the proof. □

7.2.2 *Solving the Nonzero-Sum Game Problems*

This subsection deals with solving the nonzero-sum games problem. In the nonzero-sum games, it is desired for all players to reach the Nash equilibrium by assuming that each player has the same hierarchical level as others. The definition of Nash equilibrium is given as follows.

Definition 7.4 (Vamvoudakis et al. 2017) If there exists an n-tuple of control strategies $(u_1^*, u_2^*, \ldots, u_n^*)$ satisfying the following n inequalities:

$$J_i^* \equiv J_i(u_1^*, u_2^*, \ldots, u_i^*, \ldots, u_n^*) \le J_i(u_1^*, u_2^*, \ldots, u_i, \ldots, u_n^*) \ (i = 1, 2, \ldots, n), \tag{7.70}$$

then this n-tuple of control strategies constitutes the Nash equilibrium solution of n-player finite game (7.56). And the n-tuple of quantities $(J_1^*, J_2^*, \ldots, J_n^*)$ is the Nash equilibrium outcome of n-player games (7.56) with respect to (7.57).

Now, we are in the position of solving the nonzero-sum game to find the Nash equilibrium solution. According to the dynamic programming, the following Q-function-based game Bellman equations can be derived based on Lemma 7.2.

$$z_k^T H_i z_k = x_k^T Q_i x_k + \sum_{q=1}^{n} u_{qk}^T R_q u_{qk} + z_{k+1}^T H_i z_{k+1}. \tag{7.71}$$

The optimal control policy u_{ik}^* of each player i should satisfy $\frac{\partial Q_i^*(x_k, u_{ik}, u_{-ik})}{\partial u_{ik}} = 0$ in terms of the necessary condition of optimality. Therefore, one has

$$u_i^*(k) = -K_i^* x_k, \tag{7.72}$$

where

$$K_i^* = H_{i,u_iu_i}^{-1} \left[H_{i,xu_i}^T - \sum_{r=1, r\neq i}^{n} H_{i,u_iu_r} K_r \right]. \tag{7.73}$$

Substituting K_i^* in (7.73) into (7.71) yields the optimal Q-function-based Riccati equations.

$$z_k^T H_i^* z_k = x_k^T Q_i x_k + \sum_{q=1}^{n} (u_{qk}^*)^T R_q u_{qk}^* + z_{k+1}^T H_i^* z_{k+1}, \tag{7.74}$$

where $u_{qk}^* = -K_q^* x_k$. Note that (Vamvoudakis et al. 2017) has proven that the following K_i^* $(i = 1, 2, ...n)$ guarantee system (7.56) to be stable and achieved Nash equilibrium of all players can be.

$$K_i^* = (H_{i,u_iu_i}^*)^{-1} \left[(H_{i,xu_i}^*)^T - \sum_{r=1, r\neq i}^{n} H_{i,u_iu_r}^* K_r^* \right]. \tag{7.75}$$

Remark 7.14 From (7.73) and (7.75), it can be seen that in Riccati equations (7.74) the matrices H_i^* are coupled with each other, and the values of matrices K_i^* are also coupled with each other, which make it difficult to solve Riccati equations (7.74). Therefore, the on-policy game Q-learning algorithm and off-policy game Q-learning algorithm are given in Sects. 7.2.3 and 7.2.4 to learn the optimal control laws $u_i^*(k) = -K_i^* x_k$.

7.2.3 Finding K_i^* $(i = 1, 2, \ldots, n)$ by the On-Policy Approach

In this subsection, the model-free on-policy game Q-learning algorithm is presented and the bias of solution to iterative Q-function-based Bellman equations is proved. By learning the Q-function matrices H_i^* in (7.74), the approximately optimal controller gains of multiple players can be obtained.

7.2.3.1 On-Policy Game Q-Learning Algorithm

The on-policy RL methods in Al-Tamimi et al. (2007b, a), Wei et al. (2017) and Lee et al. (2012) are extended to the case of multi-player systems, thus we present the on-policy game Q-learning algorithm.

Algorithm 7.4 On-policy game Q-learning

1: Initialization: Given the admissible controller gains for n players $K_1^0, K_2^0, \ldots, K_n^0$. Let $j = 0$ and $i = 1$, where j denotes the iteration index and i stands for player i ($i = 1, 2, \ldots, n$);

2: Policies evaluation: solve the Q-function matrices H_i^{j+1}:

$$z_k^T H_i^{j+1} z_k = x_k^T Q_i x_k + \sum_{q=1}^{n} (u_{qk}^j)^T R_q u_{qk}^j + z_{k+1}^T H_i^{j+1} z_{k+1}; \tag{7.76}$$

3: Policies update:

$$u_{ik}^{j+1} = -K_i^{j+1} x_k \tag{7.77}$$

where

$$K_i^{j+1} = (H_{i,u_iu_i}^{j+1})^{-1} \left[(H_{i,xu_i}^{j+1})^T - \sum_{r=1, r\neq i}^{n} H_{i,u_iu_r}^{j+1} K_r^{j+1} \right]; \tag{7.78}$$

4: If $i < n$, then $i = i + 1$ and go back to Step 2. Otherwise $j = j + 1$, $i = 1$, and go to Step 5;

5: Stop when $\left\| H_i^{j-1} - H_i^j \right\| \leq \varepsilon$ ($i = 1, 2, \ldots, n$) with a small constant ε ($\varepsilon > 0$). Otherwise go back to Step 2.

Remark 7.15 Algorithm 7.4 calculating Q-function matrices H_i^{j+1} yields the updated K_i^{j+1} ($H_i^{j+1} \rightarrow K_i^{j+1}$). As $j \rightarrow \infty$, H_i^{j+1} converge to H_i^* result in the convergence of K_i^{j+1} to K_i^* for all players, which can be proved in a similar way to Li et al. (2019); Vamvoudakis et al. (2017, 2012).

7.2.3.2 Bias Analysis of Solution for the On-Policy Game Q-Learning Algorithm

To satisfy the PE condition in Algorithm 7.4, probing noises are added to system (7.56). Thus, the actual control inputs applied to the system for collecting data are

$$\hat{u}_{ik}^j = u_{ik}^j + e_{ik} \tag{7.79}$$

with $e_{ik} = e_i(k)$ being probing noises and u_{ik}^j given by (7.77). Theorem 7.5 will prove the bias of solution to (7.76).

Theorem 7.5 *Rewrite Bellman equation (7.76) as*

$$x_k^T (M^j)^T H_i^{j+1} M^j x_k = x_k^T Q_i x_k + \sum_{q=1}^{n} (u_{qk}^j)^T R_q u_{qk}^j + x_{k+1}^T (M^j)^T H_i^{j+1} M^j x_{k+1}, \tag{7.80}$$

where

$$M^j = \left[I \ -(K_1^j)^T \ -(K_2^j)^T \ \ldots \ -(K_n^j)^T \right]^T.$$

Let H_i^{j+1} be the solution (7.80) with $e_{ik} = 0$ and $\hat{H}_i^{j+1}$ be the solution to (7.80) with $e_{ik} \neq 0$. Then, $H_i^{j+1} \neq \hat{H}_i^{j+1}$.

Proof Using (7.79) with $e_{ik} \neq 0$ in (7.80), Bellman equation (7.80) becomes the following:

$$x_k^T (M^j)^T \hat{H}_i^{j+1} M^j x_k = x_k^T Q_i x_k + \sum_{q=1}^{n} (u_{qk}^j + e_{qk})^T R_q \times (u_{qk}^j + e_{qk}) + x_{k+1}^T (M^j)^T \hat{H}_i^{j+1} M^j x_{k+1}, \tag{7.81}$$

where

$$x_{k+1} = A x_k + \sum_{i=1}^{n} B_i (u_{ik}^j + e_{ik}). \tag{7.82}$$

Further, (7.81) is rewritten as

$$\begin{aligned}
x_k^T (M^j)^T \hat{H}_i^{j+1} M^j x_k &= x_k^T Q_i x_k + \sum_{q=1}^{n} (u_{qk}^j + e_{qk})^T R_q (u_{qk}^j + e_{qk}) \\
&+ (A x_k + \sum_{i=1}^{n} B_i (u_{ik}^j + e_{ik}))^T (M^j)^T \hat{H}_i^{j+1} M^j (A x_k + \sum_{i=1}^{n} B_i (u_{ik}^j + e_{ik})) \\
&= x_k^T Q_i x_k + \sum_{q=1}^{n} (u_{qk}^j)^T R_q u_{qk}^j + x_{k+1}^T (M^j)^T \hat{H}_i^{j+1} M^j x_{k+1} \\
&+ 2 \sum_{q=1}^{n} e_{qk}^T R_q u_{qk}^j + \sum_{i=1}^{n} e_{ik}^T (B_i^T (M^j)^T \hat{H}_i^{j+1} M^j B_i + R_i) e_{ik} \\
&+ 2 \sum_{i=1}^{n} e_{ik}^T B_i^T (M^j)^T \hat{H}_i^{j+1} M^j x_{k+1}.
\end{aligned} \tag{7.83}$$

It can be concluded that the solution to (7.83) is not the solution to (7.80) if $e_{ik} \neq 0$. Since the solution to (7.80) is equivalent to the solution to (7.76), then one has

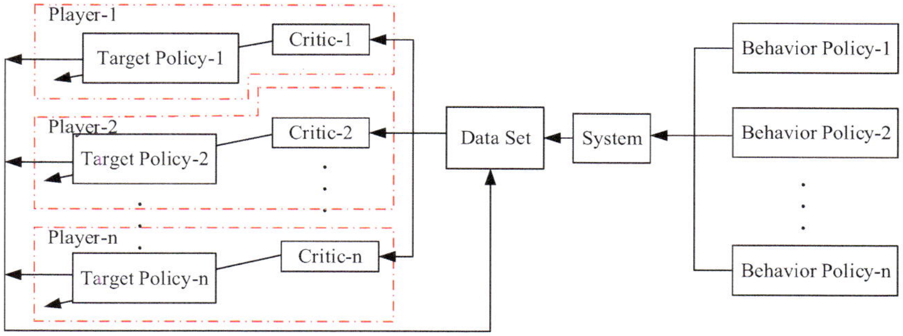

Fig. 7.6 Architecture of the off-policy game Q-learning

$H_i^{j+1} \neq \hat{H}_i^{j+1}$. Hence, adding probing noises during implementing the proposed on-policy game Q-learning Algorithm 7.4 can produce bias of solution. This completes the proof. □

Remark 7.16 It is worth noting that data are generated by $u_i^{j+1}(k) = -K_i^{j+1} x_k$ in Algorithm 7.4, which is the typical characteristic of the on-policy approach. Theorem 7.5 proves that the solution of the on-policy game Q-learning algorithm is biased. In contrast to Kiumarsi et al. (2017), we extend the proof of biased solution to iterative Bellman equations in Kiumarsi et al. (2017) to the case of multi-player nonzero-sum games.

7.2.4 *Finding K_i^* ($i = 1, 2, \ldots, n$) by the Off-Policy Approach*

In this subsection, an off-policy game Q-learning method is proposed to solve the multi-player nonzero-sum game problem. Moreover, the unbiasedness of solution to Q-learning based iterative game Bellman equation is rigorously proved even though adding probing noises ensures the PE condition. The architecture of the proposed algorithm is shown in Fig. 7.6.

7.2.4.1 Derivation of Off-Policy Game Q-Learning Algorithm

From (7.76), one has

$$\begin{aligned}(M^j)^T H_i^{j+1} M^j &= (M^j)^T \Lambda_i M^j \\ &\quad + \left(A - \sum_{i=1}^{n} B_i K_i^j\right)^T (M^j)^T H_i^{j+1} M^j \left(A - \sum_{i=1}^{n} B_i K_i^j\right), \quad (7.84)\end{aligned}$$

where

$$\Lambda_i = \text{diag}(Q_i, R_1, R_2, \ldots, R_n).$$

Adding auxiliary variables $u_{ik}^j = -K_i^j x_k$ $(i = 1, 2, \ldots, n)$ to system (7.56) yields

$$x_{k+1} = A_c x_k + \sum_{i=1}^{n} B_i(u_{ik} - u_{ik}^j), \tag{7.85}$$

where $A_c = A - \sum_{i=1}^{n} B_i K_i^j$, u_{ik} are called the behavior control policies to generate data and u_{ik}^j are called the target control policies that players need to learn. When the system trajectory is (7.85), one has

$$\begin{aligned}
&Q_i^{j+1}(x_k, u_{ik}, u_{-ik}) - x_k^T A_c^T (M^j)^T H_i^{j+1} M^j A_c x_k \\
&= x_k^T (M^j)^T H_i^{j+1} M^j x_k - \left(x_{k+1} - \sum_{i=1}^{n} B_i(u_{ik} - u_{ik}^j)\right)^T \\
&\times (M^j)^T H_i^{j+1} M^j \left(x_{k+1} - \sum_{i=1}^{n} B_i(u_{ik} - u_{ik}^j)\right) \\
&= x_k^T (M^j)^T \Lambda_i M^j x_k.
\end{aligned} \tag{7.86}$$

In view that P_i^{j+1} and H_i^{j+1} are related as shown in (7.68) and (7.69), then the following holds:

$$\begin{aligned}
&x_k^T (M^j)^T H_i^{j+1} M^j x_k - x_{k+1}^T (M^j)^T H_i^{j+1} M^j x_{k+1} + 2\left(A x_k + \sum_{i=1}^{n} B_i u_{ik}\right)^T P_i^{j+1} \\
&\times \sum_{i=1}^{n} B_i(u_{ik} - u_{ik}^j) - \sum_{i=1}^{n} (u_{ik} - u_{ik}^j)^T B_i^T P_i^{j+1} \sum_{i=1}^{n} B_i(u_{ik} - u_{ik}^j) \\
&= x_k^T (M^j)^T \Lambda_i M^j x_k.
\end{aligned} \tag{7.87}$$

Further, one has

$$\begin{aligned}
&x_k^T (M^j)^T H_i^{j+1} M^j x_k - x_{k+1}^T (M^j)^T H_i^{j+1} M^j x_{k+1} \\
&+ 2x_k^T \left[H_{i,xu_1}^{j+1} \; H_{i,xu_2}^{j+1} \; \ldots \; H_{i,xu_n}^{j+1}\right] \sum_{i=1}^{n} (u_{ik} + k_i^j x_k) \\
&+ 2\sum_{i=1}^{n} u_{ik}^T G_i^{j+1} \sum_{i=1}^{n} (u_{ik} + k_i^j x_k) - \sum_{i=1}^{n} \left(u_{ik} + k_i^j x_k\right)^T G_i^{j+1} \sum_{i=1}^{n} (u_{ik} + k_i^j x_k) \\
&= x_k^T (M^j)^T \Lambda_i M^j x_k,
\end{aligned} \tag{7.88}$$

where

$$G_i^{j+1} = \begin{bmatrix} H_{i,u_1u_1}^{j+1} - R_1 & H_{i,u_1u_2}^{j+1} & \dots & H_{i,u_1u_n}^{j+1} \\ (H_{i,u_1u_2}^{j+1})^T & H_{i,u_2u_2}^{j+1} - R_2 & \dots & H_{i,u_2u_n}^{j+1} \\ \vdots & \vdots & \ddots & \vdots \\ (H_{i,u_1u_n}^{j+1})^T & (H_{i,u_2u_n}^{j+1})^T & \dots & H_{i,u_nu_n}^{j+1} - R_n \end{bmatrix}.$$

Manipulating (7.88) can get the following form:

$$\hat{\theta}_i^j(k)\hat{L}_i^{j+1} = \hat{\rho}_{i,k}, \tag{7.89}$$

where

$$\hat{\rho}_{i,k} = x_k^T Q_i x_k + \sum_{i=1}^{n} u_{ik}^T R_i u_{ik}$$

$$\hat{L}_i^{j+1} = \left[(\text{vec}(\hat{L}_{i,rz}^{j+1}))^T, \dots, (\text{vec}(\hat{L}_{i,nn}^{j+1}))^T\right]^T, \ \hat{\theta}_i^j(k) = \left[\hat{\theta}_{i,rz}^j, \dots, \hat{\theta}_{i,nn}^j\right]$$

with $r = 0, 1, 2, \dots, n$, $z = r, r+1, r+2, \dots, n$.

$$\begin{aligned}
\hat{\theta}_{i,00}^j &= x_k^T \otimes x_k^T - x_{k+1}^T \otimes x_{k+1}^T \\
\hat{L}_{i,00}^{j+1} &= H_{i,xx}^{j+1} \\
\hat{\theta}_{i,ss}^j &= -(K_s^j x_{k+1})^T \otimes (K_s^j x_{k+1})^T + u_s^T \otimes u_s^T \\
\hat{L}_{i,ss}^{j+1} &= H_{i,u_su_s}^{j+1} \\
\hat{\theta}_{i,0s}^j &= 2x_{k+1}^T \otimes (K_s^j x_{k+1})^T + 2x_k^T \otimes u_s^T \\
\hat{L}_{i,0s}^{j+1} &= H_{i,xu_s}^{j+1} \\
\hat{\theta}_{i,st}^j &= -2(K_s^j x_{k+1})^T \otimes (K_t^j x_{k+1})^T + 2u_s^T \otimes u_t^T \\
\hat{L}_{i,st}^{j+1} &= H_{i,u_su_t}^{j+1}
\end{aligned}$$

with $s \neq t$ and $s, t = 1, 2, \dots, n$.

Thus, $K_1^{j+1}, K_2^{j+1}, \dots, K_n^{j+1}$ can be expressed as the form of $\hat{L}_i^{j+1}$

$$K_i^{j+1} = (\hat{L}_{i,ii}^{j+1})^{-1}\left((\hat{L}_{i,0i}^{j+1})^T - \sum_{r=1, r\neq i}^{n} (\hat{L}_{i,ir}^{j+1})^T K_r^j\right). \tag{7.90}$$

Theorem 7.6 *(H_i^{j+1}, K_i^{j+1}) are the solution of (7.89) and (7.90) if and only if they are the solution of (7.76) and (7.78).*

Proof From the derivation of (7.89) and (7.90), one can conclude that if (H_i^{j+1}, K_i^{j+1}) are the solution of (7.76) and (7.78), then (H_i^{j+1}, K_i^{j+1}) satisfies (7.89) and

(7.90). Next, we will prove that the solution to (7.89) and (7.90) is also the solution to (7.76) and (7.78).

It is obvious that (7.89) is equivalent to (7.87), so the solutions of both (7.89) and (7.87) are going to be the same. If (H_i^{j+1}, K_i^{j+1}) is the solution to (7.87), then it is also the solution to (7.86) based on Lemma 1. Subtracting (7.87) from (7.86) yields (7.76), then the solution to (7.87) is the same one to (7.76). Thus, (H_i^{j+1}, K_i^{j+1}) satisfied with (7.89) makes (7.76) hold resulting in (7.90) being (7.78). □

Algorithm 7.5 Off-policy game Q-learning

1: Data collection: Collect data x_k by using behavior control policies u_{ik} $(i = 1, 2, \ldots, n)$;
2: Initialization: Given initial admissible controller gains of multiple players $K_1^0, K_2^0, \ldots, K_n^0$. Let $j = 1$ and $i = 1$, where j denotes the iteration index and i stands of player i;
3: Implementing the off-policy game Q-learning: By using recursive least-square methods, $\hat{L}_i^{j+1}$ can be calculated using (7.89). And then K_i^{j+1} is updated by (7.90);
4: If $i < n$, then $i = i + 1$ and go back to Step 3. Otherwise $j = j + 1, i = 1$ and go to Step 5;
5: Stop when $\left\| K_i^j - K_i^{j-1} \right\| \leq \varepsilon$ $(i = 1, 2, \ldots, n)$, the optimal control policy is obtained. Otherwise, $i = 1$, and go back to Step 3.

Remark 7.17 If $\hat{L}_i^{j+1}$ are solved correctly, then the $u_{ik}^j = -K_i^j x_k$ can be learned by Algorithm 7.5. When $j \to \infty$, $u_{ik}^j \to u_{ik}^*$. Since the solution to Algorithm 7.5 is the same as the solution to Algorithm 7.4, and u_{ik}^j learned by Algorithm 7.4 have been proven to converge to u_{ik}^*, then u_{ik}^j found by using Algorithm 7.5 can converge to u_{ik}^*, under which the Nash equilibrium of multi-player games can be reached.

Remark 7.18 Algorithm 7.5 is indeed an off-policy Q-learning approach, since the target control policies are updated but they are not applied to the learning process. The use of arbitrary admissible behavior control policies u_{ik} to generate data and enrich data exploration is the essential feature of the off-policy learning as opposed to the on-policy learning (Kim and Lewis 2010; Jiang et al. 2017; Vamvoudakis 2017b, a).

7.2.4.2 No Bias Analysis of Solution for the Off-Policy Q-Learning Algorithm

To satisfy the condition of PE, probing noises are added into (7.76) in Algorithm 7.4. It has been proven that Algorithm 7.4 could produce bias of solution if adding probing noises into systems. The following will prove that the proposed Algorithm 7.5 does not produce deviation of the solution under the circumstance of adding probing noises.

Theorem 7.7 *Add probing noises to the behavior control policies in Algorithm 7.5. Let H_i^{j+1} be the solution to (7.86) with $e_{ik} = 0$ and $\hat{H}_i^{j+1}$ be the solution to (7.86) with $e_{ik} \neq 0$, then $\hat{H}_i^{j+1} = H_i^{j+1}$.*

Proof After probing noises are added to the behavior control policies, that is $u_{ik} + e_{ik}$, solving (7.86) is equivalent to solving the following form:

$$\begin{aligned}\hat{x}_k^T (M^j)^T \hat{H}_i^{j+1} M^j \hat{x}_k &= \hat{x}_k^T (M^j)^T \Lambda_i M^j \hat{x}_k \\ &+ \hat{x}_k^T \left(A - \sum_{i=1}^{n} B_i K_i^j\right)^T (M^j)^T \hat{H}_i^{j+1} M^j \left(A - \sum_{i=1}^{n} B_i K_i^j\right) \hat{x}_k. \end{aligned} \tag{7.91}$$

Notice that if adding probing noises into system (7.85), then it becomes

$$\hat{x}_{k+1} = A_c \hat{x}_k + \sum_{i=1}^{n} B_i (u_{ik} + e_{ik} + K_i^j \hat{x}_k). \tag{7.92}$$

In this case, (7.86) becomes

$$\begin{aligned}&\hat{x}_k^T (M^j)^T \hat{H}_i^{j+1} M^j \hat{x}_k - \left(\hat{x}_{k+1} - \sum_{i=1}^{n} B_i (u_{ik} + e_{ik} + K_i^j \hat{x}_k)\right)^T (M^j)^T \hat{H}_i^{j+1} M^j \\ &\times \left(\hat{x}_{k+1} - \sum_{i=1}^{n} B_i (u_{ik} + e_{ik} + K_i^j \hat{x}_k)\right) \\ &= \hat{x}_k^T (M^j)^T \Lambda_i M^j \hat{x}_k. \end{aligned} \tag{7.93}$$

Substituting (7.92) into (7.93), (7.93) becomes (7.91). So the solution to (7.91) is the same as (7.86). From the proof of Theorem 7.6, one can find that the solution to (7.89) is equal to that to (7.86). Therefore, it is impossible for the off-policy game Q-learning algorithm to produce deviations when adding probing noises. The unbiasedness of the solution of the off-policy game Q-learning is proved. □

Remark 7.19 Different from Kiumarsi et al. (2017), where the unbiased proof of the solution of the off-policy RL algorithm for solving the zero-sum game problem with two players was developed, while here the unbiased proof of the solution of the off-policy game Q-learning algorithm is for the nonzero-sum multi-player games.

7.2.5 Simulation Results

In this subsection, the effectiveness of the proposed off-policy game Q-learning algorithm is verified respectively for the three-player games and the five-player games.

Example 7.2 Consider the following linear DT system with three players, which play nonzero-sum game:

$$x_{k+1} = A x_k + B_1 u_1 + B_2 u_2 + B_3 u_3, \tag{7.94}$$

where

$$A = \begin{bmatrix} 0.906488 & 0.0816012 & -0.0005 \\ 0.074349 & 0.90121 & -0.000708383 \\ 0 & 0 & 0.132655 \end{bmatrix},$$

$$B_1 = \begin{bmatrix} -0.00150808 \\ -0.0096 \\ 0.867345 \end{bmatrix},\ B_2 = \begin{bmatrix} 0.00951892 \\ 0.00038373 \\ 0 \end{bmatrix},\ B_3 = \begin{bmatrix} -0.00563451 \\ -0.08962 \\ 0.356478 \end{bmatrix}.$$

Choose $Q_1 = \text{diag}(4, 4, 4)$, $Q_2 = \text{diag}(5, 5, 5)$, $Q_3 = \text{diag}(6, 6, 6)$ and $R_1 = R_2 = R_3 = 1$. Rewrite (7.74) as

$$H_i^* = \Lambda_i + \begin{bmatrix} A & B_1 & B_2 & B_3 \\ K_1A & K_1B_1 & K_1B_2 & K_1B_3 \\ K_2A & K_2B_1 & K_2B_2 & K_2B_3 \\ K_3A & K_3B_1 & K_3B_2 & K_3B_3 \end{bmatrix}^T H_i^* \begin{bmatrix} A & B_1 & B_2 & B_3 \\ K_1A & K_1B_1 & K_1B_2 & K_1B_3 \\ K_2A & K_2B_1 & K_2B_2 & K_2B_3 \\ K_3A & K_3B_1 & K_3B_2 & K_3B_3 \end{bmatrix}. \tag{7.95}$$

The model-based iterative algorithm in terms of (7.95) is used in MATLAB to obtain the real solution. Thus, the optimal Q-function matrices (H_1^*, H_2^*, H_3^*) and the optimal controller gains (K_1^*, K_2^*, K_3^*) can be obtained

$$\begin{aligned} H_1^* &= \begin{bmatrix} 26.6577 & 11.7777 & -0.0150 & -0.1100 & 0.2336 & -1.0988 \\ 11.7777 & 21.4849 & -0.0111 & -0.1434 & 0.1164 & -1.6897 \\ -0.0150 & -0.0111 & 4.0707 & 0.4622 & -0.0002 & 0.1910 \\ -0.1100 & -0.1434 & 0.4622 & 4.0223 & -0.0011 & 1.2555 \\ 0.2336 & 0.1164 & -0.0002 & -0.0011 & 1.0024 & -0.0108 \\ -1.0988 & -1.6897 & 0.1910 & 1.2555 & -0.0108 & 1.6737 \end{bmatrix} \\ H_2^* &= \begin{bmatrix} 32.0935 & 13.2638 & -0.0172 & -0.1220 & 0.2797 & -1.2318 \\ 13.2638 & 25.0310 & -0.0117 & -0.1585 & 0.1310 & -1.9340 \\ -0.0172 & -0.0117 & 5.0883 & 0.5773 & -0.0002 & 0.2383 \\ -0.1220 & -0.1585 & 0.5773 & 4.7753 & -0.0012 & 1.5664 \\ 0.2797 & 0.1310 & -0.0002 & -0.0012 & 1.0029 & -0.0121 \\ -1.2318 & -1.9340 & 0.2383 & 1.5664 & -0.0121 & 1.8244 \end{bmatrix} \\ H_3^* &= \begin{bmatrix} 37.5293 & 14.7499 & -0.0194 & -0.1340 & 0.3258 & -1.3649 \\ 14.7499 & 28.5771 & -0.0124 & -0.1736 & 0.1455 & -2.1784 \\ -0.0194 & -0.0124 & 6.1059 & 0.6925 & -0.0002 & 0.2857 \\ -0.1340 & -0.1736 & 0.6925 & 5.5283 & -0.0013 & 1.8773 \\ 0.3258 & 0.1455 & -0.0002 & -0.0013 & 1.0034 & -0.0134 \\ -1.3649 & -2.1784 & 0.2857 & 1.8773 & -0.0134 & 1.9750 \end{bmatrix} \\ K_1^* &= \begin{bmatrix} -0.2671 & -0.4385 & -0.0992 \end{bmatrix} \\ K_2^* &= \begin{bmatrix} -0.2678 & -0.1127 & -0.0006 \end{bmatrix} \\ K_3^* &= \begin{bmatrix} 0.9431 & 1.5190 & -0.0504 \end{bmatrix}. \end{aligned} \tag{7.96}$$

Table 7.1 Optimal controller gains under three probing noises

Case	On-policy Q-learning	Off-policy Q-learning
1	$K_1 = [-0.2671\ \ -0.4377\ \ -0.0992]$	$K_1 = [-0.2671\ \ -0.4385\ \ -0.0992]$
	$K_2 = [-0.2679\ \ -0.1124\ \ -0.0005]$	$K_2 = [-0.2678\ \ -0.1127\ \ -0.0006]$
	$K_3 = [0.9416\ 1.5184\ \ -0.0504]$	$K_3 = [0.9431\ 1.5190\ \ -0.0504]$
2	$K_1 = [-0.3015\ \ -0.5666\ \ -0.1009]$	$K_1 = [-0.2671\ \ -0.4385\ \ -0.0992]$
	$K_2 = [0.6741\ \ -1.7100\ \ -0.0165]$	$K_2 = [-0.2678\ \ -0.1127\ \ -0.0006]$
	$K_3 = [1.0547\ 1.9397\ \ -0.0447]$	$K_3 = [0.9431\ 1.5190\ \ -0.0504]$
3	$K_1 = [-0.4709\ \ -0.2428\ \ -0.1010]$	$K_1 = [-0.2671\ \ -0.4385\ \ -0.0992]$
	$K_2 = [-0.6146\ 0.8284\ 0.0179]$	$K_2 = [-0.2678\ \ -0.1127\ \ -0.0006]$
	$K_3 = [1.6089\ 0.8591\ \ -0.452]$	$K_3 = [0.9431\ 1.5190\ \ -0.0504]$

Three different probing noises were added to verify the unbiasedness of the off-policy game Q-learning algorithm, and compare it with the on-policy game Q-learning algorithm. To ensure PE conditions when solving (7.76) and (7.89), the following three cases of probing noises are used:

Case 1:

$$e_i = \sum_{j}^{100} 0.5 * \sin(\text{noise}_{\text{feq}}(1, j) * k). \tag{7.97}$$

Case 2:

$$e_i = \sum_{j}^{100} 6 * \sin(\text{noise}_{\text{feq}}(1, j) * k). \tag{7.98}$$

Case 3:

$$e_i = \sum_{j}^{100} 10 * \sin(\text{noise}_{\text{feq}}(1, j) * k), \tag{7.99}$$

where

$$\text{noise}_{\text{feq}}(1, j) = 500 * \text{rand}(1, 100) - 200 * \text{ones}(1, 100). \tag{7.100}$$

Table 7.1 shows the optimal controller gains when implementing Algorithms 7.4 and 7.5 under the three cases of probing noises. It can be observed that the solution when using on-policy game Q-learning Algorithm 7.4 is affected by the probing noises interference and the learned controller gains are biased. On the contrary, for all of the three probing noises, the Q-function matrices H_i^j and controller gains K_i^j always converge to the optimal values when implementing off-policy game Q-learning Algorithm 7.5.

Simulation results when utilizing the on-policy game Q-learning Algorithm 7.4: Under the probing noises Case 1, Figs. 7.7 and 7.8 show the errors between the

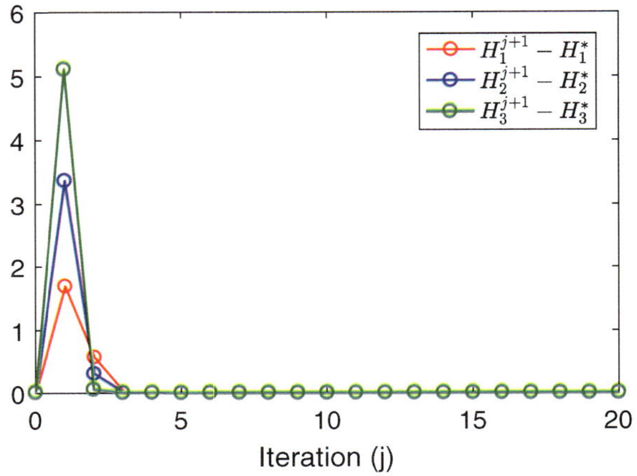

Fig. 7.7 Case 1: Convergence of H_i in on-policy game Q-learning

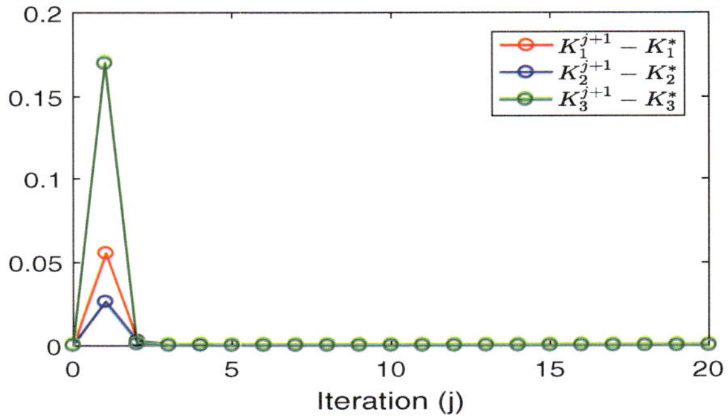

Fig. 7.8 Case 1: Convergence of K_i in on-policy game Q-learning

matrices H_i and H_i^* and the errors between the controller gains K_i and K_i^*. Figure 7.9 shows the state responses of the game system. The optimal performance J_i is plotted using the learned optimal control policy in Fig. 7.10.

Figures 7.11 and 7.12 show the errors between the matrices H_i and H_i^* and the errors between the controller gains K_i and K_i^* under the probing noises Case 2 during the implementing of on-policy game Q-learning Algorithm 7.4. Figure 7.13 shows the state responses of the game system.

Under the probing noises Case 3, Figs. 7.14 and 7.15 show the errors of matrices H_i and H_i^*, and the errors of controller gains K_i and K_i^*, Fig. 7.16 shows the state responses of the game system.

Simulation results when using the off-policy game Q-learning Algorithm 7.5: Since there is no bias of solution (H_i, K_i) even though adding probing noises, then the convergence results of H_i and K_i only under Case 1 are plotted in Figs. 7.17 and 7.18. Figure 7.19 shows the state responses of the game system. The performance J_i

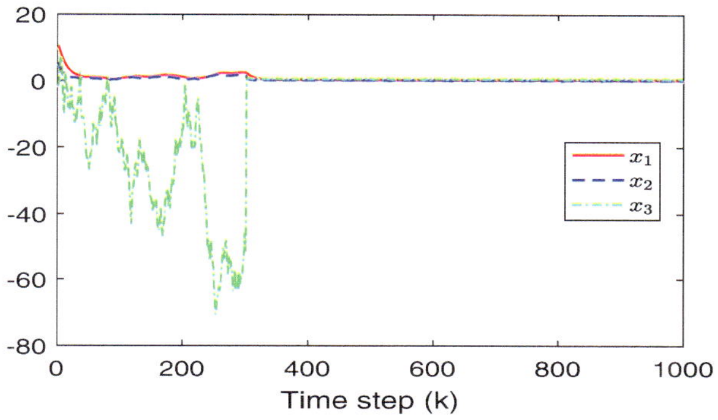

Fig. 7.9 Case 1: The system states x in on-policy game Q-learning

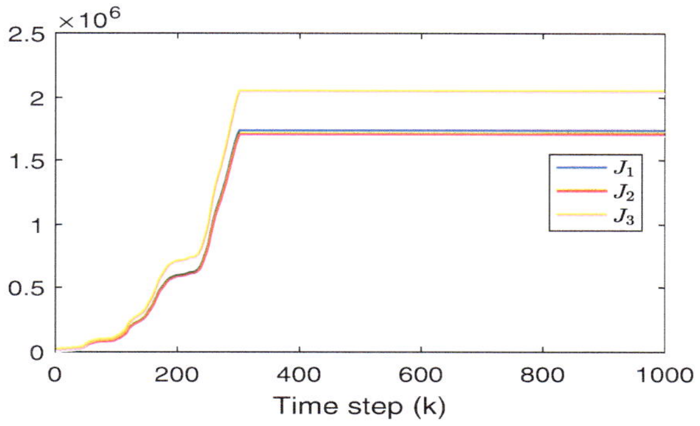

Fig. 7.10 Case 1: The optimal costs in on-policy game Q-learning

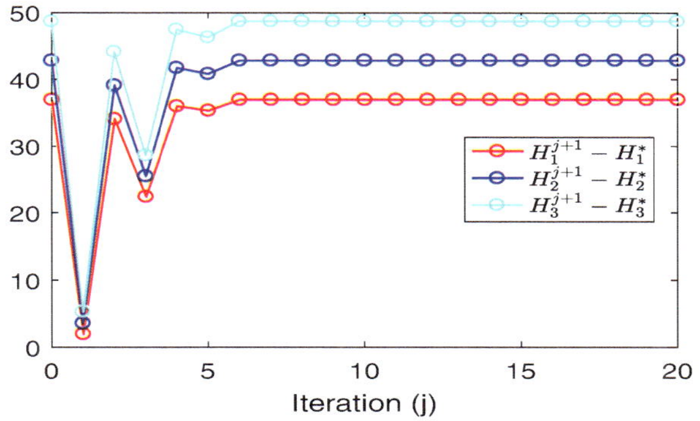

Fig. 7.11 Case 2: Convergence of H_i in on-policy game Q-learning

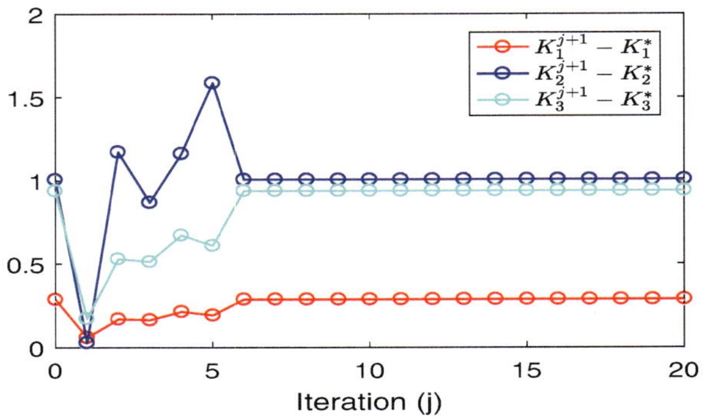

Fig. 7.12 Case 2: Convergence of K_i in on-policy game Q-learning

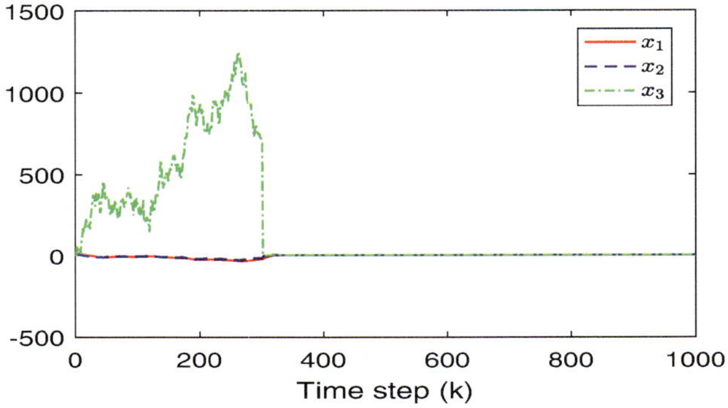

Fig. 7.13 Case 2: The system states x in on-policy game Q-learning

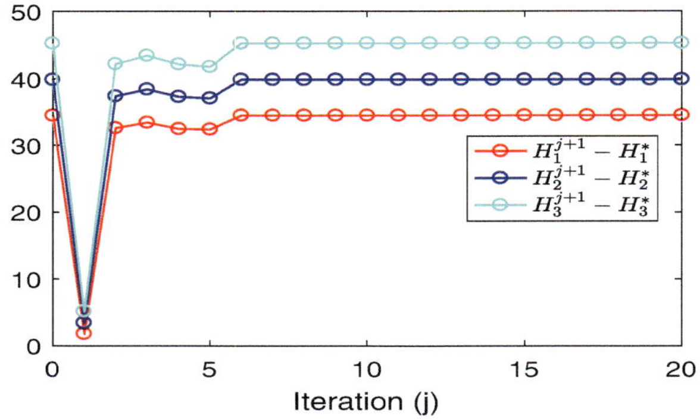

Fig. 7.14 Case 3: Convergence of H_i in on-policy game Q-learning

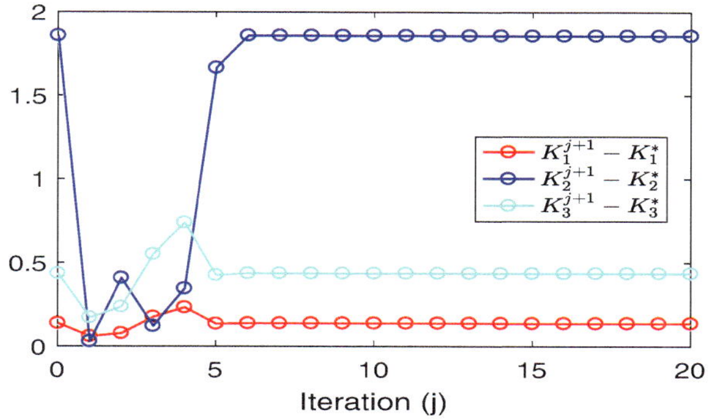

Fig. 7.15 Case 3: Convergence of K_i in on-policy game Q-learning

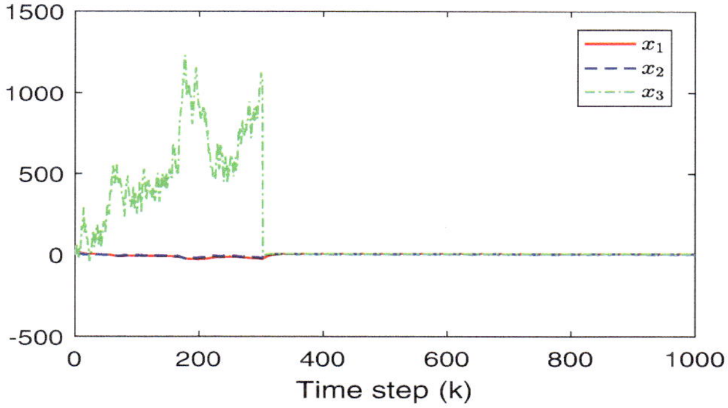

Fig. 7.16 Case 3: The system states x in on-policy game Q-learning

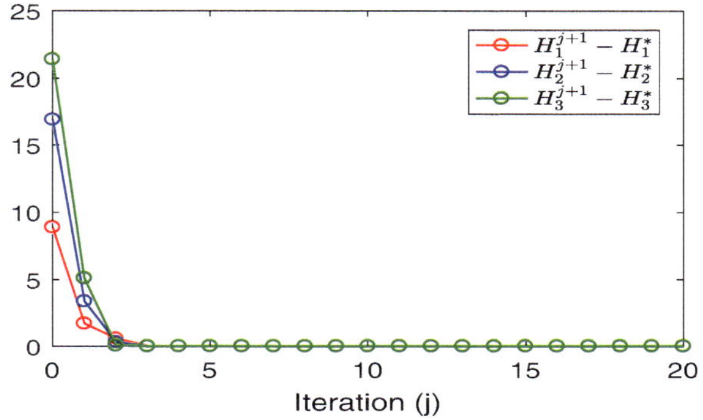

Fig. 7.17 Convergence of H_i in off-policy game Q-learning

along the system trajectories under the learned optimal control policies are plotted in Fig. 7.20.

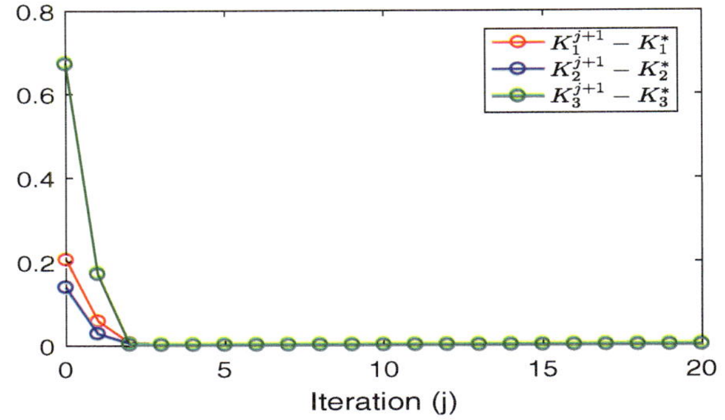

Fig. 7.18 Convergence of K_i in off-policy game Q-learning

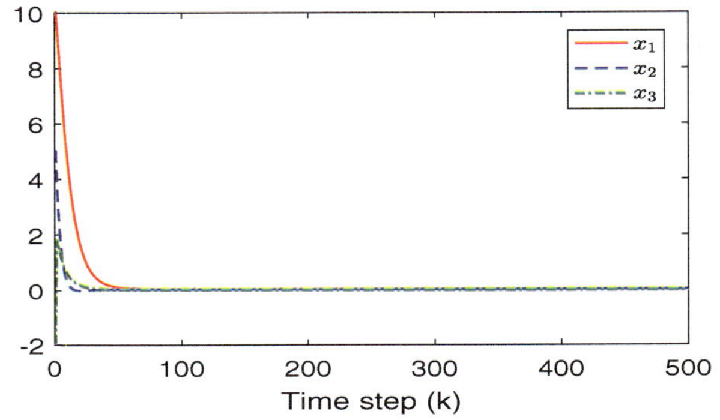

Fig. 7.19 The system states x in off-policy game Q-learning

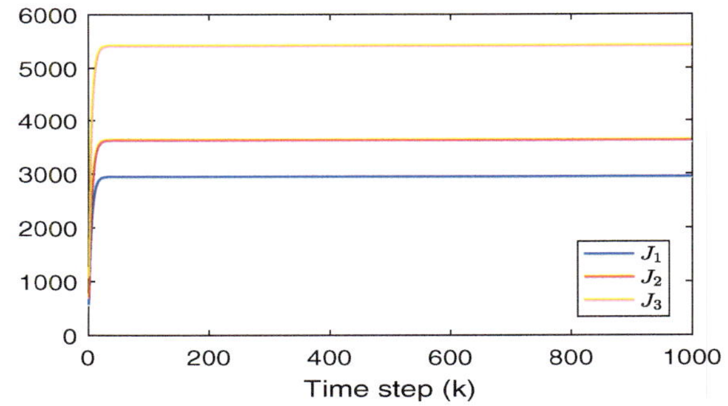

Fig. 7.20 The optimal costs in off-policy game Q-learning

Example 7.3 Consider the following linear DT system with five players, which play nonzero-sum game:

$$x_{k+1} = Ax_k + B_1u_1 + B_2u_2 + B_3u_3 + B_4u_4 + B_5u_5, \tag{7.101}$$

where

$$A=\begin{bmatrix}0.906488 & 0.0816012 & -0.0005\\ 0.074349 & 0.90121 & -0.000708383\\ 0 & 0 & 0.132655\end{bmatrix},\ B_1=\begin{bmatrix}-0.00150808\\ -0.0096\\ 0.867345\end{bmatrix}$$
$$B_2=\begin{bmatrix}0.00951892\\ 0.00038373\\ 0\end{bmatrix},\ B_3=\begin{bmatrix}-0.00563451\\ -0.08962\\ 0.356478\end{bmatrix},\ B_4=\begin{bmatrix}0.0123956\\ 0.068\\ -0.05673\end{bmatrix},\ B_5=\begin{bmatrix}-0.125\\ 0.4\\ -0.4898\end{bmatrix}.$$

Choose $Q_1=\text{diag}(4,4,4)$, $Q_2=\text{diag}(5,5,5)$, $Q_3=\text{diag}(6,6,6)$, $Q_4=\text{diag}(7,7,7)$, $Q_5=\text{diag}(3,3,3)$ and $R_1=R_2=R_3=R_4=R_5=1$. Rewrite (7.74) as

$$H_i^*=\Lambda_i+\begin{bmatrix}A & B_1 & \dots & B_5\\ K_1A & K_1B_1 & \dots & K_1B_5\\ K_2A & K_2B_1 & \dots & K_2B_5\\ K_3A & K_3B_1 & \dots & K_3B_5\\ K_4A & K_4B_1 & \dots & K_4B_5\\ K_5A & K_5B_1 & \dots & K_5B_5\end{bmatrix}^T H_i^* \begin{bmatrix}A & B_1 & \dots & B_5\\ K_1A & K_1B_1 & \dots & K_1B_5\\ K_2A & K_2B_1 & \dots & K_2B_5\\ K_3A & K_3B_1 & \dots & K_3B_5\\ K_4A & K_4B_1 & \dots & K_4B_5\\ K_5A & K_5B_1 & \dots & K_5B_5\end{bmatrix}. \tag{7.102}$$

The model-based iterative algorithm in terms of (7.102) is used in MATLAB to obtain the theoretical optimal solution. Thus, the optimal Q-function matrices (H_1^*, H_2^*, H_3^*, H_4^* and H_5^*) and the optimal controller gains (K_1^*, K_2^*, K_3^*, K_4^* and K_5^*) can be obtained

$$H_1^*=\begin{bmatrix}25.4360 & 9.7277 & -0.0168 & -0.1127 & 0.2217 & -0.9082 & 0.8767 & 0.6170\\ 9.7277 & 14.1358 & -0.0049 & -0.0663 & 0.0981 & -0.9637 & 0.8233 & 2.8781\\ -0.0168 & -0.0049 & 4.0707 & 0.4623 & -0.0002 & 0.1904 & -0.0307 & -0.2603\\ -0.1127 & -0.0663 & 0.4623 & 4.0228 & -0.0012 & 1.2483 & -0.2034 & -1.7168\\ 0.2217 & 0.0981 & -0.0002 & -0.0012 & 1.0023 & -0.0091 & 0.0089 & 0.0052\\ -0.9082 & -0.9637 & 0.1904 & 1.2483 & -0.0091 & 1.6022 & -0.1594 & -0.9780\\ 0.8767 & 0.8233 & -0.0307 & -0.2034 & 0.0089 & -0.1594 & 1.0804 & 0.3301\\ 0.6170 & 2.8781 & -0.2603 & -1.7168 & 0.0052 & -0.9780 & 0.3301 & 3.1613\end{bmatrix}$$
$$H_2^*=\begin{bmatrix}29.7586 & 10.9982 & -0.0182 & -0.1208 & 0.2561 & -1.0225 & 0.9944 & 0.6000\\ 10.9982 & 16.7580 & -0.0058 & -0.0778 & 0.1107 & -1.1192 & 0.9532 & 3.3906\\ -0.0182 & -0.0058 & 5.0883 & 0.5775 & -0.0002 & 0.2378 & -0.0383 & -0.3254\\ -0.1208 & -0.0778 & 0.5775 & 4.7758 & -0.0012 & 1.5589 & -0.2536 & -2.1457\\ 0.2561 & 0.1107 & -0.0002 & -0.0012 & 1.0027 & -0.0103 & 0.0101 & 0.0048\\ -1.0225 & -1.1192 & 0.2378 & 1.5589 & -0.0103 & 1.7443 & -0.1920 & -1.2033\\ 0.9944 & 0.9532 & -0.0383 & -0.2536 & 0.0101 & -0.1920 & 1.0941 & 0.3960\\ 0.6000 & 3.3906 & -0.3254 & -2.1457 & 0.0048 & -1.2033 & 0.3960 & 3.6386\end{bmatrix}$$

$$H_3^*=\begin{bmatrix}34.0812 & 12.2688 & -0.0196 & -0.1290 & 0.2906 & -1.1368 & 1.1121 & 0.5830\\ 12.2688 & 19.3802 & -0.0066 & -0.0893 & 0.1234 & -1.2747 & 1.0830 & 3.9032\\ -0.0196 & -0.0066 & 6.1059 & 0.6926 & -0.0002 & 0.2852 & -0.0459 & -0.3906\\ -0.1290 & -0.0893 & 0.6926 & 5.5288 & -0.0013 & 1.8694 & -0.3037 & -2.5747\\ 0.2906 & 0.1234 & -0.0002 & -0.0013 & 1.0030 & -0.0114 & 0.0112 & 0.0044\\ -1.1368 & -1.2747 & 0.2852 & 1.8694 & -0.0114 & 1.8864 & -0.2247 & -1.4285\\ 1.1121 & 1.0830 & -0.0459 & -0.3037 & 0.0112 & -0.2247 & 1.1079 & 0.4620\\ 0.5830 & 3.9032 & -0.3906 & -2.5747 & 0.0044 & -1.4285 & 0.4620 & 4.1159\end{bmatrix}$$

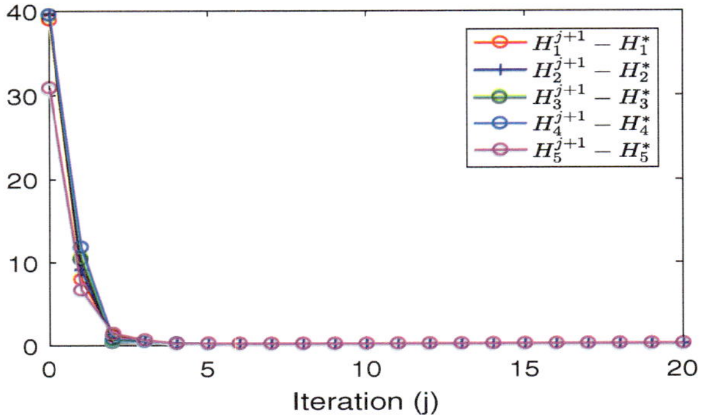

Fig. 7.21 Convergence of H_i in off-policy game Q-learning

$$H_4^* = \begin{bmatrix} 38.4038 & 13.5393 & -0.0210 & -0.1372 & 0.3250 & -1.2512 & 1.2297 & 0.5660 \\ 13.5393 & 22.0024 & -0.0074 & -0.1009 & 0.1361 & -1.4302 & 1.2129 & 4.4158 \\ -0.0210 & -0.0074 & 7.1235 & 0.8078 & -0.0002 & 0.3326 & -0.0535 & -0.4557 \\ -0.1372 & -0.1009 & 0.8078 & 6.2818 & -0.0014 & 2.1800 & -0.3539 & -3.0037 \\ 0.3250 & 0.1361 & -0.0002 & -0.0014 & 1.0034 & -0.0126 & 0.0124 & 0.0040 \\ -1.2512 & -1.4302 & 0.3326 & 2.1800 & -0.0126 & 2.0285 & -0.2573 & -1.6537 \\ 1.2297 & 1.2129 & -0.0535 & -0.3539 & 0.0124 & -0.2573 & 1.1216 & 0.5280 \\ 0.5660 & 4.4158 & -0.4557 & -3.0037 & 0.0040 & -1.6537 & 0.5280 & 4.5932 \end{bmatrix}$$

$$H_5^* = \begin{bmatrix} 21.1134 & 8.4572 & -0.0154 & -0.1045 & 0.1872 & -0.7938 & 0.7590 & 0.6340 \\ 8.4572 & 11.5136 & -0.0041 & -0.0548 & 0.0854 & -0.8082 & 0.6935 & 2.3655 \\ -0.0154 & -0.0041 & 3.0531 & 0.3471 & -0.0002 & 0.1430 & -0.0231 & -0.1951 \\ -0.1045 & -0.0548 & 0.3471 & 3.2698 & -0.0011 & 0.9378 & -0.1532 & -1.2878 \\ 0.1872 & 0.0854 & -0.0002 & -0.0011 & 1.0019 & -0.0080 & 0.0077 & 0.0056 \\ -0.7938 & -0.8082 & 0.1430 & 0.9378 & -0.0080 & 1.4600 & -0.1268 & -0.7528 \\ 0.7590 & 0.6935 & -0.0231 & -0.1532 & 0.0077 & -0.1268 & 1.0667 & 0.2641 \\ 0.6340 & 2.3655 & -0.1951 & -1.2878 & 0.0056 & -0.7528 & 0.2641 & 2.6840 \end{bmatrix}$$

$$\begin{aligned} K_1^* &= \begin{bmatrix} -0.1559 & -0.4921 & -0.1018 \end{bmatrix} \\ K_2^* &= \begin{bmatrix} -0.2372 & -0.0944 & -0.0006 \end{bmatrix} \\ K_3^* &= \begin{bmatrix} 0.8838 & 0.5288 & -0.0597 \end{bmatrix} \\ K_4^* &= \begin{bmatrix} -1.1014 & -0.7778 & 0.0084 \end{bmatrix} \\ K_5^* &= \begin{bmatrix} 0.3424 & -0.7149 & -0.0138 \end{bmatrix}. \end{aligned} \tag{7.103}$$

Simulation results when using off-policy game Q-learning Algorithm 7.5: Under the probing noises, Case 1, Figs. 7.21 and 7.22, show the convergences of the Q-function matrices H_i and the learned control gains K_i. Figure 7.23 shows the state responses of the game system. The above three figures are obtained by implementing the proposed off-policy game Q-learning algorithm. The performance J_i is plotted in Fig. 7.24 under the learned Nash equilibrium solution.

Results Analysis and Comparisons: From Table 7.1, one can find that solution deviation indeed existed when using the on-policy game Q-learning for multi-player

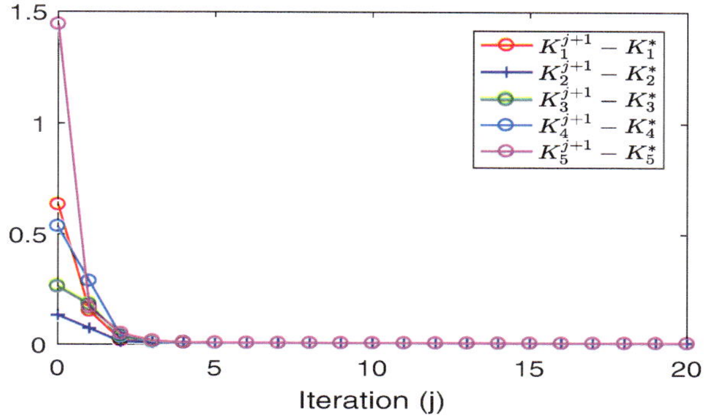

Fig. 7.22 Convergence of K_i in off-policy game Q-learning

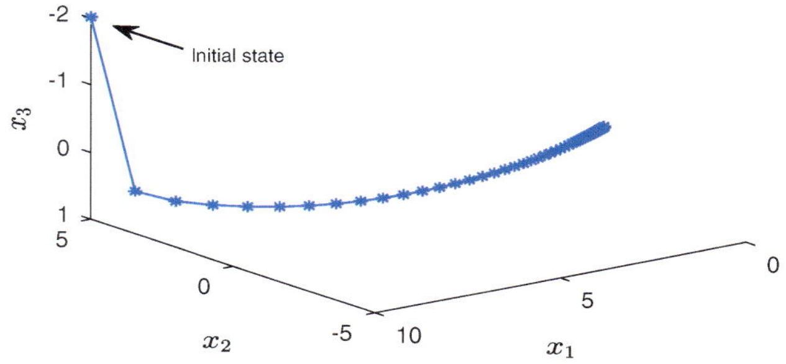

Fig. 7.23 The system states x in off-policy game Q-learning

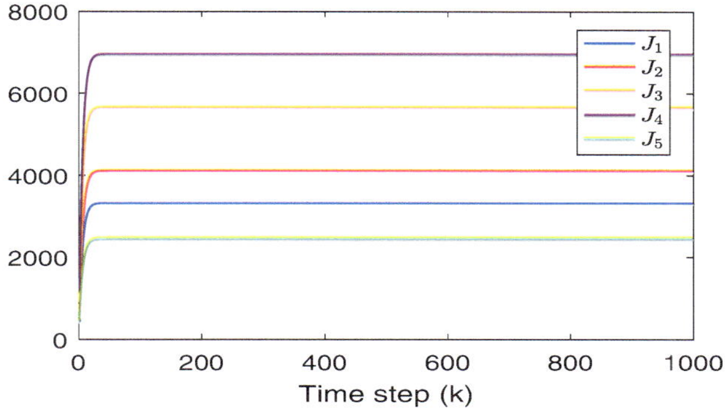

Fig. 7.24 The optimal costs in off-policy game Q-learning

games, while there is no bias of solution of Q-function matrices H_i and the optimal controller gains K_i when implementing the off-policy game Q-learning for the multi-player games. With the increasing of probing noises, the state response in Figs. 7.9, 7.13 and 7.16 and the cost in Fig. 7.10 is more remarkably affected since the system is disturbed by adding probing noises. However, as shown in Figs. 7.19 and 7.20 for the case of three-player games and Figs. 7.23 and 7.24 for the case of five-player games, the systems converge with fast velocity and without overshoot, and the costs are smaller than those using the on-policy game Q-learning algorithm.

7.3 Conclusion

In this section, an off-policy RL algorithm is presented to solve the synchronization problem of MASs in an optimal manner and using only measured data. The integral reinforcement learning idea is employed to derive off-policy Bellman equations to evaluate target policies and find improved target polices. Each agent applies a behavior policy to collect data which is used to learn the solution to the off-policy Bellman equation. This policy is different than the target policy which is not actually applied to the systems and is evaluated and learned to find the optimal policy. The proposed approach does not require any knowledge of the agent dynamics. Besides, an off-policy game Q-learning algorithm is proposed to solve multi-player nonzero-sum game problems in linear DT systems without knowing the dynamics of models. The probing noises are added into control inputs during learning solutions and the convergence and unbiasedness of the proposed off-policy game Q-learning are presented with rigorously theoretical proofs. Simulation results have demonstrated the effectiveness of the proposed method.

References

Abu-Khalaf M, Lewis FL (2005) Nearly optimal control laws for nonlinear systems with saturating actuators using a neural network hjb approach. Automatica 41(5):779–791

Al-Tamimi A, Abu-Khalaf M, Lewis FL (2007) Adaptive critic designs for discrete-time zero-sum games with application to H_∞ control. IEEE Trans Syst Man Cybern Part B (Cybern) 37(1):240–247

Al-Tamimi A, Lewis FL, Abu-Khalaf M (2007) Model-free Q-learning designs for linear discrete-time zero-sum games with application to H-infinity control. Automatica 43(3):473–481

Başar T, Olsder GJ (1998) Dynamic noncooperative game theory, 2nd edn. SIAM, PA

Bian T, Jiang Y, Jiang ZP (2014) Decentralized adaptive optimal control of large-scale systems with application to power systems. IEEE Trans Ind Electron 62(4):2439–2447

Cheng TM, Savkin AV (2011) Decentralized control of multi-agent systems for swarming with a given geometric pattern. Comput Math Appl 61(4):731–744

Dong W (2010) Distributed optimal control of multiple systems. Int J Control 83(10):2067–2079

Dunbar WB, Murray RM (2006) Distributed receding horizon control for multi-vehicle formation stabilization. Automatica 42(4):549–558

Hou ZS, Wang Z (2013) From model-based control to data-driven control: Survey, classification and perspective. Inf Sci 235:3–35

Jiang Y, Fan J, Chai T, Lewis FL, Li J (2017) Tracking control for linear discrete-time networked control systems with unknown dynamics and dropout. IEEE Trans Neural Networks Learn Syst 29(10):4607–4620

Kaya M, Alhajj R (2005) A novel approach to multiagent reinforcement learning: Utilizing olap mining in the learning process. IEEE Trans Syst Man Cybern Part C (Appl Rev) 35(4):582–590

Kim JH, Lewis FL (2010) Model-free H_∞ control design for unknown linear discrete-time systems via Q-learning with LMI, vol 46. Elsevier

Kiumarsi B, Lewis FL, Jiang Z (2017) H_∞ control of linear discrete-time systems: Off-policy reinforcement learning. Automatica 78:144–152

Lee JY, Park JB, Choi YH (2012) Integral Q-learning and explorized policy iteration for adaptive optimal control of continuous-time linear systems. Automatica 48(11):2850–2859

Lewis FL, Vrabie D, Syrmos VL (2012) Optimal control. John Wiley & Sons, New York, NY, USA

Lewis FL, Zhang H, Hengster-Movric K, Das A (2013) Cooperative control of multi-agent systems: optimal and adaptive design approaches. Springer Science & Business Media

Li J, Modares H, Chai T, Lewis FL, Xie L (2017) Off-policy reinforcement learning for synchronization in multiagent graphical games. IEEE Trans Neural Networks Learn Syst 28(10):2434–2445

Li J, Chai T, Lewis FL, Ding Z, Jiang Y (2019) Off-policy interleaved Q-learning: Optimal control for affine nonlinear discrete-time systems. IEEE Trans Neural Networks Learn Syst 30(5):1308–1320

Liu Y, Jia Y (2012) Adaptive leader-following consensus control of multi-agent systems using model reference adaptive control approach. IET Control Theory Appl 6(13):2002–2008

Luo B, Wu HN, Huang T, Liu D (2014) Data-based approximate policy iteration for affine nonlinear continuous-time optimal control design. Automatica 50(12):3281–3290

Luo B, Huang T, Wu HN, Yang X (2015) Data-driven H_∞ control for nonlinear distributed parameter systems. IEEE Trans Neural Networks Learn Syst 26(11):2949–2961

Modares H, Lewis FL (2014) Linear quadratic tracking control of partially-unknown continuous-time systems using reinforcement learning. IEEE Trans Autom Control 59(11):3051–3056

Movric KH, Lewis FL (2014) Cooperative optimal control for multi-agent systems on directed graph topologies. IEEE Trans Autom Control 59(3):769–774

Olfati-Saber R (2006) Flocking for multi-agent dynamic systems: Algorithms and theory. IEEE Trans Autom Control 51(3):401–420

Saridis GN, Lee CSG (1979) An approximation theory of optimal control for trainable manipulators. IEEE Trans Syst Man Cybern 9(3):152–159

Semsar-Kazerooni E, Khorasani K (2009) Multi-agent team cooperation: A game theory approach. Automatica 45(10):2205–2213

Starr AW, Ho YC (1969) Nonzero-sum differential games. J Optim Theory Appl 3(3):184–206

Vamvoudakis KG (2017) Q-learning for continuous-time graphical games on large networks with completely unknown linear system dynamics. Int J Robust Nonlinear Control 27(16):2900–2920

Vamvoudakis KG (2017) Q-learning for continuous-time linear systems: A model-free infinite horizon optimal control approach. Syst Control Lett 100:14–20

Vamvoudakis KG, Lewis FL (2011) Multi-player non-zero-sum games: Online adaptive learning solution of coupled hamilton-jacobi equations. Automatica 47(8):1556–1569

Vamvoudakis KG, Lewis FL, Hudas GR (2012) Multi-agent differential graphical games: Online adaptive learning solution for synchronization with optimality. Automatica 48(8):1598–1611

Vamvoudakis KG, Modares H, Kiumarsi B, Lewis FL (2017) Game theory-based control system algorithms with real-time reinforcement learning: How to solve multiplayer games online. IEEE Control Syst Mag 37(1):33–52

Vrabie D, Lewis F (2009) Neural network approach to continuous-time direct adaptive optimal control for partially unknown nonlinear systems. Neural Networks 22(3):237–246

Wang C, Wang X, Ji H (2014) A continuous leader-following consensus control strategy for a class of uncertain multi-agent systems. IEEE/CAA J Autom Sinica 1(2):187–192

Wei Q, Lewis FL, Sun Q, Yan P, Song R (2017) Discrete-time deterministic Q-learning: A novel convergence analysis. IEEE Trans Cybern 47(5):1224–1237
Xu Y, Liu W (2011) Novel multiagent based load restoration algorithm for microgrids. IEEE Trans Smart Grid 2(1):152–161
Yu C, Zhang M, Ren F (2014) Collective learning for the emergence of social norms in networked multiagent systems. IEEE Trans Cybern 44(12):2342–2355

Chapter 8
Industrial Applications of Game Reinforcement Learning

In this chapter, we are interested in how to apply game RL theory to practical industrial applications, mainly involving set-point design for optimal control of industrial process operation, particularly dual-rate rougher flotation operation, and performance optimization problems for large-scale industrial processes. To earn high economic profit viewed as one of the operational indices, we present two kinds of off-policy RL methods to learn the optimal set-points for rougher flotation operational processes without the knowledge of dynamics of unit processes and operational indices. To this end, two intractable challenges have to be faced, i.e., unknown dynamics of unit processes and operational indices and two time-scale industrial processes. Game off-policy RL algorithms are developed to find the optimal solution to industrial operational processes by using data measured in real time. Moreover, we present a RL framework for achieving plant-wide performance optimization for large-scale unknown industrial processes by integrating RL with multi-agent game theory. Illustrative examples are given to show the effectiveness of the proposed methods.

8.1 Rougher Flotation Processes and Plant-Wide Optimization Control

Industrial flow lines are composed of unit processes, such as ovens, grinders and flotation cells, operating on a fast time scale and performance measurements known as operational indices measured at a slow time scale. Operational indices may include unit process efficiency, energy consumption, makespan and output quality. Operational index optimization is an important topic in the control of industrial processes. It is desired to find optimal set-points for the unit processes to ensure that the operational indices stay within their target ranges or at their desired target values (Chai et al. 2011, 2014; Vega et al. 2014), which is called OOC. OOC of industrial

J. Li et al., *Reinforcement Learning*, Advances in Industrial Control,
https://doi.org/10.1007/978-3-031-28394-9_8

processes has received much attention. Model-based methods have been widely used to solve the OOC problem. A two-layer architecture is popular to solve the OOC problem since the unit processes operate at a fast time scale, whereas the operational indices can only be measured at a slower time scale (Würth et al. 2011). In Engell (2007), RTO is proposed to find the optimal operational indices, but it requires that the lower-layer control loop stays in its steady-state value while solving the upper-layer optimization problem. Considering dynamic characteristics and exogenous disturbances, approaches such as dynamic RTO, RTO integrated with MPC, nonlinear MPC methods and economic MPC (EMPC) have been developed to find the optimal solution of operational control problem (Vega et al. 2014; Dünnebier et al. 2005; Ellis and Christofides 2014). Recently, Chai et al. (2012) and Liu et al. (2013) presented a set-point compensation method by integrating RTO and MPC techniques for multi-rate industrial processes with transmission delay and packet losses.

Rougher flotation processes connected by several flotation cells in a serial structure are composed of flotation cell control processes, operating at a fast time scale and economic benefit measurement known as operational indices calculated at a slow time scale. Roughing is the first and basic stage of flotation processes which also include cleaning, scavenger, regrinding and classification circuits, and its primary objective is to get the valuable mineral recovery as much as possible subject to great energy consuming. Optimizing economic benefit by trading off increase in recovery rate and decrease of energy consumption, or keeping it following a desired value by forcing concentrate grades and tail grades to proper trajectories at a fast time scale is a key issue. It is desired to find optimal set-points for rougher flotation circuits to ensure that the economic operational indices stay within their target ranges or at their desired target values like most of the industrial operational processes (Chai et al. 2011, 2014; Ding et al. 2016).

Most existing literature about rougher flotation processes mainly focused on control and optimization of the basic device loops, i.e., concentrate grades and tail grades are controlled or are optimally controlled to desired set-points (Zaragoza and Herbst 1988; Thornton 1991; Xie et al. 2016; Hulbert 1995; Pulkkinen et al. 1993; Desbiens et al. 1994; Pérez-Correa et al. 1998; Duchesne et al. 2003; Ferreira and Loveday 2000; Rojas and Cipriano 2011; Maldonado et al. 2007). Optimal control (Zaragoza and Herbst 1988), adaptive control (Thornton 1991; Xie et al. 2016), multi-variable control (Hulbert 1995), expert control (Pulkkinen et al. 1993) and multi-variable predictive control (Desbiens et al. 1994; Pérez-Correa et al. 1998) were developed and adopted in rougher flotation circuits. Based on the previous work of Pérez-Correa et al. (1998), Duchesne et al. (2003), and Ferreira and Loveday (2000) and Rojas and Cipriano (2011) presented multi-variable MPC strategies with consideration of the intermediate cell grade estimates for a rougher circuit, such that concentrate grades and tail grades can follow the desired values by the optimal approach. Maldonado et al. (2007) implemented dynamic programming to minimize the Cu tailing grade in each cell given a final Cu concentrate grade by considering phenomenological models.

Actually, optimal control for achieving regulation and optimization of the local processes and an optimization procedure used to generate the set-points that maximize the economic performance function are both involved in a general industrial process control scheme (Qin and Badgwell 2003; Chai et al. 2012; Liu et al. 2013; Fan et al. 2016). Note that the set-point compensation methods (Chai et al. 2012; Liu et al. 2013; Fan et al. 2016) require the dynamics of the rougher flotation process to be accurately known, which, in general, is very difficult to obtain in practical flotation industrial environment since there exist disturbance, chemical and physical reactions between ore and chemicals (Chai et al. 2011, 2014). Therefore, the difficulty of knowing exact models of rougher flotation processes motivates us to attempt data-driven optimal control research for achieving optimality of dual-rate rougher flotation processes.

Recently, data-based methods have received considerable attention to design optimal set-points for industrial unit processes. Digital sensors are used to collect the information of the systems' states (Yin et al. 2016; Hou and Wang 2013; Cheng et al. 2015). Neural network methods and case-based-reasoning intelligent control methods have been developed to design or correct prescribed optimal set-points for large-scale complex industrial processes without requiring complete knowledge of the system dynamics (Chai et al. 2011, 2014; Dai et al. 2014; Wu et al. 2014). RL could be used to solve optimal set-point problems, since it has a capability of evaluating utility and updating control policy without requiring models of the environment (Wei et al. 2014; Lee et al. 2012; Al-Tamimi et al. 2007b). Especially off-policy RL including behavior policies and target policies is more practical and efficient technique compared with on-policy RL for dealing with optimal control problem as it can generate data of systems for enriching data exploration while the target policies are updated to find the optimal policies but not to be employed to systems (Kiumarsi et al. 2017; Modares et al. 2015). Particularly, adding probing noise into the behavior policy does not produce bias of solution when implementing policy evaluation. Therefore, the advantages of off-policy RL also motivate us to attempt off-policy learning approaches for optimal control of rougher flotation processes. How to design off-policy RL algorithms and use them to OOC problem of rougher flotation processes with completely unknown dynamics of controlled unit processes and unknown functional dependence of economic operational index generation, as well as the different time scales of operational dynamics and unit processes is our goal in this chapter.

Meanwhile, facing fierce market competition and rising raw material and labor costs, the ultimate goal of seeking economic benefits and environmental security in large-scale industries requires new plant-wide performance optimization strategies to be developed for entire industrial processes. The performance of the plant-wide industrial process is usually characterized by production efficiency, product quality, yield, energy and resource usage and production cost, and these are called production indices for the entire production line (Ding et al. 2016).

The plant-wide performance optimization problem is more complicated because a) industrial processes normally consist of multiple unit processes and each unit process has its own operation for achieving different purposes, such as ensuring unit quality of products, unit process yield and as well as minimizing unit energy consumption, etc.; b) there exists a strong coupling relationship among multiple production indices and operational indices of multiple unit processes. As a result, it is vital to select the proper target operational indices for each unit process, so that the optimum of plant-wide production index performance can be realized for large-scale industrial processes (Ding et al. 2016; Yuan et al. 2016). In addition, the nonlinear property of the dynamics of production indices and constraints on the operational indices add difficulties in solving plant-wide performance optimization.

Considerable results on plant-wide performance optimization have been reported in recent decades and this issue has been recognized as a main objective in large-scale industrial processes (Ding et al. 2016; Yuan et al. 2016; Huang et al. 2005; Wongsri and Siraworakun 2010; Ochoa et al. 2010; Oliveira et al. 2013; Xu et al. 2012). Traditionally, operators usually use their on-site experience to coordinate all of the operational indices for optimizing the plant-wide performance (Ding et al. 2016). Plant-wide performance optimization is indeed a multi-objective optimization problem that has been clearly established in Ding et al. (2016), Yuan et al. (2016), Huang et al. (2005), Wongsri and Siraworakun (2010). However, solving this multi-objective optimization problem from the perspective of compacting all variables into an augmented variable exhibits drawbacks in terms of solution quality, computational complexity and time-consuming due to high dimension and coupling relationship among variables (for example, Tennessee Eastman process has 50 state variables, 41 measurements and 12 manipulated variables (Wongsri and Siraworakun 2010)) together with nonlinear dynamics of unit processes, even though some efforts have been made in Huang et al. (2005), Wongsri and Siraworakun (2010), Ochoa et al. (2010), Oliveira et al. (2013) and Xu et al. (2012). Therefore, we want to find a general strategy to simplify the plant-wide performance optimization problem to easily find the solutions for large-scale plants in this chapter.

Decomposing the large-scale problem into several more easily solvable subproblems may overcome the shortcomings of the above methods, and the improved efficiency using the decomposing coordination methods has been shown in Ding et al. (2016), Xu et al. (2012) and Oliveira et al. (2013). The Lagrangean decomposition (LD) approach (Oliveira et al. 2013), multilevel coordinate search (MCS), the bilevel intersection coordination (IC) heuristic (Xu et al. 2012), as well as recently reported coupling relationship-based approach (Yuan et al. 2016) and integrated case-based reasoning (CBR) with the RL approach (Ding et al. 2016) have been utilized to decompose the plant-wide optimization problem into several equivalent subproblems. However, the challenges still emerge in the plant-wide performance optimization in dealing with unknown dynamic environment (unknown dynamics of unit processes), such as how to convert a complicated optimization problem into several easier solvable subproblems and how to find the optimal solution using only data. Therefore, solving the plant-wide performance optimization under unknown dynamic environment is the further aim of this chapter.

It is well known that the evolutionary algorithm, as one of supervisors, is a better way to solve multi-objective optimization problem with constraints using only data under unknown environment, while it is an off-line method that is not appropriate for industrial process control (Zhang and Tao 2016; Li et al. 2018). In real industrial processes, the relationship between changing of production indices and operational indices is time-varying and all production indices are coupled via operational indices. If conditions change, new data must be collected and all learning must be repeated. In contrast, RL is an online method that adapts to change processes by observing data measurements available in real time (Li et al. 2017; Lewis et al. 2012b; Jiang and Jiang 2012; Luo et al. 2016; Wang et al. 2017; Liu et al. 2017; Lv et al. 2019; Li et al. 2019). Moreover, the nonzero-sum multi-agent game can decompose large-scale optimization problems into simpler more easily solvable local optimality subproblems for achieving a global equilibrium of a group of interacting agents (Başar and Olsder 1998; Lewis et al. 2012a; Vamvoudakis and Lewis 2011; Johnson et al. 2014; Nash 1951; Jiang and He 2019; Qin et al. 2019). Therefore, we are driven to find a novel RL nonzero-sum graphical game method to solve the plant-wide performance optimization of industrial processes.

RL also has been used to solve multi-agent or multi-player games online in real time by using only data without requiring dynamics of systems to be known (Johnson et al. 2014; Jiang and He 2019; Qin et al. 2019; Lv and Ren 2019; Wei et al. 2018; Zhang et al. 2017, 2019; Zuo et al. 2017; Gao et al. 2019; Jiao et al. 2016; Modares et al. 2018). By carefully reviewing this literature, one can find that the results on multi-player game where all players have access to the states of augmented systems have to be achieved using RL methods for linear systems, affine nonlinear systems or general nonlinear systems with DT or CT unknown dynamics (Johnson et al. 2014; Lv and Ren 2019; Wei et al. 2018). For the multi-agent systems where each agent has own individual dynamics and is coupled with its neighbors, the reported model-free RL-based game methods are just for isomorphic linear systems (Li et al. 2017; Jiang and He 2019; Qin et al. 2019), heterogeneous linear systems (Zhang et al. 2017, 2019; Zuo et al. 2017; Gao et al. 2019; Jiao et al. 2016). However, it is still an open issue how to solve optimal control problem for general nonlinear discrete-time multi-agent systems with unknown dynamics when simultaneously considering inter-coupling characteristics, heterogeneous structure of agents, digital sampling and inherent nonlinear nature of the systems. In addition, this problem will become more difficult to be solved if control inputs of agents are constrained.

8.2 Optimal Operational Control for Industrial Process

The off-policy RL is going to be used to find the optimal set-points for the unit processes for nonlinear industrial operational processes in this section, such that the operational indices track their desired values. This is a method of data-driven optimization, where no knowledge of lower-layer control loop dynamics and operational index dynamics are required. To this end, a performance function is defined for the OOC problem in terms of the operational index tracking errors and the set-point variation effort. Using a method of linearizing industrial processes dynamics and operational index dynamics by using Taylor Series Expansion, the OOC problem is transformed to an H_∞ tracking control problem. It is shown that one could solve a DT game algebraic Riccati equation (GARE) to minimize the performance functions. The solution to the GARE equation results in the operational indices tracking their desired values while reaching Nash equilibrium. The off-policy RL algorithm is used to approximate the solutions to the DT GARE equation and learn the optimal set-points for the unit processes without any knowledge of the system dynamics. Finally, the operation control case is given to verify the effectiveness of the proposed method.

8.2.1 Problem Statement

The two-time-scale operational index control problem is first defined in terms of dynamical models for the unit processes and nonlinear functions for the operational indices. Then, the OOC problem is presented by defining a performance index to minimize the operational index tracking error by proper selection of the set-points for the unit processes.

Figure 8.1 shows the two-time-scale hierarchical architecture of the operational index optimal control problem. The hierarchical architecture of operational index control consists of two loops: (1) Lower-layer control of unit processes on a fast time scale, and (2) Upper-layer optimal control of operational indices that are measured on a slow time scale. In Fig. 8.1, the operational indices r evaluate numerically how well the unit processes are performing.

A general unit industrial process is described by

$$\begin{aligned} x(n+1) &= f(x(n), u(n)) \\ y(n) &= p(x(n)). \end{aligned} \tag{8.1}$$

This unit industrial process dynamics may describe industrial ovens, grinders and so on. Since industrial processes usually steadily operate around their operation points

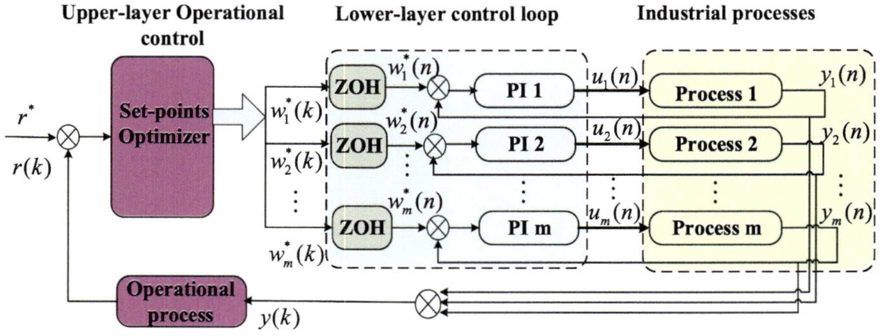

Fig. 8.1 Two-layer architecture for industrial operational control processes

in real operations, so they can be approximated as linear systems with bounded disturbance near the steady states (Chai et al. 2012; Liu et al. 2013; Rojas and Cipriano 2011). Thus, the nonlinear functions $f(x(n), u(n))$ and $p(x(n))$ are assumed to be second-order continuously differentiable such that they can be linearized by using Taylor Series Expansion.

The lower-layer control loop provides the fundamental working conditions for the industrial process in a fast time scale. The objective of the lower-layer loop control is to ensure the outputs of unit industrial process track the outputs of an upper-layer operational control loop called the set-points. To this end, a feedback PI control input is given as

$$u(n) = K_p e(n) + K_i E(n) \tag{8.2}$$

with

$$E(n+1) = E(n) - y(n) + w(n). \tag{8.3}$$

The main physical and chemical reactions in industrial process take place in the unit processes. Periodically, the performance of these unit processes can be measured in terms of the so-called operational indices which detail how well the unit processes are performing. These operational indices must follow values prescribed by industry production requirements. The operational index control loop determines set-points for the unit processes such that the operational indices follow their prescribed values. This operational control loop usually works on a slow time scale due to the fact that the reactions in the operational process always take more time compared to the basic feedback control in the device loops. Since the relationships between operational indices and the outputs of the lower-layer control loops are complicated and usually show nonlinearity, the following nonlinear function is employed to show the dynamics of the operational indices:

$$r(k+1) = g(r(k), y(k)). \tag{8.4}$$

The dynamics of operational indices expresses linear characteristics when system (8.1) steadily operating near the stable states. Thus, the nonlinear function $g(r(k), y(k))$ is naturally assumed to be a second-order continuously differentiable function to linearize it by using Taylor Series Expansion.

Remark 8.1 As shown in (8.1) and (8.4), two time scales are employed in the industrial operational processes. In the lower-layer control loop, sampled data at the time instant n is measured while the state of operation is updated at the time instant $k = N_0 n$ in the upper-layer operational control loop. It is clear that the upper-layer operational control is a slow process compared to the lower-layer control loop with fast sampling.

In the industrial processes, it is desired to design the optimal set-points to ensure the operational index is kept at a desired value or within a desired range. This can be achieved by solving the following optimization problem:

Problem 8.1 OOC problem

$$\min_{w(k)} \sum_{k=0}^{\infty} \beta^k \left[(r(k) - r^*)^T Q(r(k) - r^*) + w^T(k) R w(k)\right]. \tag{8.5}$$

Subject to

$$\begin{aligned} x(n+1) &= f(x(n), u(n)) \\ y(n) &= p(x(n)) \\ r(k+1) &= g(r(k), y(k)) \\ u(n) &= K_p e(n) + K_i E(n) \\ E(n+1) &= E(n) - y(n) + w(n) \end{aligned} \tag{8.6}$$

where Q and R are positive semi-definite matrix and positive definite matrix, respectively.

Remark 8.2 Solving Problem 8.1 is difficult due to (1) nonlinear dynamics constraints of the controlled process and the operational index; (2) The different time scales, i.e., fast sampling rate in the lower-layer control loop and slow operational velocity in the upper-layer operational control.

Remark 8.3 Note that it is essential to use the discount factor β in the performance function. This is because the set-point $w(k)$ should be a function of r and r^* with a form of $w(k) = h(r, r^*)$ for achieving the goal of r tracking r^* in Problem 8.1, which could make (8.5) unbounded as the desired operational index r^* usually does not converge to zero.

8.2.2 H_∞ *Tracking Control for Operational Processes*

This subsection presents an H_∞ tracking control formulation for the OOC Problem 8.1 that takes into account the two time-scale sampling. First, Taylor Series Expansion is used to linearize the dynamics of the unit industrial process and operational process. Then, the different time scales adopted in the industrial operational processes are dealt with by the lifting technique which models the dynamics of the lower-layer process in terms of the slow time scale of the upper-layer operational control. Finally, it is shown how to formulate the OOC problem as an H_∞ tracking control problem.

8.2.2.1 Linearization of the Operational Process and Industrial Process

Since the nonlinear functions $f(x(n), u(n))$, $p(x(n))$ and $g(r(k), y(k))$ are assumed to be second-order continuously differentiable, then problem (8.1) can be linearized using Taylor Series Expansion as

$$\begin{aligned} x(n+1) &= Gx(n) + Nu(n) + d_1(n) \\ y(n) &= Cx(n) + d_2(n) \end{aligned} \tag{8.7}$$

where $G = f_x(x_e, u_e)$, $N = f_u(x_e, u_e)$, $C = p'(x_e)$, $d_1(n) = x_e - u_e f_u(x_e, u_e) - x_e f_x(x_e, u_e) + R_1(n)$ and $d_2(n) = y_e - x_e p'(x_e) + R_2(n)$. $R_1(n) = \frac{1}{2}((x(n) - x_e)\frac{\partial}{\partial x} + ((u(n) - u_e)\frac{\partial}{\partial u})^2 f(\hbar, \bar{\lambda})$ and $R_2(n) = \frac{1}{2}p''(\varsigma)(x(n) - x_e)^2$ are the bounded residual errors, $\hbar = x_e + \theta_1(x(n) - x_e)$, $\bar{\lambda} = u_e + \theta_2(u(n) - u_e)$, $\varsigma = x_e + \theta_3(x(n) - x_e)$, and $0 \leq \theta_i \leq 1$ $(i = 1, 2, 3)$.

Similarly, (8.4) is linearized as

$$r(k+1) = Lr(k) + My(k) + \varpi(k) \tag{8.8}$$

where $L = g_r(r_e, y_e)$, $M = g_y(r_e, y_e)$ and $\varpi(k) = r_e - y_e g_y(r_e, y_e) - r_e g_r(r_e, y_e) + \rho(k)$. $\rho(k) = \frac{1}{2}((r(k) - r_e)\frac{\partial}{\partial r} + ((y(k) - y_e)\frac{\partial}{\partial y})^2 g(\kappa, \upsilon)$. κ, υ are real numbers between r_e. and $r(k)$, y_e and $y(k)$, respectively.

An augmented system composed of the linearized unit industrial process dynamics (8.7) and the tracking error dynamics (8.3) is constructed as

$$\xi(n+1) = \bar{A}\xi(n) + \bar{B}w(n) + \bar{I}\tilde{d}(n) \tag{8.9}$$

where

$$\xi(n) = \left[x^T(n)\ E^T(n)\right]^T, \ \bar{A} = \begin{bmatrix} G - NK_pC & NK_i \\ -C & I \end{bmatrix}, \ \bar{B} = \begin{bmatrix} NK_p \\ I \end{bmatrix}$$

$$\bar{I} = \begin{bmatrix} I & -NK_p \\ 0 & -I \end{bmatrix}, \ \tilde{d}(n) = \begin{bmatrix} d_1(n) \\ d_2(n) \end{bmatrix}.$$

Equation (8.8) can be rewritten as

$$r(k+1) = Lr(k) + \bar{M}\xi(k) + \varpi(k) + \hat{M}\tilde{d}(k) \tag{8.10}$$

where $\bar{M} = \left[MC\ 0\right]$, $\hat{M} = \left[0\ M\right]$.

Remark 8.4 Since practical nonlinear industrial operational processes are usually running at steady state, then their nonlinear dynamics and the dynamics of operational indices can be linearized near the steady operating points (Chai et al. 2012; Liu et al. 2013; Rojas and Cipriano 2011). Thus, the focused OOC problem in this Sect. 8.2 is in fact a local optimization problem.

It is well known that robust differential dynamic programming (DDP) also can solve Problem 8.1 to find the locally optimal policies for nonlinear systems with disturbance and model error by using second-order approximation (Mayne and David 1970). But this method is valid suppose that the dynamics of industrial operational processes are known a priori. The target here is to present a data-drive learning algorithm to solve OOC problem without the knowledge of industrial operational processes.

8.2.2.2 Lifting Technique for Two-Time-Scale OOC Problem

In the lower-layer control loop, sample data at the time instant n is measured while the state of operation is updated at the time instant $k = N_0 n$ in the upper-layer operational control loop. Then the following form holds:

$$w(k) = w(nN_0) = w(nN_0 + 1) = \cdots = w(nN_0 + N_0 + 1) \tag{8.11}$$

and (8.9) in terms of slow time scale is rewritten as

$$\begin{aligned}\xi((k+1)) &= \xi((n+1)N_0) = \xi(nN_0 + N_0)\\ &= \bar{A}^{N_0}\xi(N_0 n) + \sum_{i=0}^{N_0-1}\left(\bar{A}^i\bar{B}w(N_0 n) + \bar{A}^i\bar{I}\tilde{d}(N_0 n_0 - (N_0 - 1 - i))\right)\\ &= \tilde{A}\xi(k) + \tilde{B}w(k) + D\vartheta(k)\end{aligned} \tag{8.12}$$

where

$$D = \left[\bar{A}^0\bar{I}\ \bar{A}^1\bar{I}\ \ldots\ \bar{A}^{N_0-1}\bar{I}\right]$$
$$\vartheta = \left[\tilde{d}^T(nN_0 + (N_0-1))\ \tilde{d}^T(nN_0 + (N_0-2))\ \ldots\ \tilde{d}^T(nN_0)\right]^T.$$

8.2.2.3 Expression of OOC Problem as H_∞ Tracking Control Problem

All constraints in Problem 8.1 and the desired value of the operational index are compacted in an augmented form as

$$X(k+1) = \tilde{G}X(k) + \tilde{N}w(k) + \tilde{H}\chi(k) \tag{8.13}$$

where $X(k) = \left[\xi^T(k)\ r^T(k)\ (r^*(k))^T\right]^T$ and

$$\tilde{G} = \begin{bmatrix}\tilde{A} & 0 & 0\\ \hat{M} & L & 0\\ 0 & 0 & I\end{bmatrix},\ \tilde{N} = \begin{bmatrix}\tilde{B}\\ 0\\ 0\end{bmatrix},\ \tilde{H} = \begin{bmatrix}D & 0\\ \hat{D} & I\\ 0 & 0\end{bmatrix},\ \hat{D} = \left[0\ 0\ \ldots\ \hat{M}\right],\ \chi(k) = \begin{bmatrix}\vartheta(k)\\ \varpi(k)\end{bmatrix}.$$

Remark 8.5 The desired operational index is constant. That is, $r^*(k+1) = r^*(k)$.

Problem 8.1 in terms of the augmented form of (8.13) can be rewritten as the following Problem 8.2.

Problem 8.2 Control objective:

$$\min_{w(k)} \sum_{k=0}^{\infty} \beta^k \left(X^T(k)\tilde{Q}X(k) + w^T(k)Rw(k) \right). \tag{8.14}$$

Subject to (8.13)

where $\tilde{Q} = \tilde{I}^T Q \tilde{I}$, $\tilde{I} = [0\ I\ \ -I]$. The tracking error is $e_r(k) = r(k) - r^*(k) = \tilde{I}X(k)$.

Remark 8.6 Note that the residual error $\chi(k)$ depends on the state $X(k)$ of the augmented system (8.13). In order to find the solution to Problem 8.2, $\chi(k)$ should be considered as the disturbance corresponding to the state $X(k)$. Then, the OOC problem shown in Problem 8.2 can be formulated as an H_∞ tracking control problem.

Definition 8.1 The H_∞ tracking control problem is to find a set-point $w(k) = w(X_k)$ depending on X_k, such that

- Augmented system (8.13) satisfies the disturbance attenuation condition for a specific attenuation factor $\gamma > 0$:

$$\sum_{k=0}^{\infty} \beta^k \left(X^T(k)\tilde{Q}X(k) + w^T(k)Rw(k) \right) \leq \gamma^2 \sum_{k=0}^{\infty} \beta^k \|\chi(k)\|^2. \tag{8.15}$$

- The tracking error $e_r(k)$ converges to zero.

8.2.3 Solving the H_∞ Tracking Control Problem

To find the solution to Problem 8.2 as a zero-sum differential game problem, a new performance index is defined in terms of (8.15)

$$J(w(k), \chi(k)) = \sum_{k=0}^{\infty} \beta^k \left(X^T(k)\tilde{Q}X(k) + w^T(k)Rw(k) - \gamma^2 \|\chi(k)\|^2 \right). \tag{8.16}$$

The H_∞ control problem can be expressed as a two-player zero-sum differential game in which the control policy player seeks to minimize the value function, while the disturbance policy player desires to maximize it. The goal is to find feedback saddle point such that

$$\begin{aligned} J(w^*(k), \chi^*(k)) &= \min_{w(k)} \max_{\chi(k)} J(w(k), \chi(k)) \\ &= \min_{w(k)} \max_{\chi(k)} \sum_{k=0}^{\infty} \beta^k \left(X^T(k)\tilde{Q}X(k) + w^T(k)Rw(k) - \gamma^2 \|\chi(k)\|^2 \right). \end{aligned} \tag{8.17}$$

For an admissible control policy $w(k)$ and a disturbance policy $\chi(k)$, the value function is defined as

$$\begin{aligned} V(X(k)) &= J(w(k), \chi(k)) \\ &= \sum_{i=k}^{\infty} \beta^{i-k} \left(X^T(i)\tilde{Q}X(i) + w^T(i)Rw(i) - \gamma^2 \|\chi(i)\|^2 \right). \end{aligned} \tag{8.18}$$

Thus, the following Bellman equation can be obtained:

$$V(X(k)) = X^T(k)\tilde{Q}X(k) + w^T(k)Rw(k) - \gamma^2\|\chi(k)\|^2 + \beta V(X(k+1)). \tag{8.19}$$

It is proved that the value function for linear systems is quadratic in terms of the states of the augmented system (Kiumarsi et al. 2017). That is, $V(X(k)) = X^T(k)PX(k)$. Then, one has the Hamiltonian as

$$\begin{aligned} H(X(k), w(k), \chi(k)) &= X^T(k)\tilde{Q}\, X(k) + w^T(k)Rw(k) \\ &- \gamma^2\|\chi(k)\|^2 + \beta X^T(k+1)PX(k+1) - X^T(k)PX(k). \end{aligned} \tag{8.20}$$

Applying the stationarity conditions $H_w = 0$ and $H_\chi = 0$ give the optimal set-point and the worst case disturbance as

$$\begin{aligned} w^*(k) &= -K_1^*X(k) \\ &= -\left((R + \beta\tilde{N}^T P^*\tilde{N}) - \beta^2\tilde{N}^T P^*\tilde{N}(-\gamma^2 I + \beta\tilde{H}^T P^*\tilde{H})^{-1}\tilde{H}^T P^*\tilde{N}\right)^{-1} \\ &\times \left(\beta\tilde{N}^T P^*\tilde{G} - \beta^2\tilde{N}^T P^*\tilde{H}(-\gamma^2 I + \beta\tilde{H}^T P^*\tilde{H})^{-1}\tilde{H}^T P^*\tilde{G}\right) X(k) \end{aligned} \tag{8.21}$$

$$\begin{aligned} \chi^*(k) &= -K_2^*X(k) \\ &= -\left((-\gamma^2 I + \beta\tilde{H}^T P^*\tilde{H}) - \beta^2\tilde{H}^T P^*\tilde{N}(R + \beta\tilde{N}^T P^*\tilde{N})^{-1}\tilde{N}^T P^*\tilde{H}\right)^{-1} \\ &\times \left(\beta\tilde{H}^T P^*\tilde{G} - \beta^2\tilde{H}^T P^*\tilde{N}(R + \beta\tilde{N}^T P^*\tilde{N})^{-1}\tilde{N}^T P^*\tilde{G}\right) X(k) \end{aligned} \tag{8.22}$$

and P^* satisfies the following GARE.

$$\begin{aligned} &\tilde{Q} - P + (K_1^*)^T RK_1^* - \gamma^2(K_2^*)^T K_2^* \\ &\quad + \beta(\tilde{G} - \tilde{N}K_1^* - \tilde{H}K_2^*)^T P(\tilde{G} - \tilde{N}K_1^* - \tilde{H}K_2^*) = 0. \end{aligned} \tag{8.23}$$

Remark 8.7 Note that finding a set-point that satisfies attenuation condition (8.15) for the H_∞ tracking control problem is equivalent to maximizing and minimizing the discounted performance function (8.16) subject to augmented system (8.13) (Modares et al. 2015; Luo and Wu 2013).

Remark 8.8 Note that the disturbance $\chi(k)$ is in fact the residual error generated by $\tilde{G}X(k) + \tilde{N}w(k)$ estimating the nonlinear dynamics of $X(k+1)$ as shown in (8.13) and it depends on the state $X(k)$ of augmented systems (8.13) and cannot be chosen arbitrarily.

Remark 8.9 Since the direct solution of the GARE is difficult, iterative algorithms are used to solve it. PI as a class of RL algorithms (Kiumarsi et al. 2017; Modares et al. 2015; Wei et al. 2017; Narayanan and Jagannathan 2017), Algorithm 8.1, is used to obtain the optimal control input and disturbance offline and online.

Algorithm 8.1 Model-based offline RL algorithm

1: Initialize the system with stabilizing control gains K_1^0 and K_2^0 and set $j = 0$. Given $0 \le \beta \le 1$ and $\gamma > 0$;

2: Evaluate policies by solving P^{j+1}:

$$
\begin{aligned}
P^{j+1} &= \tilde{Q} + (K_1^j)^T R K_1 - \gamma^2 (K_2^j)^T K_2^j \\
&\quad + \beta\left(\tilde{G} - \tilde{N}K_1^j - \tilde{H}K_2^j\right)^T P^{j+1}\left(\tilde{G} - \tilde{N}K_1^j - \tilde{H}K_2^j\right);
\end{aligned} \tag{8.24}
$$

3: Find improved set-point and disturbance policy gains:

$$
\begin{aligned}
K_1^{j+1} &= -\left((R + \beta\tilde{N}^T P^{j+1}\tilde{N}) - \beta^2\tilde{N}^T P^{j+1}\tilde{N}(-\gamma^2 I + \beta\tilde{H}^T P^{j+1}\tilde{H})^{-1}\tilde{H}^T P^{j+1}\tilde{N}\right)^{-1} \\
&\quad \times \left(\beta\tilde{N}^T P^{j+1}\tilde{G} - \beta^2\tilde{N}^T P^{j+1}\tilde{H}(-\gamma^2 I + \beta\tilde{H}^T P^{j+1}\tilde{H})^{-1}\tilde{H}^T P^{j+1}\tilde{G}\right)
\end{aligned} \tag{8.25}
$$

$$
\begin{aligned}
K_2^{j+1} &= -\left((-\gamma^2 I + \beta\tilde{H}^T P^{j+1}\tilde{H}) - \beta^2\tilde{H}^T P^{j+1}\tilde{N}(R + \beta\tilde{N}^T P^{j+1}\tilde{N})^{-1}\tilde{N}^T P^{j+1}\tilde{H}\right)^{-1} \\
&\quad \times \left(\beta\tilde{H}^T P^{j+1}\tilde{G} - \beta^2\tilde{H}^T P^{j+1}\tilde{N}(R + \beta\tilde{N}^T P^{j+1}\tilde{N})^{-1}\tilde{N}^T P^{j+1}\tilde{G}\right);
\end{aligned} \tag{8.26}
$$

4: Stop when $\left\| P^{j+1} - P^j \right\| \le \epsilon$ with a small constant ϵ $(\epsilon > 0)$.

In order to have a unique feedback saddle point in the class of strictly feedback stabilizing policies, the inequalities in (8.27) and (8.28) should be satisfied (Başar and Bernhard 2008).

$$
-\gamma^2 I + \beta\tilde{H}^T P\tilde{H} < 0 \tag{8.27}
$$

$$
R + \beta\tilde{N}^T P\tilde{N} > 0. \tag{8.28}
$$

Remark 8.10 The convergence of Algorithm 8.1 has been proved in Luo and Wu (2013) and Al-Tamimi et al. (2007a). That means that the sequence P^j obtained by iterative (8.24)–(8.26) can converge to the solution of the GARE equation (8.23).

Note that Algorithm 8.1 requires the knowledge of the whole industrial operational process dynamics including the upper-layer operational control loop and the lower-layer control loop. To obviate this requirement, a model-free approach is presented based on Algorithm 8.1 next.

8.2.4 Off-Policy Reinforcement Learning Algorithm

An off-policy RL algorithm is used to find the optimal set-point using only measured data shown in Fig. 8.2.

The augmented system (8.13) can be written as

$$X(k+1) = \tilde{G}_c X(k) + \tilde{N}(K_1^j X(k) + w(k)) + \tilde{H} K_2^j \chi(k) \tag{8.29}$$

where $w^j(k) = -K_1^j X(k)$ and $\chi^j(k) = -K_2^j X(k)$ are the target policies which are under evaluation and $\tilde{G}_c = \tilde{G} - \tilde{N} K_1^j - \tilde{H} K_2^j$. Using Taylor Series Expansion for the value function $V^{j+1}(X(k)) = X^T(k) P^{j+1} X(k)$ yields

$$\begin{aligned} & V^{j+1}(X(k)) - \beta V^{j+1}(X(k+1)) \\ & \quad = X^T(k) P^{j+1} X(k) - \beta X^T(k) \tilde{G}_c^T P^{j+1} \tilde{G}_c X(k) - \beta X^T(k) \tilde{G}_c^T P^{j+1} X(k+1) \\ & \quad + \beta X^T(k) \tilde{G}_c^T P^{j+1} \tilde{G}_c X(k) - \beta \left(K_1^j X(k) + w(k) \right)^T \tilde{N}^T P^{j+1} X(k+1) \\ & \quad - \beta \left(K_2^j X(k) + \chi(k) \right)^T \tilde{H}^T P^{j+1} X(k+1). \end{aligned} \tag{8.30}$$

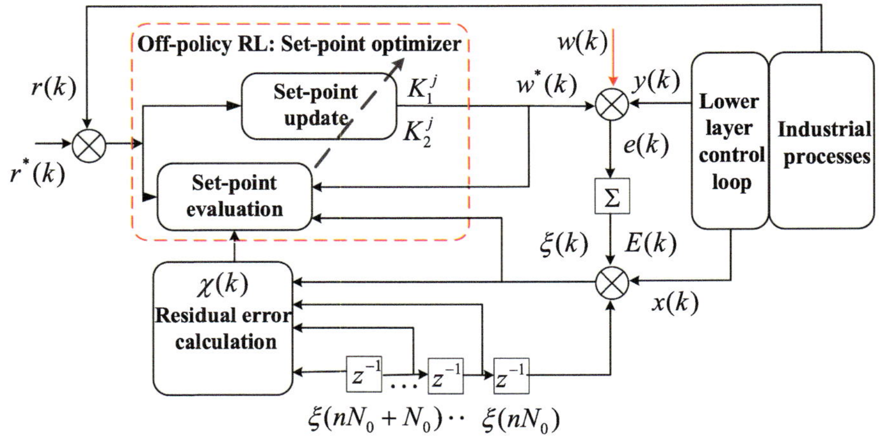

Fig. 8.2 Off-policy RL scheme by using zero-sum games

By considering Bellman equation (8.19), one has

$$X^T(k)P^{j+1}X(k) - \beta X^T(k)\tilde{G}_c^T P^{j+1}\tilde{G}_c X(k) = \Pi_k^j \tag{8.31}$$

where

$$\Pi_k^j = X^T(k)\tilde{Q}X(k) + X^T(k)(K_1^j)^T R K_1^j X(k) - \gamma^2 X^T(k)(K_2^j)^T K_2^j X(k).$$

Substituting (8.31) in (8.30) yields

$$\begin{aligned}
&V^{j+1}(X(k)) - \beta V^{j+1}(X(k+1)) \\
&\quad = \Pi_k^j - \beta X^T(k)\tilde{G}_c^T P^{j+1}X(k+1) + \beta X^T(k)\tilde{G}_c^T P^{j+1}\tilde{G}_c X(k) \\
&\quad + \beta X^T(k)\tilde{G}_c^T P^{j+1}\tilde{G}_c X(k) - \beta\left(K_1^j X(k) + w(k)\right)^T \tilde{N}^T P^{j+1}X(k+1) \\
&\quad - \beta\left(K_2^j X(k) + \chi(k)\right)^T \tilde{H}^T P^{j+1}X(k+1).
\end{aligned} \tag{8.32}$$

Remark 8.11 From (8.32), one can easily see that the residual error $\chi(k)$ is required to be known and measured when solving the value function V^{j+1} and calculating the set-point and disturbance policy gains K_1^j and K_2^j by using data-driven approach. But $\chi(k)$ is in fact the residual error of the systems generated by linearizing the operational processes and cannot be collected a priori. Hence Proposition 8.1 is proposed to replace $\chi(k)$ with the measured data.

Proposition 8.1 *The residual error $\chi(k)$ can be expressed as*

$$\chi(k) = S_1\eta(k) + S_2 w(k) \tag{8.33}$$

with

$$\begin{aligned}
&\eta(k) = \left[\xi^T(nN_0)\ \xi^T(nN_0+1)\ \dots\ \xi^T(nN_0+N_0)\ r^T(k)\ r^T(k+1)\right]^T \\
&S_1 = \begin{bmatrix} 0 & 0 & \dots & -W_{\bar{I}}\bar{A} & W_{\bar{I}} & 0 & 0 \\ 0 & 0 & \dots & W_{\bar{I}} & 0 & 0 & 0 \\ \vdots & \vdots & \iddots & \vdots & \vdots & \vdots & \vdots \\ -W_{\bar{I}}\bar{A} & W_{\bar{I}} & \cdots & 0 & 0 & 0 & 0 \\ \hat{M}\bar{A} - \bar{M} & -\hat{M}W_{\bar{I}} & \dots & 0 & 0 & -L & I \end{bmatrix},\ S_2 = \begin{bmatrix} -W_{\bar{I}}\bar{B} \\ -W_{\bar{I}}\bar{B} \\ \vdots \\ -W_{\bar{I}}\bar{B} \\ \hat{M}\bar{B} \end{bmatrix}.
\end{aligned} \tag{8.34}$$

Proof Since the matrix $\bar{I}^T\bar{I}$ is invertible based on the definition of $\bar{I}$, then (8.9) can be transformed as

$$\tilde{d}(n) = W_{\bar{I}}(\xi(n+1) - \bar{A}\xi(n) - \bar{B}w(n)) \tag{8.35}$$

where $W_{\bar{I}} = (\bar{I}^T\bar{I})^{-1}\bar{I}^T$. According to (8.35), thus the following expression holds

$$\begin{aligned}
\tilde{d}(nN_0) &= W_{\bar{I}}\left(\xi(nN_0+1) - \bar{A}\xi(nN_0) - \bar{B}w(nN_0)\right) \\
\tilde{d}(nN_0+1) &= W_{\bar{I}}\left(\xi(nN_0+2) - \bar{A}\xi(nN_0+1) - \bar{B}w(nN_0+1)\right) \\
&\vdots \\
\tilde{d}(nN_0+N_0-1) &= W_{\bar{I}}\left(\xi(nN_0+N_0) - \bar{A}\xi(nN_0+N_0-1) - \bar{B}w(nN_0+N_0-1)\right).
\end{aligned} \tag{8.36}$$

By (8.10), one has

$$\gamma(k) = r(k+1) - Lr(k) - \bar{M}\xi(k) - \hat{M}\tilde{d}(k). \tag{8.37}$$

Combining with (8.35) yields the following form:

$$\begin{aligned}
\gamma(k) = r(k+1) &- Lr(k) + (\hat{M}\bar{A} - \bar{M})\xi(nN_0) \\
&- \hat{M}W_{\bar{I}}\xi(nN_0+1) + \hat{M}\bar{B}w(k).
\end{aligned} \tag{8.38}$$

Integrating (8.36) and (8.38) yields (8.33). This completes the proof. □

Remark 8.12 Note that computing S_1 and S_2 requires knowledge of $\bar{A}$ and $\bar{B}$ in (8.34), which are unknown. In the following, it is shown how to identify S_1 and S_2 using observed data.

By using (8.32) and following the procedure in Kiumarsi et al. (2017), one gets the following off-policy Bellman equation:

$$\begin{aligned}
&\left((X^T(k)\otimes X^T(k)) - \beta(X^T(k+1)\otimes X^T(k+1))\right)\mathrm{vec}(L_1^{j+1}) \\
&+\beta\left(X^T(k)\otimes X^T(k+1)\right)\mathrm{vec}(L_2^{j+1}) + \beta\left(w^T(k)\otimes X^T(k+1)\right)\mathrm{vec}(L_3^{j+1}) \\
&-\beta\left(X^T(k)\otimes X^T(k)\right)\mathrm{vec}(L_4^{j+1}) + 2\beta\left(X^T(k)\otimes (K_1^j X(k))^T\right)\mathrm{vec}(L_5^{j+1}) \\
&+2\beta\left(X^T(k)\otimes (K_2^j X(k))^T\right)\mathrm{vec}(L_6^{j+1}) - \beta\left((K_1^j X(k))^T\otimes (K_1^j X(k))^T\right)\mathrm{vec}(L_7^{j+1}) \\
&-2\beta\left((K_1^j X(k))^T\otimes (K_2^j X(k))^T\right)\mathrm{vec}(L_8^{j+1}) \\
&-\beta\left((K_2^j X(k))^T\otimes (K_2^j X(k))^T\right)\mathrm{vec}(L_9^{j+1}) \\
&+\beta\left(\eta^T(k)\otimes X^T(k+1)\right)\mathrm{vec}(L_{10}^{j+1}) + \beta\left(w^T(k)\otimes X^T(k+1)\right)\mathrm{vec}(L_{11}^{j+1}) \\
&= \Pi_k^j
\end{aligned} \tag{8.39}$$

where

$$L_1^{j+1} = P^{j+1},\ L_2^{j+1} = \tilde{G}^T P^{j+1},\ L_3^{j+1} = \tilde{N}^T P^{j+1},\ L_4^{j+1} = \tilde{G}^T P^{j+1}\tilde{G}$$
$$L_5^{j+1} = \tilde{G}^T P^{j+1}\tilde{N},\ L_6^{j+1} = \tilde{G}^T P^{j+1}\tilde{H},\ L_7^{j+1} = \tilde{N}^T P^{j+1}\tilde{N},\ L_8^{j+1} = \tilde{N}^T P^{j+1}\tilde{H}$$
$$L_9^{j+1} = \tilde{H}^T P^{j+1}\tilde{H},\ L_{10}^{j+1} = S_1^T \tilde{H}^T P^{j+1},\ L_{11}^{j+1} = S_2^T \tilde{H}^T P^{j+1}.$$

Further, (8.39) can be expressed as

$$\bar{H}^j(k)L^{j+1} = \rho^j \tag{8.40}$$

where

$$L^{j+1} = \left[(\text{vec}(L_1^{j+1}))^T\ (\text{vec}(L_2^{j+1}))^T \ \ldots\ (\text{vec}(L_{11}^{j+1}))^T\right]^T$$
$$\rho^j = \Pi_k^j,\ \bar{H}^j(k) = \left[\bar{H}_1^j\ \bar{H}_2^j\ \ldots\ \bar{H}_{11}^j\right]$$

and

$$\bar{H}_1^j = \left(X^T(k) \otimes X^T(k)\right) - \beta\left(X^T(k+1) \otimes X^T(k+1)\right)$$
$$\bar{H}_2^j = \beta\left(X^T(k) \otimes X^T(k+1)\right),\ \bar{H}_3^j = \beta\left(w^T(k) \otimes X^T(k+1)\right)$$
$$\bar{H}_4^j = -\beta\left(X^T(k) \otimes X^T(k)\right),\ \bar{H}_5^j = 2\beta\left(X^T(k) \otimes (K_1^j X(k))^T\right)$$
$$\bar{H}_6^j = 2\beta\left(X^T(k) \otimes (K_2^j X(k))^T\right),\ \bar{H}_7^j = -\beta\left((K_1^j X(k))^T \otimes (K_1^j X(k))^T\right)$$
$$\bar{H}_8^j = -2\beta\left((K_1^j X(k))^T \otimes (K_2^j X(k))^T\right),\ \bar{H}_9^j = -\beta\left((K_2^j X(k))^T \otimes (K_2^j X(k))^T\right)$$
$$\bar{H}_{10}^j = \beta\left(\eta^T(k) \otimes X^T(k+1)\right),\ \bar{H}_{11}^j = \beta\left(w^T(k) \otimes X^T(k+1)\right).$$

The Bellman equation (8.40) can be solved using the data from data sets $\bar{H}_i^j$ and ρ^j based on the following recursive least squares (RLS) method:

$$\begin{aligned} L^{j+1}(k+1) &= L^{j+1}(k) + K(k)(\rho^j(k) - \bar{H}^j(k)L^{j+1}(k)) \\ P(k) &= (I - K(k)\bar{H}^j(k))P(k-1) \\ K(k) &= \frac{P(k-1)\left(\bar{H}^j(k)\right)^T}{1+\bar{H}^j(k)P(k-1)\left(\bar{H}^j(k)\right)^T}. \end{aligned} \tag{8.41}$$

The unknown parameters in L^{j+1} can be found by using at least μ data set which is equal to the number of unknown parameters of the off-policy Bellman equation. That is,

$$\begin{aligned} \mu &= (1 + 2m + 4q + 4s)(m + q + 2s) + \frac{q}{2}(1 + q) \\ &+ \left(m + 2q + 2s + \frac{1 + N_0(m + q) + s}{2}\right)(N_0(m + q) + s). \end{aligned} \tag{8.42}$$

Using the solution L_1^{j+1} to L_{11}^{j+1} calculated from RLS (8.41), K_1^{j+1} and K_2^{j+1} can be obtained as

$$
\begin{aligned}
K_1^{j+1} &= \left((R + \beta L_7^{j+1}) - \beta^2 L_8^{j+1}(-\gamma^2 I + \beta L_9^{j+1})^{-1} (L_8^{j+1})^T\right)^{-1} \\
&\quad \times \left(\beta (L_5^{j+1})^T - \beta^2 L_8^{j+1}(-\gamma^2 I + \beta L_9^{j+1})^{-1} (L_6^{j+1})^T\right) \qquad (8.43)\\
K_2^{j+1} &= \left(-\gamma^2 I + \beta L_9^{j+1} - \beta^2 (L_8^{j+1})^T (R + \beta L_7^{j+1})^{-1} L_8^{j+1}\right)^{-1} \\
&\quad \left(\beta (L_6^{j+1})^T - \beta^2 (L_8^{j+1})^T (R + \beta L_7^{j+1})^{-1} (L_5^{j+1})^T\right). \qquad (8.44)
\end{aligned}
$$

Algorithm 8.2 Off-policy RL for optimal operational control

1: Data collection: Collecting real system data $(x(n), E(n), r(k))$ from the controlled process and operational process for the sample sets H_i^j and ρ^j using a behavior set-point $w(k)$;
2: Initialization: Start with the initial stabilizing gains K_1^0 and K_2^0 and set $j = 0$;
3: Implementing RL: By RLS (8.41), L_i^{j+1} $(i = 1, 2, \ldots 11)$ can be found using the collected data in Step 1 to update K_1^j and K_2^j in terms of (8.43) and (8.44);
4: Let $j = j + 1$. If $\left\| K_1^{j+1} - K_2^j \right\| \leq l_1$ and $\left\| K_2^{j+1} - K_2^j \right\| \leq l_2$ (l_1 and l_2 are small positive numbers), then stop the iteration and the optimal set-point has been obtained. Otherwise, go back to Step 3 and continue.

Theorem 8.1 *The set-point* $w^j(k) = -K_1^j X(k)$ *and the worst disturbance* $\chi^j(k) = -K_2^j X(k)$ *found by Algorithm 8.2 converge to the Nash equilibrium solution* $w^*(k)$ *and* $\chi^*(k)$ *as* $j \to \infty$, *i.e.,* $\lim_{j\to\infty} w^j(k) = w^*(k)$ *and* $\lim_{j\to\infty} \chi^j(k) = \chi^*(k)$.

Proof If system (8.13) is persistently excited ensuring

$$
\varepsilon_0 I \leq \frac{1}{\mu} \sum_{k=1}^{\mu} \left(H^j(k)\right)^T H^j(k) \leq \varepsilon_1 I
$$

where ε_0 and ε_1 are positive integers with $\varepsilon_0 \leq \varepsilon_1$, then L^j in RLS (8.41) can converge to the real value, thus the $w^j(k) = -K_1^j X(k)$ and $\chi^j(k) = -K_2^j X(k)$ will converge to the optimal set-point $w^*(k)$ and the worst disturbance $\chi^*(k)$ since the solution of Algorithms 8.1 and 8.2 is equivalent (Kiumarsi et al. 2017) and the convergence of solution of Algorithm 8.1 has been proved in Luo and Wu (2013) and Al-Tamimi et al. (2007a). This completes the proof. □

Remark 8.13 In off-policy RL, two different types of policies are used: the behavior set-point which is used to generate data for learning and the target set-point and target disturbance policy which are evaluated and updated. This is in contrast to on-policy RL (Al-Tamimi et al. 2007a; Mu et al. 2016) that requires the learning data to be generated by the same set-point and the same disturbance policy as the ones under evaluation. This not only greatly increases the information exploration ability, but also avoids the set-points and the disturbance to be adjusted during the learning process, which is more acceptable in real industrial operations since the disturbance is in fact the residual error generated by linearizing nonlinear industrial operational processes.

Remark 8.14 Inspired by Kiumarsi et al. (2017), a data-driven off-policy RL method is developed to learn the optimal set-point for the industrial operation processes in Algorithm 8.2. This algorithm is implemented without requiring the knowledge of dynamics of the upper-layer operational control and lower-layer control loop, which is different from model-based OOC for industrial operational processes (Vega et al. 2014; Würth et al. 2011; Engell 2007; Dünnebier et al. 2005; Ellis and Christofides 2014; Chai et al. 2012; Liu et al. 2013). In the data collection step, a specific behavior set-point is applied to the industrial operational process and it is tracked by the output of lower-layer control loop, then the data $(x(n), E(n), r(k))$ can be collected (the state $x(n)$ and the operational indices $r(k)$ are required to be measurable) and the residual error $\chi(k)$ does not need to be calculated in terms of (8.33). In the off-policy RL learning, the set-point $w^j(k) = -K_1^j X(k)$ and the disturbance policy $\chi^j(k) = -K_2^j X(k)$ are evaluated and updated using the measured data generated by the specific behavior set-point, while they are not applied to the industrial operational processes.

Remark 8.15 Practical industrial processes are usually steadily operated near the equilibrium points under the control inputs obtained by operators based on their rich experience. Thus, the initial stabilizing gains K_1^0 and K_2^0 can be obtained easily. Besides, H_∞ control method can be used to yield initial stabilizing control policies for systems (Modares and Lewis 2014).

Remark 8.16 The parameters of PI controller in the lower-layer control loop are turned by using Refined Ziegler–Nichols methods. This method is extensively applied in industrial operational processes (Chai et al. 2011, 2014; Dai et al. 2014; Pomerleau et al. 2000).

8.2.5 Simulation Results

Here, the proposed off-policy RL algorithm is applied to the rougher flotation operational process composed of only one cell.

Example 8.1 Under some natural assumptions (e.g., constant airflow, two mineralogical classes, perfect mixing and bilateral material transfer), the following single-cell flotation model according to mass balance in froth and pulp phases respectively is established, which has been fully investigated in Chai et al. (2012), Liu et al. (2013) and Rojas and Cipriano (2011):

$$\frac{dM_p^i}{dt} = -\left(k_p^i + \frac{q_T}{\left(1-\varepsilon_g\right)Ah_p}\right)M_p^i + k_e^i M_e^i + q_a X_a^i$$
$$\frac{dM_e^i}{dt} = -\left(k_e^i + \frac{q_c}{\left(1-\varepsilon_g\right)A\left(H-h_p\right)}\right)M_e^i + k_p^i M_p^i. \tag{8.45}$$

The concentrate grade is

$$L_{cg} = \frac{M_e^1 g_{cp}^1 + M_e^2 g_{cp}^2}{M_e^1 + M_e^2} L_{cu}. \tag{8.46}$$

The tail grade is

$$L_{tg} = \frac{M_p^1 g_{cp}^1 + M_p^2 g_{cp}^2}{M_p^1 + M_p^2} L_{cu} \tag{8.47}$$

where M_e^i and M_p^i are the foam quality and the slurry quality, respectively. q_a and h_p are respectively the feed flow rate ($m^3/$ min) and the pulp liquid level height. $i = 1, 2$ represent the mineral species 1 (mainly chalcopyrite) and the mineral species 2 (mainly gangue). The cell size is $53.2 \times 3.2m^3$ and values of other parameters are listed in Table 8.1.

In this simulation, we choose $x = \left[M_p^1\ M_p^2\ M_e^1\ M_e^2\right]^T$, $u = \left[h_p\ q_a\right]$, $y = \left[L_{cg}\ L_{tg}\right]^T$. $w = \left[L_{cg}\ L_{tg}\right]^T$ denotes the set-points of concentrate grade and tail grade. For the real rougher flotation operational process, it is desired for economic benefit to track an ideal value on a slow time scale by forcing the output of lower-layer device loop to the set-points on a fast time scale. Actually, the economic benefit depends on not only $y = \left[L_{cg}\ L_{tg}\right]^T$, but also control consumption, thus the economic benefit has the form of $r_k = M_r y_k + N_r u_k$, where $M_r = [100\ \ -20]$, $N_r = [0.01\ 0.1]$. Under the operational condition listed in Table 8.1, the economic objective is set to be 20 referring to Chai et al. (2012) and Liu et al. (2013).

The external disturbance signal and measurement noise are incorporated and assumed as $0.05e^{-0.001t}\sin(100v(t)t)$ to simulate the real experimental scenario ($v(t) = [v_1(t)\ v_2(t)\ v_3(t)\ v_4(t)]^T$ is a vector with $v_i(t) \in [0\ 1]$, $(i = 1, 2, 3, 4)$. Under the operational conditions shown in Table 8.1, an equilibrium point for the flotation process of chalcopyrite is

$$\left[h_p\ q_a\right] = \left[2.8\ 17\right] \tag{8.48}$$
$$\left[M_p^1\ M_p^2\ M_e^1\ M_e^2\right] = \left[16.8\ 829.48\ 4.56\ 0.105\right]. \tag{8.49}$$

Linearizing (8.45)–(8.47) at the equilibrium point yields the following system:

Table 8.1 Parameters and values

Parameter	Physical meaning	Value
k_p^i	Flotation rate	$k_p^1 = 17.9\,\text{min}^{-1}$ $k_p^2 = 0.04\,\text{min}^{-1}$
k_e^i	Drainage rate	$k_e^1 = 63.6\,\text{min}^{-1}$ $k_e^2 = 316\,\text{min}^{-1}$
q_T	Week	$9.3\text{m}^3/\text{min}$
ε_g	Stagnation constant	0
A	Cross-sectional area	53.2m^2
X_a^i	Mineral species concentration	$X_a^1 = 0.1549$ $X_2^a = \frac{g_{cp}^1 - g_a}{g_a - g_{cp}^2} X_a^1 = 3.0484$
g_{cp}^1	Brass grade in brass pulp	0.417
g_{cp}^2	The grade of brass in gangue pulp	0.0034
L_{cu}	Chalcopyrite brass grade	0.412
q_c	Concentrate pulp flow	$7.392\text{m}^3/\text{min}$
H	Total height	3.2m
g_a	Feed mineral grade	0.0234

$$\dot{x} = \begin{bmatrix} -17.9624 & 0 & 65.6 & 0 \\ 0 & -0.1024 & 0 & 316 \\ 17.9 & 0 & -65.9474 & 0 \\ 0 & 0.04 & 0 & -316.247 \end{bmatrix} x + \begin{bmatrix} 0.3746 & 0.1549 \\ 18.3790 & 3.0484 \\ -3.96 & 0 \\ -0.0903 & 0 \end{bmatrix} u$$

$$y = \begin{bmatrix} 0 & 0 & 0.0832 & -3.6501 \\ 0 & 0 & 0.0203 & -0.0004 \end{bmatrix} x. \tag{8.50}$$

First, Algorithm 8.2 is implemented to learn the optimal set-point gain. The sampling interval of the lower-layer device loop is chosen as 1 min, and the updating period of the upper-layer operational control loop is 30 min. The initial gains K_1^0 and K_2^0 can be chosen randomly but they must guarantee the operational process to be stable. K_1^0 is given in (8.51), and K_2^0 is not shown here due to its high dimension.

$$K_1^0 = \begin{bmatrix} -0.1331 & 0.0027 & 0.4590 & -20.1262 & -0.2396 & 0.2671 & 0 & -0.0080 \\ -0.1187 & 0.0024 & 0.4095 & -17.9528 & -0.2136 & 0.2383 & 0 & -0.0071 \end{bmatrix}. \tag{8.51}$$

Algorithm 8.2 is implemented and Fig. 8.3 demonstrates the norms of L^j, K_1^j and K_2^j at each iteration. One can easily see that Algorithm 8.2 achieves the convergence at the 14th iteration, thus the optimal set-point gain can be obtained as shown in (8.52).

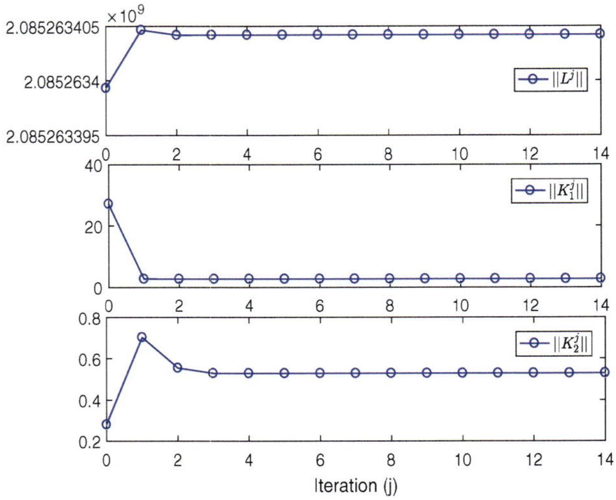

Fig. 8.3 Convergence results of unknown parameters

$$K_1^* = \begin{bmatrix} 0.0100 & -0.0002 & -0.0344 & 1.5104 & 0.0179 & -0.0201 & 0 & 0.0006 \\ -0.0134 & 0.0003 & 0.0463 & -2.0295 & -0.0241 & 0.0270 & 0 & -0.0008 \end{bmatrix}. \tag{8.52}$$

Then, the obtained optimal set-point $w_k^* = -K_1^* X_k$ is as the reference input of the lower-layer control loop, and it is followed by the output of the lower-layer control loop under the designed PI controller. Figures 8.4 and 8.5 respectively show the tracking results of the set-points and the economic operational index for three cases, i.e., the updating periods of the upper-layer operational control loop are (a): 5 min, (b): 30 min and (c): 100 min. From Figs. 8.4 and 8.5, one can find that it takes longer time to follow the desired economic benefit for the upper-layer operational control loop with the slow updating period than that with the fast updating period. The fast updating rate of upper-layer operational control loop also increases computational load. Proper updating time scales are indeed chosen according to the requirements of practical industrial operations.

To confirm the contributions of the proposed off-policy RL algorithm, the comparisons with the fixed set-point method and the intelligent control method (Chai et al. 2011, 2014) are made under the same initial operational scenarios. The tracking results of set-points and economic benefit are shown in Fig. 8.6. As shown in Fig. 8.6, the better tracking results are obtained by using the proposed off-policy RL algorithm, since the operators' experience-based fixed set-points are not the optimal set-points. The intelligent control method corrects the prescribed set-points according to the errors between the economic benefit and the desired value based on operators' experience. Then, it is unclear whether it is the desired economic benefit

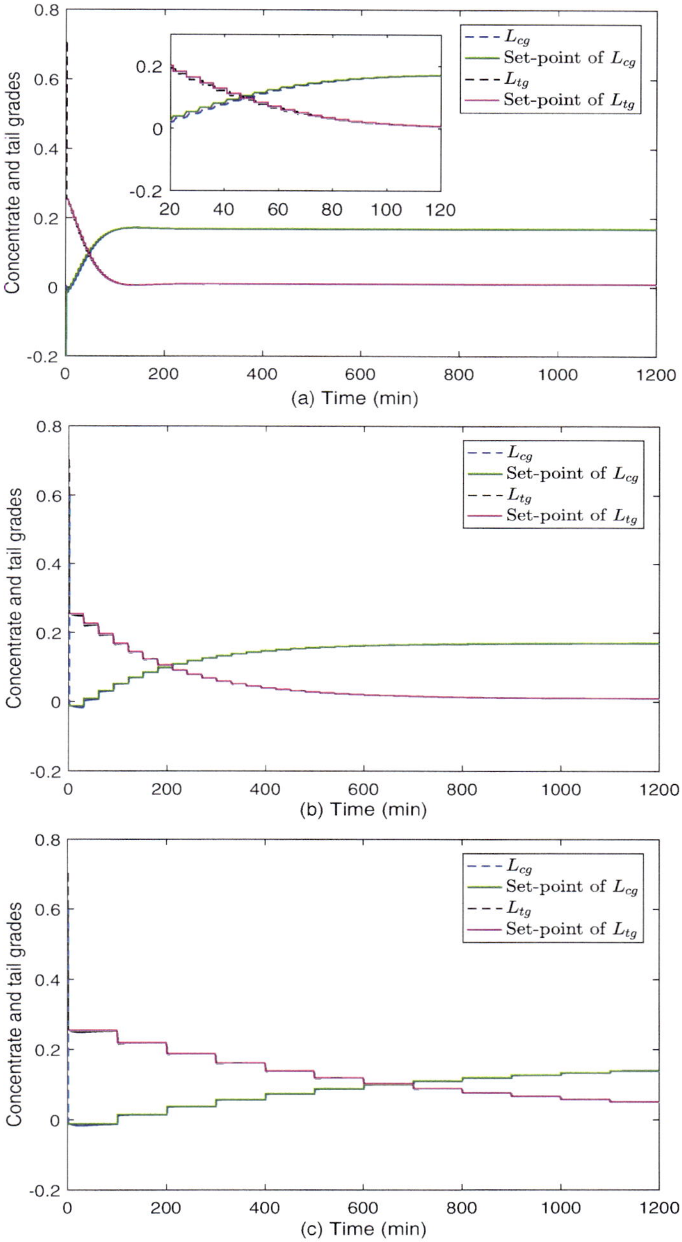

Fig. 8.4 Tracking performance of the optimal set-points **a** Case a; **b** Case b; **c** Case c

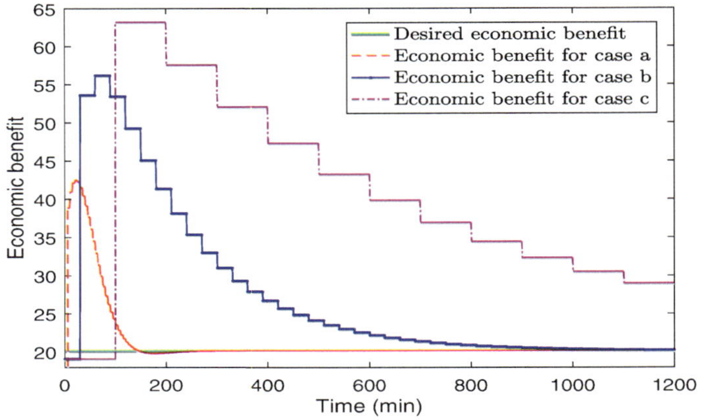

Fig. 8.5 Tracking performance of the economic benefit

or the desired set-point which is chosen based on the operators' experience. From Fig. 8.6, one can see that the desired economic benefit cannot be tracked well when using the intelligent control method even though this method can converge faster than the proposed off-policy RL method.

8.3 Optimal Set-Point Design for Rougher Flotation Processes with Multiple Cells

Rougher flotation, composed of multiple cells operating at a fast time scale and economic performance measurements known as operational indices measured at a slower time scale, is very basic and the first concentration stage for flotation plants. In this section, Q-learning is used to solve optimal control problems with multiple subsystems. Thus, an off-policy Q-learning algorithm is presented here to learn optimal set-points for rougher flotation processes with multiple cells using only measured data, such that economic benefit tracks the desired value by forcing the concentrate grades and tail grades of the lower-layer control loops to the corresponding set-points. To this end, first, the OOC for dual-rate rougher flotation processes with multiple cells is formulated. Second, H_∞ tracking control problem is developed to optimally prescribe the set-points for the rougher flotation processes. Then, a zero-sum game off-policy Q-learning algorithm is proposed to find the optimal set-points by using measured data.

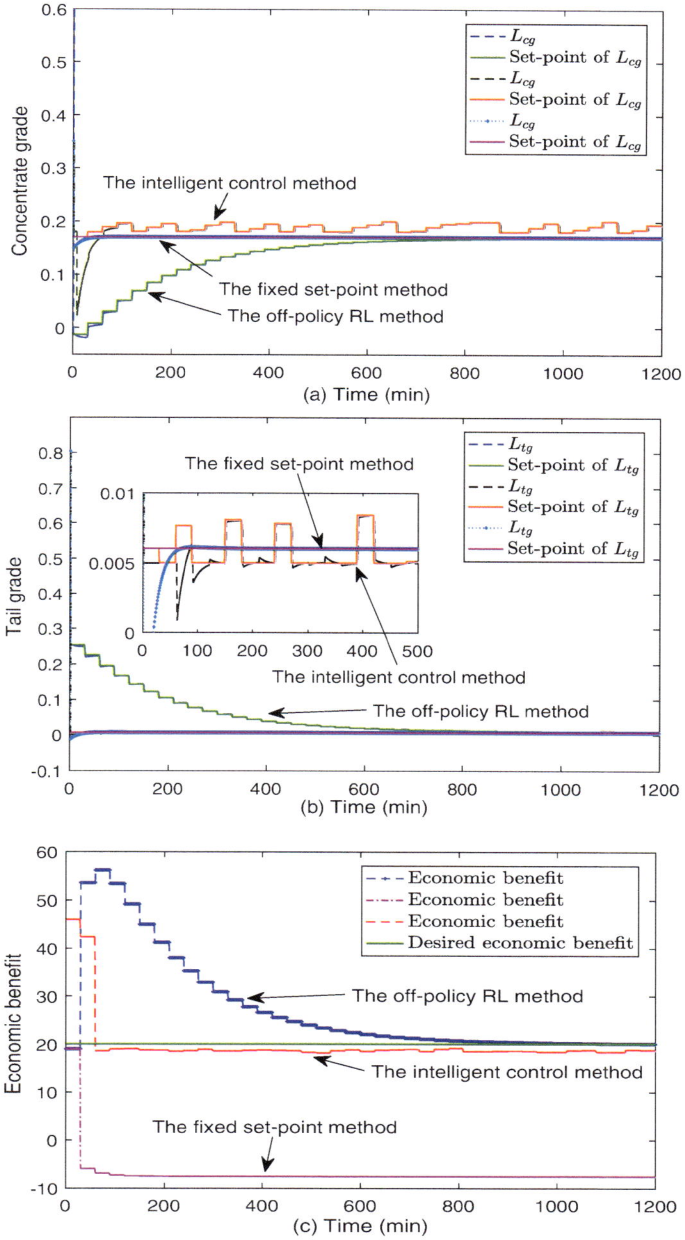

Fig. 8.6 Tracking performances by using the off-policy RL method, the fixed set-point method and the intelligent control method for different updating scales of operational control

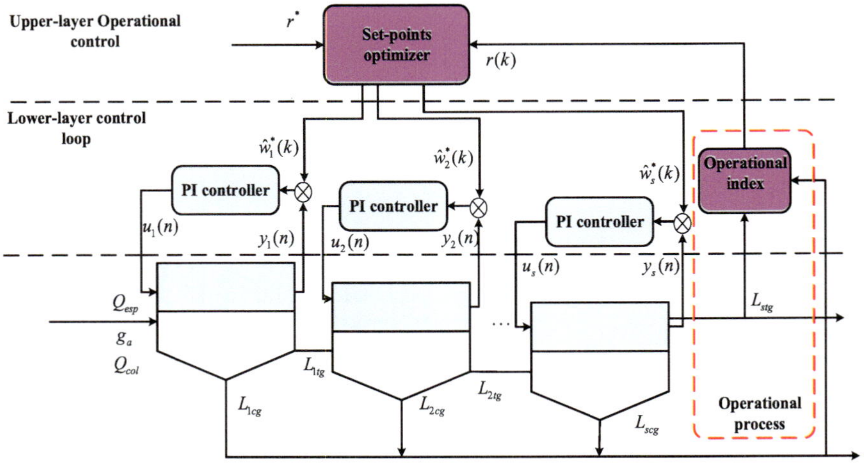

Fig. 8.7 Two-layer architecture for optimal operational control of rougher flotation processes

8.3.1 Problem Statement

The OOC problem of dual-rate rougher flotation process is formulated, along with lower-layer rougher flotation cells with fast sampling and upper-layer economic operational control with slow updating as shown in Fig. 8.7.

A rougher flotation process consists of several cells, whose nonlinear dynamics can be expressed as

$$\begin{aligned} \dot{x}_l(t) &= f_l(x_l(t), u_l(t)) \\ y_l(t) &= p_l(x_l(t)) \end{aligned} \tag{8.53}$$

where $f_l(x_l(t), u_l(t))$ and $p_l(x_l(t))$ are assumed to be second-order continuously differentiable.

The objective of the lower-layer loop control is to ensure the control output to steadily track the set-points input by the operational control layer. To this end, a digital output feedback PI controller is employed as follows:

$$u_l(n) = K_{lP} e_l(n) + K_{lI} E_l(n) \tag{8.54}$$

where $E_l(n) = \sum_{i=1}^{n-1} e_l(i)$ is the summation of the tracking errors, and its dynamic can be expressed in the following form:

$$E_l(n+1) = E_l(n) - y_l(n) + w_l(n). \tag{8.55}$$

Since economic benefit is closely related to the concentrate grade, the tail grade and the control input of rougher flotation process, and their relationship usually shows nonlinear feature, then the following nonlinear function is employed to show the dynamics of the economic benefit:

$$r(k+1) = g(r(k), L_{lcg}(k), L_{stg}(k), u(k)) \tag{8.56}$$

where the nonlinear function $g(r(k), L_{lcg}(k), L_{stg}(k), u(k))$ is usually second-order continuously differentiable.

In order to render the economic benefit to a desired value by the optimal approach, the goal here is to design optimal set-points for achieving the optimal control of the economic benefit by minimizing the following performance index:

$$J = \sum_{k=0}^{\infty} \beta^k ((r(k) - r^*)^T Q(r(k) - r^*) + w^{\mathrm{T}}(k) R w(k)) \tag{8.57}$$

where Q and R are positive semi-definite matrix and positive definite matrix, respectively.

Remark 8.17 Naturally, the discount factor β is introduced into (8.57) since the set-point $w(k)$ usually depends on the desired operational index r^* and $r^* \neq 0$.

Problem 8.3 is given to clearly formulate the OOC problem for rougher flotation circuits.

Problem 8.3 Control objective:

$$\min_{w_k} \sum_{k=0}^{\infty} \beta^k ((r(k) - r^*)^{\mathrm{T}} Q(r(k) - r^*) + w^{\mathrm{T}}(k) R w(k)) \tag{8.58}$$

subject to (8.53)–(8.56)

Remark 8.18 It is quite hard to solve Problem 8.3 due to (1) nonlinear dynamics constraints of the rougher flotation cells and the economic benefit index. (2) The dual rates, i.e., fast sampling rate in the lower-layer control loops and slow operational velocity in the upper-layer economic benefit operational control.

8.3.2 H_∞ Tracking Control for Rougher Flotation Processes

To solve Problem 8.3, the rougher flotation process operating at steady state is first linearized using the Taylor Series Expansion; then, the lifting technique (Chai et al. 2012; Liu et al. 2013; Fan et al. 2016) is employed for dealing with the dual-rate sampling, and further Problem 8.3 is transformed to H_∞ tracking control problem for finding the optimal solution.

8.3.2.1 Linearization of Rougher Flotation Process

Since the dynamics of rougher flotation process are approximately linear near steady-state points (Chai et al. 2012; Liu et al. 2013; Fan et al. 2016), then (8.53) and (8.56) are rewritten using Taylor Series Expansion as

$$\begin{aligned} \dot{x}_l(t) &= G_l x_l(t) + N_l u_l(t) + d_{l1}(t) \\ y_l(t) &= C_l x_l(t) + d_{l2}(t) \end{aligned} \tag{8.59}$$

$$r(k+1) = Lr(k) + M\hat{y}(k) + Su(k) + \gamma(k) \tag{8.60}$$

where

$$\begin{aligned} \hat{y}(k) &= \left[L_{1cg}(k)\ L_{2cg}(k)\ \ldots\ L_{scg}(k)\ L_{stg}(k)\right] \\ G_l &= \tfrac{\partial f_l(x_{le},u_{le})}{\partial x},\ N_l = \tfrac{\partial f_l(x_{le},u_{le})}{\partial u},\ C_l = p_l'(x_{le}) \\ d_{l1}(t) &= x_{le} - u_{le}\tfrac{\partial f_l(x_{le},u_{le})}{\partial u} - x_{le}\tfrac{\partial f_l(x_{le},u_{le})}{\partial x} + R_{l1}(t) \\ d_{l2}(t) &= y_{le} - x_{le}\tfrac{\partial p_l(x_{le})}{\partial x} + R_{l2}(t) \\ L &= \tfrac{\partial g(r_e,\hat{y}_e,u_e)}{\partial r},\ M = \tfrac{\partial g(r_e,\hat{y}_e,u_e)}{\partial \hat{y}},\ S = \tfrac{\partial g(r_e,\hat{y}_e,u_e)}{\partial u} \\ u_e &= [u_{1e}^T\ u_{2e}^{\mathrm{T}}\ \cdots\ u_{se}^{\mathrm{T}}]^{\mathrm{T}} \\ \hbar_l &= x_{le}+\theta_{l1}(x_l(t) - x_{le}). \end{aligned}$$

$R_{l1}(t)$ and $R_{l2}(t)$ are the bounded residual errors, and

$$\begin{aligned} R_{l1}(t) &= \tfrac{1}{2}\left((x_l(t) - x_{le})\tfrac{\partial}{\partial x} + (u_l(t) - u_{le})\tfrac{\partial}{\partial u}\right)^2 f_l(\hbar_l, \bar{\lambda}_l) \\ R_{l2}(t) &= \tfrac{1}{2}p_l''(\varsigma_l)(x_l(t) - x_{le})^2 \\ u_e &= \left[u_{1e}^T\ u_{2e}^T\ \ldots\ u_{se}^T\right]^T,\ \hbar = x_{le} + \theta_{l1}(x_l(t) - x_{le}) \\ \bar{\lambda}_l &= u_{le} + \theta_{l2}(u_l(t) - u_{le}),\ \varsigma_l = x_{le} + \theta_{l3}(x_l(t) - x_{le})\ (0 \le \theta_{li} \le 1,\ i = 1, 2, 3) \\ \gamma(k) &= r_e - y_e M - r_e L + \rho(k) - u_e S \\ \rho(k) &= \tfrac{1}{2}\left((r(k) - r_e)\tfrac{\partial}{\partial r} + (\hat{y}(k) - \hat{y}_e)\tfrac{\partial}{\partial \hat{y}} + (u(k) - u_e)\tfrac{\partial}{\partial u}\right)^2 g(\kappa, \upsilon, \tau). \end{aligned}$$

κ, υ, τ are real numbers between r_e and $r(k)$, $\hat{y}_e$ and $\hat{y}_k$, u_e and $u(k)$, respectively.

Remark 8.19 Since systems (8.53) and (8.56) are linearized in terms of Taylor Series Expansion due to their steady operation at steady state, the solution of the focused OOC problem in Sect. 8.1 is essentially suboptimal.

8.3.2.2 Lifting Technique for Dual-Rate Rougher Flotation Processes

Since digital controller (8.54) is employed, then system (8.59) is in fact the following discrete-time system:

$$\begin{aligned} x_l(n+1) &= \hat{G}_l x_l(n) + \hat{N}_l u_l(n) + \hat{d}_{l1}(n) \\ y_l(n) &= C_l x_l(n) + d_{l2}(n) \end{aligned} \tag{8.61}$$

where $\hat{G}_l = e^{G_l T_0}$, $\hat{N}_l = \int_0^{T_0} e^{G_l \tau} d\tau N_l$, $\hat{d}_{l1}(n) = \int_0^{T_0} e^{G_l \tau} d_{l1}(\tau) d\tau$. T_0 is the sampling period of the rougher flotation cells.

An augmented system is constructed by defining a compact form $\xi_l(n) = \left[x_l^T(n)\ E_l^T(n)\right]^T$ as

$$\xi_l(n+1) = \bar{A}_l \xi_l(n) + \bar{B}_l w_l(n) + \bar{I}_l \tilde{d}_l(n) \tag{8.62}$$

where

$$\bar{A}_l = \begin{bmatrix} \hat{G}_l - \hat{N}_l K_P C & \hat{N}_l K_I \\ -C_l & I \end{bmatrix},\ \bar{B}_l = \begin{bmatrix} \hat{N}_l K_P \\ I \end{bmatrix},\ \bar{I}_l = \begin{bmatrix} I & -\hat{N}_l K_P \\ 0 & -I \end{bmatrix},\ \tilde{d}_l(n) = \begin{bmatrix} \hat{d}_{l1}(n) \\ d_{l2}(n) \end{bmatrix}.$$

Further, defining $\xi(n) = \left[\xi_1^T(n)\ \xi_2^T(n)\ \ldots\ \xi_s^T(n)\right]^T$ yields the following compact form for the rougher flotation process:

$$\xi(n+1) = \bar{A}\xi(n) + \bar{B}w(n) + \bar{I}\tilde{d}(n) \tag{8.63}$$

where

$$w(n) = \left[w_1^T(n)\ w_2^T(n)\ \ldots\ w_s^T(n)\right]^T,\ \tilde{d}(n) = \left[\tilde{d}_1^T(n)\ \tilde{d}_2^T(n)\ \ldots\ \tilde{d}_s^T(n)\right]^T$$
$$\bar{A} = \mathrm{diag}(\bar{A}_1, \bar{A}_2, \ldots, \bar{A}_s),\ \bar{B} = \mathrm{diag}(\bar{B}_1, \bar{B}_2, \ldots, \bar{B}_s),\ \bar{I} = \mathrm{diag}(\bar{I}_1, \bar{I}_2, \ldots, \bar{I}_s).$$

Equation (8.60) can be rewritten as

$$\begin{aligned} r(k+1) = Lr(k) + (\bar{M} - \bar{S} + S\bar{K}_I)\xi(k) + S\bar{K}_p w(k) \\ + \gamma(k) + (\hat{M} - \hat{S})\tilde{d}(k) \end{aligned} \tag{8.64}$$

where

$$\begin{gathered} \bar{M} = M\hat{I}\bar{C},\ \bar{C} = \mathrm{diag}(\bar{C}_1, \bar{C}_2, \ldots, \bar{C}_s),\ \bar{C}_j = [C_j\ 0] \\ \hat{M} = M\hat{I}\bar{\Pi},\ \bar{\Pi} = \mathrm{diag}(\Pi, \Pi, \ldots, \Pi),\ \Pi = [0\ I] \\ \bar{S} = S\bar{K}_p\bar{C},\ \bar{K}_P = \mathrm{diag}(K_{1P}, K_{2P}, \ldots, K_{sP}),\ \hat{S} = S\bar{K}_p\bar{\Pi} \\ \bar{K}_I = \mathrm{diag}([0\ K_{1I}], [0\ K_{2I}], \ldots, [0\ K_{sI}]),\ \hat{I} = \mathrm{diag}(I,\ 1). \end{gathered}$$

It is well known that data are measured at the time instant n in the lower-layer flotation cell, while the economic benefit value is updated at the time instant $k = N_0 n$ in the upper-layer operational control loop. Then the following form holds:

$$w(k) = w(nN_0) = w(nN_0 + 1) = \cdots = w(nN_0 + N_0 - 1) \tag{8.65}$$

and (8.63) is rewritten in terms of slow time scale as

$$\begin{aligned}\xi(k+1) &= \xi((n+1)N_0) = \xi(nN_0 + N_0)\\ &= \tilde{A}\xi(k) + \tilde{B}w(k) + D\vartheta(k)\end{aligned} \tag{8.66}$$

where

$$\tilde{A} = \bar{A}^{N_0},\ \tilde{B} = \sum_{i=0}^{N_0-1} \bar{A}^i \bar{B},\ D = \left[\bar{A}^0 \bar{I}, \bar{A}^1 \bar{I}, \ldots, \bar{A}^{N_0-1}\bar{I}\right]$$
$$\vartheta(k) = \left[\tilde{d}^T(nN_0 + (N_0 - 1))\ \tilde{d}^T(nN_0 + (N_0 - 2))\ \ldots \tilde{d}^T(nN_0)\right]^T.$$

Let $X(k) = \left[\xi(k)^T\ r(k)^T\ r(k)^*\right]^T$, one has

$$X(k+1) = \tilde{G}X(k) + \tilde{N}w(k) + \tilde{H}\chi(k) \tag{8.67}$$

where

$$\tilde{G} = \begin{bmatrix} \tilde{A} & 0\ 0 \\ \bar{M} - \bar{S} + S\bar{K}_I & L\ 0 \\ 0 & 0\ I \end{bmatrix},\ \tilde{N} = \begin{bmatrix} \tilde{B} \\ S\bar{K}_p \\ 0 \end{bmatrix},\ \tilde{H} = \begin{bmatrix} D & 0 \\ \hat{D} & I \\ 0 & 0 \end{bmatrix}$$
$$\hat{D} = \left[0\ 0 \ldots \hat{M} - \hat{S}\right],\ \chi(k) = \left[\vartheta^T(k)\ \gamma^T(k)\right]^T.$$

Remark 8.20 Here, the desired economic benefit is constant, i.e., $r^*(k+1) = r^*(k)$.

Note that the residual error $\chi(k)$ is dependent on the state $X(k+1)$ of augmented system (8.67) by the definition of $\chi(k)$. For solving Problem 8.3, $\chi(k)$ should be considered as the disturbance corresponding to the state $X(k)$. Thus, the OOC problem shown in Problem 8.3 can be formulated as the H_∞ tracking control problem below.

Problem 8.4 Find a set-point $w(k) = w(X_k)$ satisfying

- the attenuation condition below for a specific attenuation factor $\gamma > 0$:

$$\sum_{k=0}^{\infty} \beta^k (X(k)^T \tilde{Q} X(k) + w^T(k) R w(k)) \leq \gamma^2 \sum_{k=0}^{\infty} \beta^k \|\chi(k)\|^2; \tag{8.68}$$

- the tracking error $e_r(k)$ ($e_r(k) = r(k) - r^*$) converges to zero.

8.3.3 On-Policy Q-Learning Based on Zero-Sum Game

It is shown how to find the solution of Problem 8.4 by using zero-sum game-based Q-learning method. Define a new performance index in terms of (8.68):

$$J(w(k), \chi(k)) = \sum_{k=0}^{\infty} \beta^k \left(X(k)^T \tilde{Q} X(k) + w^T(k) R w(k) - \gamma^2 \|\chi(k)\|^2 \right). \quad (8.69)$$

Actually, solving H_∞ tracking control problem is equivalent to maximizing and minimizing the cost function (8.69) by using zero-sum game approach (Al-Tamimi et al. 2007b; Kiumarsi et al. 2017), that is,

$$\begin{aligned} J(w^*(k), \chi^*(k)) &= \min_{w(k)} \max_{\chi(k)} J(w(k), \chi(k)) \\ &= \min_{w(k)} \max_{\chi(k)} \left(\sum_{k=0}^{\infty} \beta^k (X(k)^T \tilde{Q} X(k) + w^T(k) R w(k) - \gamma^2 \|\chi(k)\|^2) \right). \end{aligned} \quad (8.70)$$

Define the value function as

$$V^*(X(k)) = \min_{w(k)} \max_{\chi(k)} \sum_{k=0}^{\infty} \beta^k (X(k)^T \tilde{Q} X(k) + w^T(k) R w(k) - \gamma^2 \|\chi(k)\|^2). \quad (8.71)$$

Combining with $w^*(k) = -K_1^* X(k)$ and $\chi^*(k) = -K_2^* X(k)$ yields $V^*(X_k) = X^T(k) P X(k)$ $(P > 0)$ by referring to Al-Tamimi et al. (2007b) and Kiumarsi et al. (2014). By (8.71), the action-dependent optimal Q-function is defined below

$$\begin{aligned} Q^*(X(k), w(k), \chi(k)) = X(k)^T \tilde{Q} X(k) + w^T(k) R w(k) - \gamma^2 \chi(k)^T \chi(k) \\ + \beta V^*(X(k+1)) \end{aligned} \quad (8.72)$$

and one has

$$\begin{aligned} V^*(X(k)) &= \min_{w(k)} \max_{\chi(k)} Q^*(X(k), w(k), \chi(k)) \\ &= Q^*(X(k), w^*(k), \chi^*(k)). \end{aligned} \quad (8.73)$$

Thus, Proposition 8.2 is naturally derived by referring to Al-Tamimi et al. (2007b).

Proposition 8.2 *If we set* $z(k) = \left[X^T(k)\ w^T(k)\ \chi^T(k)\right]^T$, $w(k) = -K_1 X(k)$, $\chi(k) = -K_2 X(k)$, *thus the following form holds:*

$$Q^*(X(k), w(k), \chi(k)) = z^T(k) H z(k). \quad (8.74)$$

H is denoted as

$$\begin{bmatrix} H_{XX} & H_{Xw} & H_{X\chi} \\ H_{Xw}^T & H_{ww} & H_{w\chi} \\ H_{X\chi}^T & H_{w\chi}^T & H_{\chi\chi} \end{bmatrix} = \begin{bmatrix} \beta \tilde{G}^T P \tilde{G} + \tilde{Q} & \beta \tilde{G}^T P \tilde{N} & \beta \tilde{G}^T P \tilde{H} \\ \beta (\tilde{G}^T P \tilde{N})^T & \beta \tilde{N}^T P \tilde{N} + R & \beta \tilde{N}^T P \tilde{H} \\ \beta (\tilde{G}^T P \tilde{H})^T & \beta (\tilde{N}^T P \tilde{H})^T & -\gamma^2 I + \beta \tilde{H}^T P \tilde{H} \end{bmatrix} \quad (8.75)$$

and

$$P = (M^*)^T H M^* \tag{8.76}$$

where $M^* = \begin{bmatrix} I & -(K_1^*)^T & -(K_2^*)^T \end{bmatrix}^T$.

By Proposition 8.2, one has the Q-function-based game Bellman equation

$$\begin{aligned} z^T(k)Hz(k) &= X^T(k)\tilde{Q}X(k) + w^T(k)Rw(k) \\ &\quad - \gamma^2\chi^T(k)\chi(k) + \beta z^T(k+1)Hz(k+1). \end{aligned} \tag{8.77}$$

Implementing $\frac{\partial Q^*(X(k),w(k),\chi(k))}{\partial w(k)} = 0$ and $\frac{\partial Q^*(X(k),w(k),\chi(k))}{\partial \chi(k)} = 0$ yields the optimal set-point and the worst disturbance as

$$\begin{aligned} w^*(k) &= -K_1^* X(k) \\ &= -\left(H_{ww} - H_{w\chi}H_{\chi\chi}^{-1}H_{w\chi}^T\right)^{-1}\left(H_{Xw}^T - H_{w\chi}H_{\chi\chi}^{-1}H_{X\chi}^T\right)X(k) \\ \chi^*(k) &= -K_2^* X(k) \\ &= -\left(H_{\chi\chi} - H_{w\chi}^T H_{ww}^{-1}H_{w\chi}\right)^{-1}\left(H_{X\chi}^T - H_{w\chi}^T H_{ww}^{-1}H_{Xw}^T\right)X(k) \end{aligned} \tag{8.78}$$

where the matrix H satisfies (8.77). The PI Algorithm 8.3 is used to learn the optimal set-point and the worst disturbance.

Algorithm 8.3 Model-free Q-learning algorithm

1: Initialization: Given stabilizing set-point and disturbance policy gains K_1^0 and K_2^0, $0 < \beta \leq 1$ and $\gamma > 0$. Let $j = 0$, where j denotes iteration index;

2: Policy evaluation by solving Q-function matrix H^{j+1}:

$$\begin{aligned} z^T(k)H^{j+1}z(k) &= X^T(k)\tilde{Q}X(k) + (w^j)^T(k)Rw^j(k) \\ &\quad - \gamma^2(\chi^j(k))^T\chi^j(k) + \beta z^T(k+1)H^{j+1}z(k+1); \end{aligned} \tag{8.79}$$

3: Policy Update:

$$w^{j+1}(k) = -K_1^{j+1}X_k, \quad \chi^{j+1}(k) = -K_2^{j+1}X_k$$

where

$$K_1^{j+1} = \left(H_{\chi\chi}^{j+1} - (H_{w\chi}^{j+1})^T(H_{ww}^{j+1})^{-1}H_{w\chi}^{j+1}\right)^{-1}\left((H_{X\chi}^{j+1})^T - (H_{w\chi}^{j+1})^T(H_{ww}^{j+1})^{-1}(H_{Xw}^{j+1})^T\right) \tag{8.80}$$

$$K_2^{j+1} = \left(H_{\chi\chi}^{j+1} - (H_{w\chi}^{j+1})^T(H_{ww}^{j+1})^{-1}H_{w\chi}^{j+1}\right)^{-1}\left((H_{X\chi}^{j+1})^T - (H_{w\chi}^{j+1})^T(H_{ww}^{j+1})^{-1}(H_{Xw}^{j+1})^T\right); \tag{8.81}$$

4: Stop when $\left\| H^{j+1} - H^j \right\| \leq \varepsilon$ with a small constant $\varepsilon(\varepsilon > 0)$.

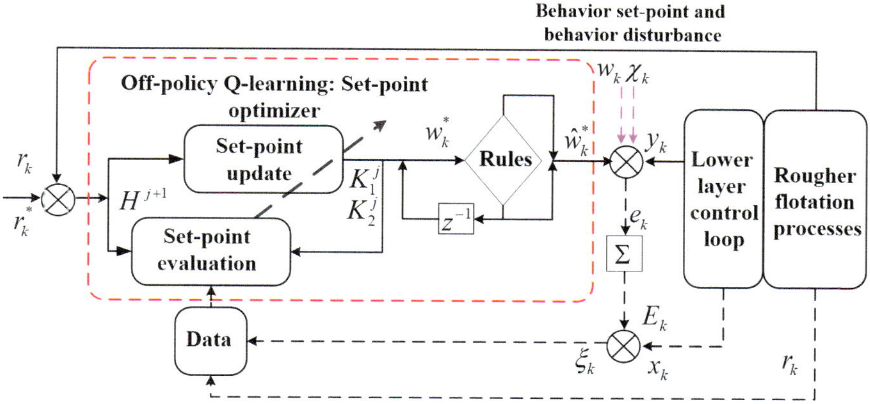

Fig. 8.8 Off-policy Q-learning scheme by using zero-sum games

Remark 8.21 Note that Algorithm 8.3 is in fact an on-policy Q-learning approach wherein the disturbance $\chi(k)$ needs to be updated and act the rougher flotation cells using $\chi_k^{j+1} = -K_2^{j+1} X_k$, while it essentially is the residual error generated by linearizing the nonlinear dynamics and cannot be specified. Moreover, as pointed out in Kiumarsi et al. (2017), adding probing noise to the set-point results in a bias in solving the real value of H^{j+1}. Compared with on-policy Q-learning, off-policy Q-learning is more practical technique for handling optimal control problem as it can overcome the two shortcomings generated by on-policy Q-learning. Hence the sequels will devote to designing off-policy Q-learning algorithm for achieving optimal operation of rougher flotation processes.

8.3.4 *Off-Policy Q-Learning Algorithm*

An off-policy Q-learning algorithm is presented for learning the optimal set-point and the worst disturbance using only data, and then rules used for selecting the optimum set-point are proposed as shown in Fig. 8.8.

Q-function-based Lyapunov equation is given as follows by using (8.79):

$$(M^j)^T H^{j+1} M^j = (M^j)^T \Pi M^j + \beta \tilde{G}_c^T (M^j)^T H^{j+1} M^j \tilde{G}_c \tag{8.82}$$

where $M^j = \begin{bmatrix} I & -(K_1^j)^T & -(K_2^j)^T \end{bmatrix}^T$, $\Pi = \mathrm{diag}(\tilde{Q}, R, -\gamma^2 I)$.

Introducing the auxiliary variables $w^j(k) = -K_1^j X(k)$ and $\chi^j(k) = -K_2^j X(k)$ into augmented system (8.67) yields

$$X(k+1) = \tilde{G}_c X(k) + \tilde{N}(K_1^j X(k) + w(k)) + \tilde{H}(K_2^j X(k) + \chi(k)) \tag{8.83}$$

where $\tilde{G}_c = \tilde{G} - \tilde{N}K_1^j - \tilde{H}K_2^j$. Along the trajectory of (8.83), one has

$$
\begin{aligned}
&(Q^*)^{j+1}(X(k), w^j(k), \chi^j(k)) - \beta X^T(k)\tilde{G}_c^T(M^j)^T H^{j+1}M^j\tilde{G}_c X(k) \\
&\quad = X^T(k)(M^j)^T H^{j+1}M^j X(k) - \beta(X(k+1) - \tilde{N}(K_1^j X(k) + w(k)) \\
&\quad\quad - \tilde{H}(K_2^j X(k) + \chi(k))^T(M^j)^T H^{j+1}M^j(X(k+1) \\
&\quad\quad - \tilde{N}(K_1^j X(k) + w(k)) - \tilde{H}(K_2^j X(k) + \chi(k))) \\
&\quad = X^T(k)(M^j)^T \Pi M^j X(k).
\end{aligned}
\tag{8.84}
$$

Since P^{j+1} and H^{j+1} have the same relationship as that shown in (8.76), then the off-policy Q-function game Bellman equation can be obtained.

$$
\begin{aligned}
&X^T(k)(M^j)^T H^{j+1}M^j X(k) - \beta X^T(k+1)(M^j)^T H^{j+1}M^j X(k+1) \\
&\quad + 2\beta X^T(k+1)P^{j+1}\tilde{N}(K_1^j X(k) + w(k)) \\
&\quad + 2\beta X^T(k+1)P^{j+1}\tilde{H}(K_2^j X(k) + \chi(k)) \\
&\quad - \beta(K_1^j X(k) + w(k))^T \tilde{N}^T P^{j+1}\tilde{N}(K_1^j X(k) + w(k)) \\
&\quad - 2\beta(K_1^j X(k) + w(k))^T \tilde{N}^T P^{j+1}\tilde{H}(K_2^j X(k) + \chi(k)) \\
&\quad - \beta(K_2^j X(k) + \chi(k))^T \tilde{H}^T P^{j+1}\tilde{H}(K_2^j X(k) + \chi(k)) \\
&\quad = X^T(k)(M^j)^T \Pi M^j X(k).
\end{aligned}
\tag{8.85}
$$

Consider that the relationship between H^{j+1} and P^{j+1} is same as that shown in (8.75), (8.85) can be rewritten as

$$
H^j(k)L^{j+1} = \rho_k^j \tag{8.86}
$$

where

$$
\begin{aligned}
\rho_k^j &= X^T(k)\tilde{Q}X(k) + w^T(k)Rw(k) - \gamma^2\chi^T(k)\chi(k) \\
L_1^{j+1} &= H_{XX}^{j+1},\ L_2^{j+1} = H_{Xw}^{j+1},\ L_3^{j+1} = H_{X\chi}^{j+1} \\
L_4^{j+1} &= H_{ww}^{j+1},\ L_5^{j+1} = H_{w\chi}^{j+1},\ L_6^{j+1} = H_{\chi\chi}^{j+1} \\
L^{j+1} &= \left[(\mathrm{vec}(L_1^{j+1}))^T\ (\mathrm{vec}(L_2^{j+1}))^T\ \ldots\ (\mathrm{vec}(L_6^{j+1}))^T\right]^T \\
H^j(k) &= \left[H_1^j\ H_2^j\ \ldots\ H_6^j\right] \\
H_1^j &= (X^T(k) \otimes X^T(k)) - \beta(X^T(k+1) \otimes X^T(k+1)) \\
H_2^j &= 2X^T(k) \otimes w^T(k) + 2\beta X^T(k+1) \otimes (K_1^j X(k+1))^T \\
H_3^j &= 2X^T(k) \otimes \chi^T(k) + 2\beta X^T(k+1) \otimes (K_2^j X(k+1))^T \\
H_4^j &= -\beta(K_1^j X(k+1)))^T \otimes (K_1^j(X(k+1))^T + w^T(k) \otimes w^T(k) \\
H_5^j &= -2\beta(K_1^j X(k+1))^T \otimes (K_2^j(X(k+1))^T + 2\chi^T(k) \otimes w^T(k) \\
H_6^j &= -\beta(K_2^j X(k+1))^T \otimes (K_2^j X(k+1))^T + \chi^T(k) \otimes \chi^T(k).
\end{aligned}
$$

If L^{j+1} is estimated, then K_1^{j+1} and K_2^{j+1} can be calculated as

$$K_1^{j+1} = \left(L_4^{j+1} - L_5^{j+1}(L_6^{j+1})^{-1}(L_5^{j+1})^T\right)^{-1}\left((L_2^{j+1})^T - L_5^{j+1}(L_6^{j+1})^{-1}(L_3^{j+1})^T\right) \tag{8.87}$$

$$K_2^{j+1} = \left(L_6^{j+1} - (L_5^{j+1})^T(L_4^{j+1})^{-1}L_5^{j+1}\right)^{-1}\left((L_3^{j+1})^T - (L_5^{j+1})^T(L_4^{j+1})^{-1}(L_2^{j+1})^T\right). \tag{8.88}$$

Algorithm 8.4 Off-policy Q-learning algorithm

1: Data collection: Collect system data $(x(n), E(n), r(k))$ from the rougher flotation operational process using a behavior set-point $w(k)$ and a behavior disturbance $\chi(k)$ and store them in the sample sets H_i^j and ρ^j. Given $\beta > 0$ and $\gamma > 0$;

2: Initiation: Choose the initial gains K_1^0 and K_2^0, such that system (8.67) can be stabilized. Let $j = 0$;

3: Implementing Q-learning: L_i^{j+1} $(i = 1, 2, \ldots, 6)$ are estimated in terms of (8.86) using the collected data in Step 1, and then K_1^j and K_2^j are updated in terms of (8.87) and (8.88);

4: If $\left\|K_1^j - K_1^{j-1}\right\| \leq l_1$ and $\left\|K_2^j - K_2^{j-1}\right\| \leq l_2$ (l_1 and l_2 are small positive numbers), then stop the iteration and the optimal set-points have been obtained. Otherwise, let $j = j + 1$ and go back to Step 3.

Remark 8.22 In Algorithm 8.4, a specific behavior set-point and a specific behavior disturbance are applied to generate data of rougher flotation operational process while the behavior set-point is tracked by the output of lower-layer control loops, thus the data $(x_l(n), E_l(n), r(k))$ can be firstly collected. Then the set-point $w^j(k) = -K_1^j X(k)$ and the disturbance policy $\chi^j(k) = -K_2^j X(k)$ are evaluated and updated using the collected data, but they are not applied to the rougher flotation operational processes, which is different from the on-policy RL (Wei et al. 2014; Lee et al. 2012; Al-Tamimi et al. 2007b) where the target set-point and target disturbance under evaluation are applied to systems to collect data.

Remark 8.23 The initial stabilizing gains K_1^0 and K_2^0 can be obtained easily based on operator's rich experience. Besides, H_∞ control method can be used to yield an initial stabilizing control policies for systems (Modares and Lewis 2014).

To accurately estimate the real value of L^{j+1} in (8.86) by using the RLS or batch least squares (BLS), the behavior set-point should be in the form of the $w(k) + e(k)$ ($e(k)$ is a probing noise) to guarantee the PE (Al-Tamimi et al. 2007a; Kiumarsi et al. 2014; Modares and Lewis 2014). Lemma 8.1 is presented to show no bias result from adding probing noise in Algorithm 8.4.

Lemma 8.1 *Let H^{j+1} be the solution to (8.84) with the behavior set-point $w(k)$, then it is equivalent to the solution to (8.84) with the behavior set-point $w(k) + e(k)$ $(e(k) \neq 0)$.*

Proof If the behavior set-point is $w(k) + e(k)$, then the collected data $X(k+1)$ is in fact $\hat{X}(k+1)$ with the form of

$$\hat{X}(k+1) = \tilde{G}_c X(k) + \tilde{N}(K_1^j X(k) + w(k) + e(k)) + \tilde{H}(K_2^j X(k) + \chi(k)). \tag{8.89}$$

Substituting (8.89) and $w(k) + e(k)$ into (8.84) yields

$$\begin{aligned} X^T(k)(M^j)^T H^{j+1} M^j X(k) \\ &- \beta \left(\hat{X}(k+1) - \tilde{N}(K_1^j X(k) + w(k) + e(k)) - \tilde{H}(K_2^j X(k) + \chi(k)\right)^T \\ &\times (M^j)^T H^{j+1} M^j \left(\hat{X}(k+1) - \tilde{N}(K_1^j X(k) + w(k) + e(k))\right) \\ &= X^T(k)(M^j)^T \Pi M^j X(k). \end{aligned} \tag{8.90}$$

By (8.89) and (8.83), (8.90) becomes (8.84). Hence, adding the probing noise during learning in the proposed off-policy Q-learning algorithm cannot produce bias. This completes the proof. □

Theorem 8.2 *(H^{j+1}, K_1^j, K_2^j) is the solution of (8.79)–(8.81) if and only if it is the solution of (8.86)–(8.88).*

Proof It is easily concluded that if (H^{j+1}, K_1^j, K_2^j) is the solution of (8.79)–(8.81), then it can satisfy (8.86)–(8.88) from the above derivation. Next the fact that the solution of (8.86)–(8.88) is also the solution of (8.79)–(8.81) will be shown.

Note that (8.86) is equivalent to (8.85) by checking them. Thus, the solution of (8.86) can make (8.85) hold. Subtracting (8.85) from (8.84), one has

$$\begin{aligned} X^T(k)(M^j)^T H^{j+1} M^j X(k) - \beta X^T(k) \tilde{G}_c^T (M^j)^T H^{j+1} M^j \tilde{G}_c X(k) \\ = X^T(k)(M^j)^T \Pi M^j X(k). \end{aligned} \tag{8.91}$$

Due to $X(k+1) = \tilde{G}_c X(k)$ by using $w^{j+1}(k) = -K_1^{j+1} X(k)$, and $\chi^{j+1}(k) = -K_2^{j+1} X(k)$ and the definition of $z(k)$, the solution of (8.91) is equivalent to that of (8.79). Moreover, (8.87) and (8.88) are the same as (8.80) and (8.81). This completes the proof. □

From Theorem 8.2, it is easy to know that $w^j(k)$ and $\chi^{j+1}(k)$ learned by Algorithm 8.4 can converge to the Nash equilibrium solution $w^*(k)$ and $\chi^*(k)$ as $j \to \infty$, i.e., $\lim_{j\to\infty} w^j(k) = w^*(k)$, $\lim_{j\to\infty} \chi^j(k) = \chi^*(k)$ since the convergence of solutions of Algorithm 8.3 has been proven in Luo and Wu (2013) and Al-Tamimi et al. (2007a).

Remark 8.24 One can notice that the existing model-free OOC methods (Chai et al. 2011, 2014; Dai et al. 2014; Wu et al. 2014) depended on the operator's experience or presented the neural-network-based set-points design on the premise that the optimal performance indices are known a priori. The proposed off-policy RL method in this part can learn the optimal set-points for achieving OOC of operational processes for completely unknown dynamics of controlled unit processes and unknown function of operational index.

Remark 8.25 In contrast to the on-policy Q-learning method (Wei et al. 2014; Lee et al. 2012; Al-Tamimi et al. 2007b; Kiumarsi et al. 2014), the arbitrary behavior policy is introduced in the off-policy Q-learning method, which is used to generate data of systems for enriching data exploration while the target policies, especially disturbance policy, are updated to find the optimal policy but not to be employed to systems. Particularly, adding probing noise into the behavior policy would not produce bias of solution when implementing policy evaluation (Kiumarsi et al. 2017). Actually, Algorithm 8.4 can also be applied into other stages of flotation processes such as cleaning, scavenger, regrinding and classification circuits, and therefore, the optimal operation of the whole flotation processes can be achieved.

8.3.5 Optimum Set-Point Selector

In practical rougher flotation operational processes, the set-points followed by the concentrate grades and the tail grades are usually bounded, that is $w_l \in [w_{l\,\min},\ w_{l\,\max}]$, where $w_{l\,\max}$ and $w_{l\,\min}$ are non-negative real vectors with $w_{l\,\min} \leq w_{l\,\max}$. To satisfy the constraint of set-points, the following rules are presented to get the optimum set-point $\hat{w}_k^*$:

Rule 1. If $w_{\min} \leq w_k^* \leq w_{\max}$, then $\hat{w}_k^* = w_k^*$;

where

$$w_{\max} = \left[w_{1\max}^T\ w_{2\max}^T\ \cdots\ w_{s\max}^T\right]^T,\ w_{\min} = \left[w_{1\min}^T\ w_{2\min}^T\ \cdots\ w_{s\min}^T\right]^T.$$

Rule 2. If $w_k^* < w_{\min}$ or $w_k^* > w_{\max}$, then $\hat{w}_k^* = w_{k-1}^*$.

Theorem 8.3 *The set-point derived by the above rules can guarantee $\hat{w}_k^{j+1}$ to converge to the optimal set-points, that is,*

$$\lim_{k\to\infty} \hat{w}_k^* = w_k^*. \tag{8.92}$$

Proof By Rule 1 and Rule 2, one has

$$\hat{w}_k^* = w_k^* + \alpha(w_{k-1}^* - w_k^*) \tag{8.93}$$

where $\alpha = 0$ or 1. Thus,

$$\left\|\hat{w}_k^* - w_k^*\right\| \leq \left\|w_{k-1}^* - w_k^*\right\|. \tag{8.94}$$

Since w_k^* can guarantee the stability of the whole operational process (8.67), then $w_{k-1}^* - w_k^* \to 0$ as $k \to \infty$. By (8.94), (8.92) holds. This completes the proof. □

Remark 8.26 Refined Ziegler–Nichols method can design the PI controller parameters for the lower-layer control loop (Chai et al. 2011, 2014, 2012; Liu et al. 2013).

8.3.6 Simulation Results

The proposed off-policy Q-learning algorithm is verified in the rougher flotation operational process with two cells. Moreover, the dual rates operation is analyzed and proper comparisons are made to show the advantages of the proposed method.

Example 8.2 The flotation models according to mass balance in froth and pulp phases are given below (Rojas and Cipriano 2011; Chai et al. 2012; Liu et al. 2013)

$$\begin{aligned}\frac{dM_{lp}^i}{dt} &= f_{lp}^i = -\left(k_p^i + \frac{q_{lT}}{\left(1-\varepsilon_g\right)A_l h_{lp}}\right)M_{lp}^i + k_e^i M_{le}^i + q_{la}X_{la}^i\\ \frac{dM_{le}^i}{dt} &= f_{le}^i = -\left(k_e^i + \frac{q_{lc}}{\left(1-\varepsilon_g\right)A_l\left(H_l - h_{lp}\right)}\right)M_{le}^i + k_p^i M_{lp}^i.\end{aligned} \tag{8.95}$$

The concentrate grade and the tail grade in cell l, respectively, are

$$L_{lcg} = \frac{M_{le}^1 g_{lcp}^1 + M_{le}^2 g_{lcp}^2}{M_{le}^1 + M_{le}^2} L_{cu},\ L_{ltg} = \frac{M_{lp}^1 g_{lcp}^1 + M_{lp}^2 g_{lcp}^2}{M_{lp}^1 + M_{lp}^2} L_{cu}. \tag{8.96}$$

The general concentrate grade and the recovery, respectively, are

$$L_{cg} = \frac{\sum_{i=1}^2 \sum_{l=1}^2 M_{le}^i g_{lcp}^i}{\sum_{i=1}^2 \sum_{l=1}^2 M_{le}^i} L_{cu} = \frac{(g_a - L_{tg})L_{cg}}{(L_{cg} - L_{tg})g_a} \times 100 \tag{8.97}$$

where $l = 1, 2$. The sizes of two cells are both $53.2 \times 3.2m^3$ and values of other parameters are listed in Table 8.2. The economic benefit has the form of $r_k = M_r \hat{y}_k + N_r u_k$, where $M_r = [100\ 100\ 10]^T$, $N_r = [0.01\ 0.1\ 0.01\ 0.1]$. The economic benefit objective is 10, $Q = 800$, $R = \text{diag}(50, 50, 50, 50)$, $\beta = 0.9$, $\gamma = 5$, $w_{l\max} = [1\ 1]^T$ and $w_{lmin} = [0\ 0]^T$.

The equilibrium points of the rougher flotation process under the operational parameters in Table 8.2 are

$$\begin{aligned}\left[h_{1p}\ q_{1a}\right] = \left[2.8\ 17\right],\ \left[h_{2p}\ q_{2a}\right] = \left[2.5\ 12\right]\\ \left[M_{1p}^1\ M_{1p}^2\ M_{1e}^1\ M_{1e}^2\right] = \left[16.8\ 824.266\ 4.56\ 0.104\right]\\ \left[M_{2p}^1\ M_{2p}^2\ M_{2e}^1\ M_{2e}^2\right] = \left[20.84\ 1300\ 5.67\ 0.1645\right].\end{aligned}$$

First, we implement Algorithm 8.4 to learn the optimal set-point. Set the sampling interval of the lower-layer device loops and the updating period of the upper-layer operational control loop to be 1 min and 30 min, respectively. Figure 8.9 demonstrates the convergence results of H^{j+1}, K_1^j and K_2^j. The learned optimal set-point gain K_1^* is

Table 8.2 Parameters and values

Parameter	Physical meaning	Value
k_p^i	Flotation rate	$k_p^1 = 17.9 \quad k_p^2 = 0.04\ (\text{min}^{-1})$
k_e^i	Drainage rate	$k_e^1 = 65.6 \quad k_e^2 = 316\ (\text{min}^{-1})$
$q_{1T}\ q_{2T}$	Week	9.3 6.3 (m^3/min)
ε_g	Stagnation constant	0
$A_1 = A_2$	Cross-sectional area	53.2m^2
L_{cu}	Chalcopyrite brass grade	0.412
X_a^i	Mineral species concentration	$X_a^1 = 0.1549 \quad X_2^a = \frac{g_{cp}^1 - g_a}{g_a - g_{cp}^2} X_a^1 = 3.0484$
$g_{1cp}^1\ g_{2cp}^1$	Brass grade in brass pulp a	0.417 0.62
$g_{1cp}^2\ g_{2cp}^2$	Brass grade in gangue pulp	0.0034 0.0034
$q_{1c}\ q_{2c}$	Concentrate pulp flow	7.392 8.103 (m^3/min)
$H_1 = H_2$	Total height	3.2m
g_a	Feed mineral grade	0.0234

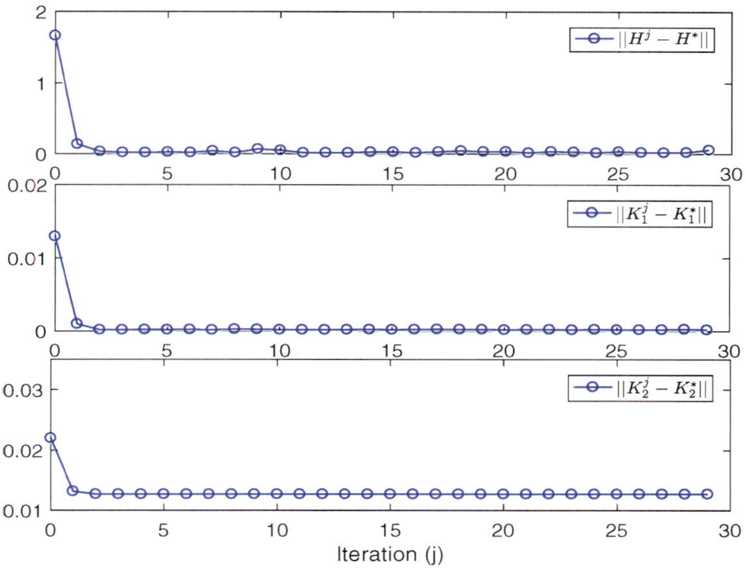

Fig. 8.9 Convergence results of parameters

presented in (8.98). The optimal set-point can be further approximated by combining with Rule 1 and Rule 2.

In the real operation of rougher flotation process, it is hard to avoid external disturbance and measurement errors. Hence, the external disturbance signal and measurement noise are assumed as $0.05e^{-0.001t}\cos(100\nu(t)t)(\nu(t){=}[\nu_1(t)\ \nu_2(t)\ \nu_3(t)\nu_4(t)]^T$

is a vector with $v_i(t) \in [0, 1]$ $(i = 1, 2, 3, 4)$ and put into cell 1 and cell 2, respectively.

$$K_1^* = \begin{bmatrix} K_{11}^* & K_{12}^* & K_{13}^* \end{bmatrix} \tag{8.98}$$

where

$$K_{11}^* = \begin{bmatrix} -0.0002 & 0.0000 & 0.0038 & -0.1671 \\ 0.0002 & -0.0000 & -0.0044 & 0.1919 \\ -0.0001 & 0.0000 & 0.0024 & -0.1048 \\ 0.0002 & -0.0000 & -0.0043 & 0.1869 \end{bmatrix}$$
$$K_{12}^* = \begin{bmatrix} -0.0114 & 0.0127 & -0.0001 & 0.0000 & 0.0054 \\ 0.0131 & -0.0146 & 0.0001 & -0.0000 & -0.0062 \\ -0.0071 & 0.0080 & -0.0001 & 0.0000 & 0.0034 \\ 0.0127 & -0.0142 & 0.0001 & -0.0000 & -0.0060 \end{bmatrix}$$
$$K_{13}^* = \begin{bmatrix} -0.1858 & -0.0167 & 0.0296 & 0 & -0.0004 \\ 0.2134 & 0.0192 & -0.0339 & 0 & 0.0004 \\ -0.1164 & -0.0104 & 0.0185 & 0 & -0.0002 \\ 0.2076 & 0.0186 & -0.0329 & 0 & 0.0004 \end{bmatrix}.$$

Then, three cases of the updating periods of the upper-layer operational control loop (a): 10 min, (b): 30 min, and (c): 120 min are taken into account (see Figs. 8.10 and 8.11a) for showing the tracking results of the set-points and the economic operational index. In Fig. 8.11a, one can find that the longer it takes to follow the desired economic benefit, the slower updating period of upper-layer operational process is. The fast updating rate of upper-layer operational control loop also can bring increasing computational load. In general, proper updating rate of the upper-layer operational process should be chosen according to the requirements of practical industrial operations. Moreover, the results shown in Figs. 8.10 and 8.11 also indicate that the proposed off-policy Q-learning method has good performance and robustness to both disturbance and measurement errors.

The comparisons with the fixed set-point method and the intelligent control method (Chai et al. 2011, 2014) are made under the same initial operational scenario listed in Table 8.2. The fixed set-points of cell 1 and cell 2 are respectively chosen as $[0.0117\ 0.01]^T$ and $[0.1274\ 0.01]^T$. The set-points in the intelligent control method are adjusted according to the error between the desired economic benefit and the real value. The tracking results of economic benefit and calculated recovery are respectively shown in Fig. 8.11b, c, where one can see that the better tracking result and high recovery rate are obtained by using the proposed off-policy RL algorithm, since the operators' experience-based fixed set-points maybe not the optimal set-points. Moreover, the increments of set-points in the intelligent control method are chosen also based on the operators' experience, then the improper increments could produce negative effects on tracking the desired economic benefit and mineral recovery.

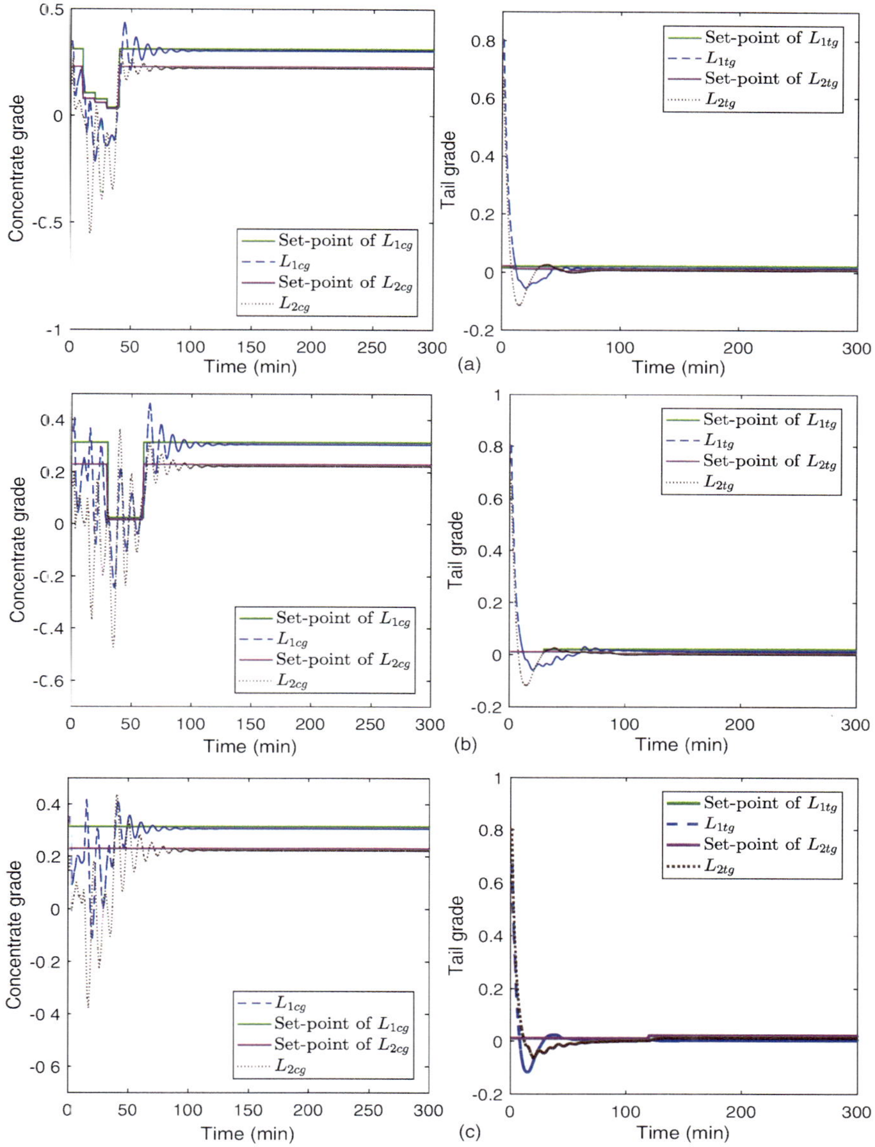

Fig. 8.10 Tracking performance of the optimal set-points **a** Case a; **b** Case b; **c** Case c

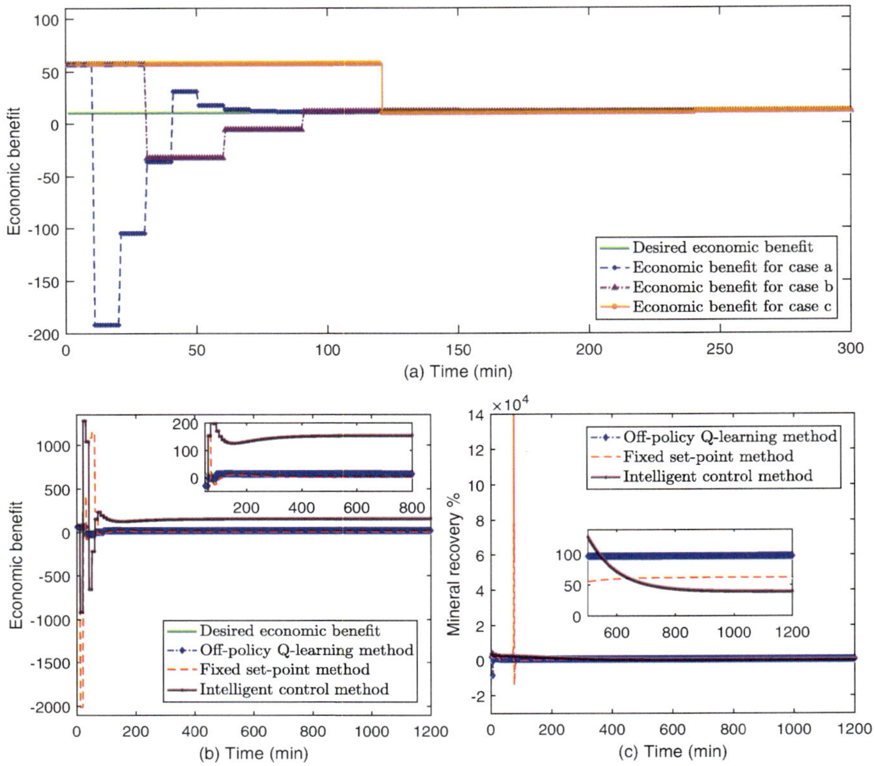

Fig. 8.11 Tracking performance of the economic benefit and calculated recovery

8.4 Plant-Wide Optimization Using Game Reinforcement Learning

This section presents a novel technique to achieve plant-wide performance optimization for large-scale unknown industrial processes by integrating the RL method with multi-agent game theory. A main advantage of the proposed technique is that plant-wide optimal performance is achieved by a distributed approach where multiple agents solve simplified local nonzero-sum optimization problems so that a global Nash equilibrium is reached. To this end, first, the plant-wide performance optimization problem is reformulated by decomposition into local optimization subproblems for each production index in a multi-agent framework. Then the nonzero-sum graphical game theory is utilized to compute the operational indices for each unit process with the purpose of reaching the global Nash equilibrium, resulting in production indices following their prescribed target values. The stability and the global Nash equilibrium of this multi-agent graphical game solution are rigorously proved. Reinforcement learning methods are then developed for each agent to solve the nonzero-sum graphical game problem using data measurements available in the system in real time. The plant dynamics do not have to be known. The difficulties we will face are how to model the plant-wide performance optimization as a multi-agent optimization problem and how to solve this problem with heterogeneous dynamics of agents and constraints using nonzero-sum game theory and RL for industrial processes by model-free approaches.

8.4.1 Problem Statement

A multi-agent framework is established for the plant-wide performance optimization problem of industrial processes.

8.4.1.1 Graph Theory

Generally, a graph is denoted by $\mathcal{G} = (\mathcal{V}, \mathcal{E})$ wherein $\mathcal{V} = \{v_1, v_2, \ldots, v_N\}$ and $\mathcal{E} \subseteq \mathcal{V} \times \mathcal{V}$ represent a set of vertices and a set of edges or arcs, respectively. The connectivity matrix of graph $\mathcal{G}$ is $E = [e_{ij}]$ with $e_{ij} > 0$ if $(v_j, v_i) \in \mathcal{E}$, and otherwise $e_{ij} = 0$. $(v_j, v_i) \in \mathcal{E}$ indicates that there exists an edge between the vertex v_j and v_i in an undirected graph and nodes v_i and v_j are neighbors. A simple graph means there are no repeated edges or self-loops $(v_i, v_i) \in \mathcal{E}$ for $\forall i$, which indicates $e_{ii} = 0$. Denote the set of neighbors of node v_i as $N_i = \{v_j : (v_j, v_i) \in \mathcal{E}\}$. A sequence of edges $(v_{i_1}, v_{i_2}), (v_{i_2}, v_{i_3}), \ldots, (v_{i_{j-1}}, v_{i_j})$ means a path between the node v_{i_1} and v_{i_j}, where $(v_{i_{l-1}}, v_{i_l}) \in \mathcal{E}$ for $l \in \{2, \ldots, j\}$. A graph is called strongly connected if there exists a path for every two distinct vertices (Movric and Lewis 2014; Vamvoudakis et al. 2012).

8.4.1.2 Plant-Wide Performance Optimization Description

In practical industrial plants, the scheduling department takes charge of decisions or allocation of overall plant-wide production indices denoted here by p_m ($m = 1, 2 \ldots \vartheta$ and ϑ is a positive integer). Production indices are usually product quality, yield, energy and raw material consumption of the overall plant (Ding et al. 2016; Yuan et al. 2016; Huang et al. 2005). For given target production indices p_m^*, decision-making is performed in the technical department in order to find appropriate operational indices to make the production indices follow the target production indices. The obtained operational indices become the target operational indices r_m^* for unit processes in the production line. Unit operational indices may be unite process yield, quality and energy consumption, etc. Figure 8.12 shows the detailed flowchart of the overall plant.

The actual production indices p_m of industrial plants are dynamic and interact with each other caused by the operational indices of each unit s ($s = 1, 2, \ldots, n$).

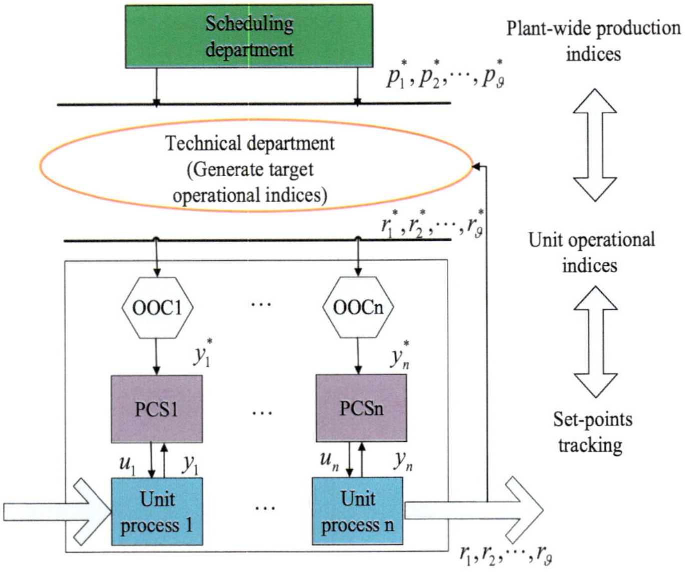

Fig. 8.12 Production line for processing production plant ($u_1, \ldots, u_n$ and $y_1, \ldots, y_n$ denote control inputs and outputs of unit processes, respectively. $y_1^*, \ldots, y_n^*$ are set-points for all of unit processes)

Since there exists coupling relationship among production indices, then all of these production indices might be thought of as nodes distributed on communication graph $\mathcal{G}$ with node dynamics

$$p_m(k+1) = f_m\left(p_m(k), r_m(k), \sum_{j \in N_m} e_{mj} a_j r_j(k), \pi(k), q(k)\right) \quad (m = 1, 2, ..., \vartheta) \tag{8.99}$$

where $r_m(k)$ represents the operational index m associated with node m, $\pi(k)$ and $q(k)$ denote average unit process throughput per hour and process raw material quality or quantity, respectively. At each working condition, $\pi(k)$ and $q(k)$ almost remain constant. k means the sampling time instant to measure the operational indices and production indices. $f_m(\cdot)$ stand for the dynamics of the nonlinear systems generating production indices p_m and are smooth. a_j denotes the appropriate dimensional matrices with ϑ block matrices, where the jth block element is identity matrix and the others are zero matrices. $\sum_{j \in N_m} e_{mj} a_j r_j(k)$ in fact includes the operational indices associated with the neighbors of node m.

System (8.99) clearly shows which operational indices can influence which production indices. Actually, the dynamics of (8.99) is with multiple control inputs from node m and its neighbors if the operational indices here are supposed to be control inputs. Our objective shall be to optimize the plant-wide performance expressed as minimizing the sum of errors between production indices and the target production indices subject to (8.99). Hence, the performance indices are defined as follows (Ding et al. 2016):

$$\begin{aligned}
&\min_{r(k)} \sum_{k=0}^{\infty} \|p_1(k) - p_1^*\|_{Q_1}^2 \\
&\min_{r(k)} \sum_{k=0}^{\infty} \|p_2(k) - p_2^*\|_{Q_2}^2 \\
&\vdots \\
&\min_{r(k)} \sum_{k=0}^{\infty} \|p_\vartheta(k) - p_\vartheta^*\|_{Q_\vartheta}^2
\end{aligned} \tag{8.100}$$

where Q_m $(m = 1, 2, \ldots, \vartheta)$ are positive definite matrices with appropriate dimensions. $r(k)$ is a decision variable that consists of all of $r_m(k)$.

In practical industrial processes, the operational index $r_m(k)$ is always constrained by production technology and security of operation. Here $r_m(k)$ is constrained with $r_{m,\min} \leq r_m \leq r_{m,\max}$, $r_{m,\max}$ and $r_{m,\min}$ are the upper bound and lower bound, respectively. Thus, the plant-wide optimization problem is formulated as Problem 8.5 below.

Problem 8.5

$$\begin{aligned}
&\min_{r(k)} \sum_{k=0}^{\infty} \|p_1(k) - p_1^*\|_{Q_1}^2 \\
&\min_{r(k)} \sum_{k=0}^{\infty} \|p_2(k) - p_2^*\|_{Q_2}^2 \\
&\vdots \\
&\min_{r(k)} \sum_{k=0}^{\infty} \|p_\vartheta(k) - p_\vartheta^*\|_{Q_\vartheta}^2
\end{aligned} \tag{8.101}$$

s.t. (8.99) and

$$r_{m,\min} \leq r_m \leq r_{m,\max} \; (m = 1, 2, \ldots, \vartheta). \tag{8.102}$$

Notice that Problem 8.5 is a multi-objective optimization problem distributed on a communication graph where each node interplays with its neighbors. Moreover, one can notice that the dynamics of the production indices on each node is nonlinear and the operational indices are constrained. If we define an augmented vector of all of production indices by compacting this problem, then a high-dimensional optimization problem is derived below

$$\min_{r(k)} \sum_{k=0}^{\infty} \left\| p(k) - p^* \right\|_{\tilde{Q}}^2 . \tag{8.103}$$

Subject to

$$p(k+1) = f(p(k), r(k), \pi(k), q(k)) \tag{8.104}$$

where

$$\begin{aligned} p(k) &= \begin{bmatrix} p_1(k) \\ p_2(k) \\ \vdots \\ p_\vartheta(k) \end{bmatrix}, \ f = \begin{bmatrix} f_1(k) \\ f_2(k) \\ \vdots \\ f_\vartheta(k) \end{bmatrix}, \ r(k) = \begin{bmatrix} r_1(k) \\ r_2(k) \\ \vdots \\ r_\vartheta(k) \end{bmatrix} \\ \tilde{Q} &= \operatorname{diag}(Q_1, Q_2, \ldots, Q_m), \ r_{\min} \le r \le r_{\max} \\ p^* &= \begin{bmatrix} p_1^* \\ p_2^* \\ \vdots \\ p_\vartheta^* \end{bmatrix}, \ r_{\min} = \begin{bmatrix} r_{1,\min} \\ r_{2,\min} \\ \vdots \\ r_{\vartheta,\min} \end{bmatrix}, \ r_{\max} = \begin{bmatrix} r_{1,\max} \\ r_{2,\max} \\ \vdots \\ r_{\vartheta,\max} \end{bmatrix}. \end{aligned} \tag{8.105}$$

$\tilde{Q}$ is a positive semi-definite matrix.

The fact that modern large-scale industrial processes are typical multi-variable control systems (Wongsri and Siraworakun 2010) makes it extremely difficult to solve Problem 8.5 using this compact high-dimensional way, since this method is high computational loaded and time-consuming (Yuan et al. 2016; Oliveira et al. 2013; Xu et al. 2012). Moreover, it is worth pointing out that the graphical connectivity of nodes (production indices) cannot be shown and utilized in this expression of optimization problem.

8.4.1.3 Nonzero-Sum Graphical Game Formulation

The basic idea is to find an easy and efficient approach to solve Problem 8.5. So what we shall do here is decompose Problem 8.5 into several easier to solve subproblems and to formulate these as a nonzero-sum multi-agent graphical game.

For the plant-wide optimization Problem 8.5, it is desired for each production index to follow its own target value by an optimal approach using cooperative control strategy or decision-making of operational indices. Notice that the performance of each production index is determined not only its own control policy, but also its neighbors' control policies. In this sense, Problem 8.5 is a graphical game. Thus, each production index could be viewed as an agent located on a node in a graphical game that is struggling to minimize the error from its own target value by decision-making of the operational indices. Problem 8.6 is thus proposed.

Problem 8.6 For each agent, the objective function is to minimize the following performance index:

$$J_m = \sum_{k=0}^{\infty} \gamma^k \left(\| p_m(k) - p_m^* \|_{Q_m}^2 + 2 \int_0^{r_m(k)} \varpi_m(s) ds \right) \tag{8.106}$$

s.t.

$$\begin{aligned} p_m(k+1) &= f_m(p_m(k), r_m(k), r_{-m}(k), \pi(k), q(k)) \\ r_{m,\min} &\le r_m \le r_{m,\max} \end{aligned} \tag{8.107}$$

where γ is a discount factor with $0 < \gamma \le 1$, and $\varpi_m(s) = \tanh^{-T}(\bar{U}_m^{-1} s)\bar{U}_m S_m$. $r_{-m}(k)$ stands for the vector of the control inputs $\{r_j | j \in N_m\}$ of the neighbors of node m, that is $r_{-m}(k) = \sum_{j \in N_m} e_{mj} a_j r_j(k)$. $r_m = [r_{m1}, r_{m2}, \ldots, r_{m\kappa_m}]^T$ and $|r_{mi}| \le \bar{r}_{mi}$, $(i = 1, 2, \ldots, \kappa_m)$. $\bar{r}_{mi}$ can be determined by $r_{m,\max}$ and $r_{m,\min}$. $\bar{U}_m = \text{diag}\{\bar{r}_{m1}, \bar{r}_{m2}, \ldots, \bar{r}_{m\kappa_m}\}$, S_m are the positive definite matrices with appropriate dimensions. $\int_0^{r_m(k)} \varpi_m(s) ds$ is positive definite if $r_m(k) \ne 0$ and it is zero if $r_m(k) = 0$ (Kiumarsi and Lewis 2015).

Remark 8.27 Problem 8.6 is indeed a nonzero-sum multi-agent graphical game problem, in which each agent in graph $\mathcal{G}$ needs to decide its own control policy for achieving its own objective, while its performance has to be affected by its neighbors' control policies.

Remark 8.28 Different from Ding et al. (2016), Yuan et al. (2016), Oliveira et al. (2013), Xu et al. (2012) as well as the high-dimensional optimization problem mentioned in the above (Huang et al. 2005; Wongsri and Siraworakun 2010; Ochoa et al. 2010), the plant-wide optimization problem is decomposed into subproblems under multi-agent framework. Each subproblem corresponding to each production index is subject to a reduced-order subsystem. From this point, complexity of computation is reduced resulting in less time-consuming.

8.4.2 *Nonzero-Sum Graphical Game for Solving Multi-objective Optimization Problem*

Using nonzero-sum graphical game methods, multi-agent optimization Problem 8.6 is solved. The idea of Nash equilibrium is used to solve the nonzero-sum graphical game, so that the optimal operational indices are found to achieve the production indices tracking their targets.

8.4.2.1 Nonzero-Sum Graphical Game

In Problem 8.6, one can find that the performance of each agent m is affected not only by the policy $r_m(k)$ of itself and but also by other neighbors' control policies $r_{-m}(k)$. To achieve Nash equilibrium, the idea of nonzero-sum game is employed below.

Assume the control polices $r_m(k)$ to be state dependent polices, the value function for node m ($m = 1, 2, ..., \vartheta$) corresponding to (8.106) is defined as

$$V_m(W_m(k), p_m^*) = \sum_{\tau=k}^{\infty} \gamma^{\tau-k} \left(\| p_m(\tau) - p_m^* \|_{Q_m}^2 + 2 \int_0^{r_m(\tau)} \varpi_m(s) ds \right) \quad (8.108)$$

where $W_m(k)$is called an associated vector constructed for node m using the local measured information $p_m(k)$ and $\{p_j(k)|j \in N_m\}$, that is,

$$W_m(k) = \left\{ p_j(k) | j \in \{m \cup N_m\} \right\}. \quad (8.109)$$

It is desired for agent m to minimize

$$V_m^*(W_m(k), p_m^*) = \min_{r_m} \sum_{\tau=k}^{\infty} \gamma^{\tau-k} \left(\| p_m(\tau) - p_m^* \|_{Q_m}^2 + 2 \int_0^{r_m(\tau)} \varpi_m(s) ds \right). \quad (8.110)$$

Thus, the following Bellman equations can be obtained:

$$V_m(W_m(k), p_m^*) = \| p_m(k) - p_m^* \|_{Q_m}^2 + 2 \int_0^{r_m(k)} \varpi_m(s) ds + \gamma V_m(W_m(k+1), p_m^*). \quad (8.111)$$

Define the Hamiltonians for all agents

$$\begin{aligned} H_m(W_m, \Delta V_m, r_m, r_{-m}) = {} & \| p_m(k) - p_m^* \|_{Q_m}^2 + 2 \int_0^{r_m(k)} \varpi_m(s) ds \\ & + \gamma V_m(W_m(k+1), p_m^*) - V_m(W_m(k), p_m^*). \end{aligned} \quad (8.112)$$

Based on the necessary condition for optimality, we take the derivative of H_m along the trajectory of $p_m(k+1)$, then it follows:

$$\frac{\partial H_m}{\partial r_m(k)} = 2(\bar{U}_m S_m)^T \tanh^{-1}\left(\bar{U}_m^{-1} r_m(k)\right)\big|_{r_m(k)=r_m^*}$$
$$+ \gamma \frac{\partial p_m^T(k+1)}{\partial r_m(k)} \times \frac{\partial V_m(W_m(k+1), p_m^*)}{\partial p_m(k+1)}\big|_{r_m(k)=r_m^*} = 0. \quad (8.113)$$

Thus, one has

$$r_m^*(k) = \bar{U}_m \tanh\left(-\frac{1}{2}\gamma(\bar{U}_m S_m)^{-T} \frac{\partial p_m^T(k+1)}{\partial r_m(k)} \frac{\partial V_m(W_m(k+1), p_m^*)}{\partial p_m(k+1)}\right). \quad (8.114)$$

Let all operational indices be (8.114) and substitute them into (8.111), the coupled cooperative game DT HJB equations are obtained as follows:

$$H_m(W_m, \frac{\partial V_m^*}{\partial p_m}, r_m^*, r_{-m}^*) = \left\| p_m(k) - p_m^* \right\|_{Q_m}^2 + 2\int_0^{r_m^*} \varpi_m(s)ds$$
$$+ \gamma V_m(W_m^*(k+1), p_m^*) - V_m(W_m(k), p_m^*) = 0 \quad (8.115)$$

where the elements $p_j(k+1)$ $(\forall j \in \{m \cup N_m\})$ in the associated vector $W_m^*(k+1)$ have the dynamics of $p_j(k+1) = f_j\left(p_j(k), r_j^*(k), r_{-j}^*(k), \pi(k), q(k)\right)$ constructed for node j.

Lemma 8.2 is given to prove that the operational indices (8.114) satisfy the constraint condition of operational indices.

Lemma 8.2 *The operational indices $r_m^*(k)$ make the following inequality hold.*

$$r_{m,\min} \le r_m^*(k) \le r_{m,\max}. \quad (8.116)$$

Proof For Problem 8.6 with the added term $\int_0^{r_m(k)} \varpi_m(s)ds$ in the performance indices, since tanh(.) is bounded by $[-1,\ 1]$, then by (8.114) and the definition of $\bar{U}_m$, one has

$$\left|r_{mi}^*(k)\right| \le \bar{r}_{mi} \ (i = 1, 2, \ldots, m\kappa_m). \quad (8.117)$$

Therefore, $r_m^*(k)$ naturally dropped into the range of $[r_{m,\min},\ r_{m,\max}]$ of $r_m(k)$. This completes the proof. □

8.4.2.2 Global Nash Equilibrium

Now we are in the position to prove that the control policies defined as (8.114) with V_m^* that are the solution to the coupled cooperative game DT HJB equations guarantee all agents with interior interactions to reach a global Nash equilibrium. Here, a Nash equilibrium definition is presented first.

Definition 8.2 (Li et al. 2019; Vamvoudakis et al. 2012) A global Nash equilibrium solution for a ϑ player game is given by a $\vartheta-$ tuple of policies $\{r_1^*, r_2^*, \ldots, r_\vartheta^*\}$ if it satisfies

$$J_m^* \triangleq J_m(p_m(0), r_m^*, r_{\mathcal{G}-m}^*) \leq J_m(p_m(0), r_m, r_{\mathcal{G}-m}^*) \tag{8.118}$$

for all m $(m \in \mathcal{V})$ and $\forall r_m, r_{\mathcal{G}-m}$, where $r_{\mathcal{G}-m}^* = \{r_j^* | j \neq m, j \in \mathcal{V}\}$ and $r_{\mathcal{G}-m} = \{r_j | j \neq m, j \in \mathcal{V}\}$. The $\vartheta-$ tuple of game values $\{J_1^*, J_2^*, \ldots, J_\vartheta^*\}$ is said to be a Nash equilibrium outcome of the $\vartheta-$ player game.

Note that how to rigorously prove the stability and Nash equilibrium for coupled and heterogeneous nonlinear agents with constraints has not been reported, even though some results on optimal control using RL for linear or affine nonlinear agents have been available (Li et al. 2017; Jiang and He 2019; Qin et al. 2019; Zhang et al. 2017, 2019; Zuo et al. 2017; Gao et al. 2019; Jiao et al. 2016; Modares et al. 2018). Extend the theory in Kiumarsi and Lewis (2015) for single affine nonlinear optimal tracking problem to the graphical game, in the sequel, Theorem 8.4 is provided to prove the stability and Nash equilibrium of all agents in a graph using nonzero-sum graphical game theory.

Assumption 8.1 The graph $\mathcal{G}$ is strongly connected.

Theorem 8.4 *(Stability and Nash Equilibrium Solution): Let the graph $\mathcal{G}$ be structured as $\mathcal{G} = (\mathcal{V}, \mathcal{E})$, V_m^* (all $m \in \mathcal{V}$) be smooth solutions to coupled cooperative game DTHJB equations (8.115) and the control policies r_m^* as*

$$r_m^*(k) = \bar{U}_m \tanh\left(-\frac{1}{2}\gamma(\bar{U}_m S_m)^{-T} \frac{\partial p_m^T(k+1)}{\partial r_m(k)} \frac{\partial V_m^*(W_m(k+1), p_m^*)}{\partial p_m(k+1)}\right) \tag{8.119}$$

then, the following conclusions hold:

1) *The dynamics of agents (8.107) are stable and if $\gamma = 1$ for the case of the desired production indices being zeros, agents (8.107) are asymptotically stable;*
2) *The control polices $\{r_1^*, r_2^*, \ldots, r_\vartheta^*\}$ make all agents be in the global Nash equilibrium with the optimal performance $J_m^*(p_m(0)) = V_m^*(W_m(0))$.*

Proof 1) Stability: Suppose that the value functions $V_m^*(\cdot)$ make coupled cooperative game DTHJB equations (8.115) hold, we choose a Lyapunov function for each agent m as

$$V_m(W_m(k), p_m^*) = \gamma^k V_m^* \left(W_m(k), p_m^*\right). \tag{8.120}$$

Thus

$$\begin{aligned} V_m(W_m^*(k+1), p_m^*) &- V_m(W_m(k), p_m^*) \\ &= \gamma^k \left(\gamma V_m^*(W_m^*(k+1), p_m^*) - V_m^*(W_m(k), p_m^*)\right) \\ &= -\gamma^k \left(\| p_m(k) - p_m^* \|_{Q_m}^2 + 2\int_0^{r_m^*} \varpi_m(s)ds \right) < 0. \end{aligned} \tag{8.121}$$

Equation (8.121) shows dynamics (8.107) are stable and the tracking errors $e_m = p_m(k) - p_m^*$ are bounded. If $\gamma = 1$ and $p_m^* = 0$ as $k \to \infty$, then $p_m(k) \to 0$ holds by Barbalat's Extension (Kiumarsi and Lewis 2015), which shows that all agents can asymptotically track the target production indices by using the optimal policies r_m^*. For the case of $0 < \gamma < 1$ and $p_m^* \neq 0$, the tracking error $e_m = p_m(k) - p_m^*$ can be made as small as desired by choosing a discount γ approaching one and the proper matrix Q_m (Kiumarsi and Lewis 2015; Kim and Lewis 2010).

2) Optimality: For the arbitrary operational polices r_m, r_{-m} ($\forall m \in \mathcal{V}$) and initial condition $p_m(0)$, one can present the following form based on performance (8.106) and Hamiltonians (8.112):

$$\begin{aligned} V_m(W_m(0), r_m, r_{-m}) &= \sum_{k=0}^{\infty} \gamma^k \left(\| p_m(k) - p_m^* \|_{Q_m}^2 + 2\int_0^{r_m(k)} \varpi_m(s)ds \right) \\ &= \sum_{k=0}^{\infty} \gamma^k H_m(W_m, \Delta V_m, r_m, r_{-m}) + V_m(W_m(0)). \end{aligned} \tag{8.122}$$

Since

$$\begin{aligned} H_m(W_m, \Delta V_m, r_m, r_{-m}) &= H_m(W_m, \Delta V_m, r_m^*, r_{-m}^*) + 2\int_{r_m^*(k)}^{r_m(k)} \varpi_m(s)ds \\ &+ \gamma V_m(W_m(k+1), p_m^*) - \gamma V_m(W_m^*(k+1), p_m^*) \end{aligned} \tag{8.123}$$

substituting (8.123) into (8.122) and considering $H_m\left(W_m, \frac{\partial V_m^*}{\partial p_m}, r_m^*, r_{-m}^*\right) = 0$ yield

$$V_m(W_m(0), r_m, r_{-m}) = M_m + V_m^*(W_m(0)) \tag{8.124}$$

where

$$M_m = \sum_{k=0}^{\infty} \gamma^k \left(2 \int_{r_m^*}^{r_m(k)} \varpi_m(s)ds + \gamma V_m^*(W_m(k+1), p_m^*) \right.$$
$$\left. -\gamma V_m^* \left(W_m^*(k+1), p_m^* \right) \right). \tag{8.125}$$

Consider (8.119), one has

$$2(\bar{U}_m S_m)^T \tanh^{-1}(\bar{U}_m^{-1} r_m^*(k)) = -\gamma \frac{\partial V_m^*(W_m(k+1), p_m^*(k+1))}{\partial r_m(k)} \tag{8.126}$$

where $\frac{\partial V_m^*(p_m(k+1), p_m^*)}{\partial r_m(k)}$ is done along the trajectory of $p_m(k+1)$. By integrating (8.126), one gets

$$2\varpi r_m^*(r_m(k) - r_m^*)^T = \gamma V_m^*(W_m^*(k+1), p_m^*) - \gamma V_m^*(W_m(k+1), p_m^*). \tag{8.127}$$

By (8.127), M_m becomes as

$$M_m = \sum_{k=0}^{\infty} \gamma^k \left[2 \int_{r_m^*}^{r_m(k)} \varpi_m(s)ds - 2\varpi_m(r_m^*) \left(r_m(k) - r_m^* \right)^T \right]$$
$$= 2 \sum_{k=0}^{\infty} \sum_{j=1}^{m\kappa_m} \gamma^k l_M^j. \tag{8.128}$$

Using mean value theorem for the integrals, there exist a $\tilde{r}_{mj}(k) \in \left[r_{mj}^*, r_{mj}(k) \right]$ such that

$$\int_{r_{mj}^*}^{r_{mj}(k)} \varpi_j(s_j)ds_j = \varpi_j(\tilde{r}_{mj}(k)) \left(r_{mj}(k) - r_{mj}^* \right). \tag{8.129}$$

Then, one has $M_m > 0$ if $r_m \neq r_m^*$ and $M_m = 0$ if $r_m = r_m^*$. Thus, (8.124) yields

$$V_m^*(W_m(0), r_m^*, r_{-m}^*) \leq V_m(W_m(0), r_m, r_{-m}^*). \tag{8.130}$$

Equation (8.130) holds. According to (8.106), one has

$$V_m^*(W_m(0), r_m^*, r_{-m}^*) = V_m^*(W_m(0), r_m^*, r_{-m}^*, r_z) \forall z \notin \{m \cup N_m\}$$

so that

$$J_m^*(W_m(0), r_m^*, r_{\mathcal{G}-m}^*) \leq J_m(W_m(0), r_m, r_{\mathcal{G}-m}^*) \forall m \in \mathcal{V}. \tag{8.131}$$

Therefore, the global Nash equilibrium is reached. If the graph $\mathcal{G}$ is strong connected, which means there is a path from every node z to every node m, then

$$J_m(W_m(0), r_z^*, r_{\mathcal{G}-z}^*) \neq J_m(W_m(0), r_z, r_{\mathcal{G}-z}^*) \tag{8.132}$$

for all $z, m \in \mathcal{V}$. This indicates that changes of every agent will eventually influence all agents in the graph $\mathcal{G}$ if the graph is strongly connected. This completes the proof.

□

Remark 8.29 To get the Nash equilibrium solution r_m^*, we have to solve the value functions in (8.115). Solving this problem is quite hard due to the coupled nonlinear equations with partial derivative and integral terms, especially under unknown dynamics of unit processes and production indices environments. An RL-based algorithm is designed in the next section to overcome these difficulties.

8.4.3 Model Free Solution to Nonzero-Sum Graphical Game

By combining RL and NNs approximation methods, a RL-based graphical game algorithm is finally proposed to learn the Nash equilibrium points, which drive the production indices to follow the desired targets by an approximately optimal approach.

8.4.3.1 Reinforcement Learning for Nonzero-Sum Graphical Game

Referring to the existing RL methods for solving optimization problem of multi-agent systems (Li et al. 2017; Vamvoudakis and Lewis 2011; Jiang and He 2019; Zhang et al. 2017, 2019; Zuo et al. 2017; Gao et al. 2019; Jiao et al. 2016; Modares et al. 2018; Vamvoudakis et al. 2012; Mu et al. 2017), Algorithm 8.5 is developed to learn the optimal operational indices that make all of agents reach the Nash equilibrium and are controlled within the required range.

Here, if we use the single RL approach to solve the Problem 8.5 subject to (8.104), then (8.133) and (8.134) will be respectively replaced through following the RL technique like (Kiumarsi and Lewis 2015; Zhang et al. 2009) by

$$V^{(i+1)}(p(k), p^*) = \gamma V^{(i)}(p(k+1), p^*) + \|p(k) - p^*\|_{\bar{Q}}^2 + 2\int_0^{r^{(i)}} \varpi(s)ds \tag{8.135}$$

and

$$r^{(i+1)}(k) = \bar{U}\tanh\left(-\frac{1}{2}\gamma(\bar{U}S)^{-T}\frac{\partial p(k+1)}{\partial r(k)}\frac{\partial V^{(i+1)}(p(k+1), p^*)}{\partial p(k+1)}\right) \tag{8.136}$$

where

Algorithm 8.5 Value iteration: learning the operational indices

1: Initialize the iterative functions $V_m^{(0)}(\cdot) = 0$ and further calculate the initial operational policies $r_m^{(0)}$ by (8.134) for all m agents and set $i = 0$;

2: **For** $M = 1 : \vartheta$:

2.1: Evaluate policies: by solving $V_m^{(i+1)}$ using $V_m^{(i)}$, $r_m^{(i)}$ and $r_{-m}^{(i)}$:

$$V_m^{(i+1)}(W_m(k), p_m^*) = \gamma V_m^{(i)}(W_m(k+1), p_m^*) + \left\| p_m(k) - p_m^* \right\|_{Q_m}^2 + 2\int_0^{r_m^{(i)}} \varpi_m(s)ds \tag{8.133}$$

where $p_j(k+1) = f_j\left(p_j(k), r_j^{(i)}(k), r_{-j}^{(i)}(k), \pi(k), q(k)\right)$ $(j \in \{m \cup N_m\})$ and i denotes the iteration index;

2.2: Update ϑ-tuple of control policies:

$$r_m^{(i+1)}(k) = \bar{U}_m \tanh\left(-\frac{1}{2}\gamma(\bar{U}_m S_m)^{-T} \frac{\partial p_m(k+1)}{\partial r_m(k)} \frac{\partial V_m^{(i+1)}(W_m(k+1), p_m^*)}{\partial p_m(k+1)}\right) \tag{8.134}$$

End for;

3: If $\left\| V_m^{(i)} - V_m^{(i+1)} \right\| > \varepsilon$ $(\varepsilon > 0)$ for any agent m, then set $i = i + 1$ and go back to step 2; Otherwise, stop and obtain the approximate optimal control policies $r_m^{(i+1)}$.

$$\varpi(s) = \tanh^{-T}(\bar{U}^{-1}s)\bar{U}S$$

$$V^{(i+1)}(p(k), p^*) = \sum_{\tau=0}^{\infty} \gamma^{\tau-k}\left(\|p(\tau) - p^*\|_Q^2 + 2\int_0^{r^{(i)}(\tau)} \varpi(s)ds\right). \tag{8.137}$$

S is a positive definite matrix with appropriate dimensions. $\bar{U} = \text{diag}\{\bar{U}_1, \bar{U}_2, \ldots, \bar{U}_\vartheta\}$.

Remark 8.30 From the definitions of production indices $p(k)$ and operational indices $r(k)$ in (8.105), it is clear that the computations of (8.135) and (8.136) is more complex than (8.133) and (8.134), since the dimensions of $p(k)$ and $r(k)$ are larger than $p_m(k)$ and $r_m(k)$, respectively. In addition, the interior correlation among production indices cannot be clearly shown or used if adopting (8.135) and (8.136).

Remark 8.31 In the graph $\mathcal{G} = (\mathcal{V}, \mathcal{E})$, if $V_m^{(0)} = 0$ and all agents m $(m \in \mathcal{V})$ update their control policies simultaneously at each iteration i using Algorithm 8.5, then the sequences $\{V_m^{(i)}\}_{i=0}^{\infty}$ can converge monotonically to the Nash equilibrium solution V_m^* of (8.115) resulting in $r_m^{(i)}$ converging to r_m^* given by (8.114). The proof of this convergence is similar to Kiumarsi and Lewis (2015), Wei et al. (2014), Mu et al. (2017).

Remark 8.32 Notice that nonzero-sum game theory is also used for optimization problem of multi-agent systems (Li et al. 2017; Jiang and He 2019; Qin et al. 2019; Zhang et al. 2017, 2019; Zuo et al. 2017; Gao et al. 2019; Jiao et al. 2016; Modares et al. 2018). Li et al. (2017), Jiang and He (2019) and Qin et al. (2019) require the dynamics of agents to be isomorphic and linear, and the constraints of control inputs are not taken into account in Zhang et al. (2017, 2019), Zuo et al. (2017), Gao et al. (2019) and Jiao et al. (2016) with heterogeneous linear dynamics of agents and in Modares et al. (2018) with heterogeneous affine nonlinear continuous-time systems. However, Algorithm 8.5 is developed for more general and practical case, since it can find the control policies for solving optimal control problem of multi-agents with the features of general nonlinearity, heterogeneity, inter-coupling and control constraints.

Remark 8.33 Notice that what (Zhang et al. 2009) focused is to find the approximate optimal control policy for single affine nonlinear systems with control constraints through solving single DTHJB equation using iteration dynamic programming method, while approximate Nash equilibrium solution of coupled and heterogeneous multi-agents with control constraints is founded in this paper by solving multiple coupled DTHJB equations (8.115).

8.4.3.2 Neural Network-Based Approximate Solution for Designing Optimal Operational Index

We developed Algorithm 8.6 wherein neural networks are used to approximate the production indices $p_m(k+1)$, the operational indices $r_m^{(i)}$ and the value functions $\hat{V}_m^{(i+1)}$ by function approximation approach. The followings are given for approximating the dynamics of the production indices $p_m(k+1)$:

$$\hat{p}_m(k+1) = \hat{w}_{pm}^T \sigma(v_{pm}^T X_{pm}(k)) \tag{8.138}$$

where $\hat{w}_{pm}$ and v_{pm}, respectively, are the weights of the hidden layer to the output layer and the weights of the input layer to the hidden layer, and $X_{pm}(k) = \left[p_m^T(k)\; r_m^T(k)\; r_{-m}^T(k)\; \pi(k)\; q(k)\right]^T$. $\sigma\left(v_{pm}^T X_{pm}(k)\right)$ is the activation function vector, $[\sigma(z)]_q = (e^z - e^{-z})/(e^z + e^{-z})$ and q is the number of neurons in the hidden layer.

The critic NNs are given by the form of the three-layer neural networks

$$\hat{V}_m^{(i+1)}(W_m(k), p_m^*) = (w_{vm}^{(i+1)})^T \sigma((v_{vm}^{(i+1)})^T Z_{vm}(k)) \tag{8.139}$$

where $Z_{vm}(k) = \left[p_j^T(k)\; (p_m^*)^T\right]^T$ $(j \in \{m \cup N_m\})$, $w_{vm}^{(i+1)}$ and $v_{vm}^{(i+1)}$ respectively are the weights of the hidden layer to the output layer and the weights of the input layer to the hidden layer for each critic NN.

The actor NNs are given below

$$\hat{r}_m^{(i+1)}(k) = \left(w_{rm}^{(i+1)}\right)^T \sigma\left((v_{rm}^{(i+1)})^T Z_{rm}(k)\right) \tag{8.140}$$

where $Z_{rm}(k) = \left[p_j^T(k)\ (p_m^*)^T\right]^T$ $(j \in \{m \cup N_m\})$. $w_{rm}^{(i+1)}$ and$v_{rm}^{(i+1)}$ respectively denote the weights of the hidden layer to the output layer and the weights of the input layer to the hidden layer for each actor NN.

To train the NNs (8.138)–(8.140), the gradient descent algorithm (Wang et al. 2017; Mu et al. 2017; Zhang et al. 2009) can be used to update the weights of NNs. Now, Algorithm 8.6 is developed to summarize the nonzero-sum-based RL graphical game with actor–critic structure for optimizing the plant-wide performance of industrial processes.

Algorithm 8.6 Data driven decision-making for the operational indices online in real time

---

1: Training the NNs of the production indices (for all agents in the graph $\mathcal{G} = (\mathcal{V}, \mathcal{E})$:

(1) Initialize weight vectors $\hat{w}_{pm}, v_{pm}$ $(m \in \mathcal{V})$;
(2) Train weights in terms of gradient descent algorithms using the measured data $(p_m(k), r_m(k), r_{-m}(k), \pi(k), q(k))$ from the practical operation of industrial process until $e_{pm} = \left|p_m - \hat{p}_m\right| \leq \varepsilon$ $(\varepsilon > 0)$;
(3) Get the trained weights $\hat{w}_{pm}$ and v_{pm};

2: Approximate critic and actor NNs:

(1) Initialize iterative functions $V_m^{(0)}(\cdot) = 0$ for all agents m $(m \in \mathcal{V})$ and further calculate the initial operational policies $r_m^{(0)}$ by (8.134) for all agents. Let the iteration index $i = 0$;
(2) **For** $m = 1 : \vartheta$:

(2.1) Train the critic network of agent m: Calculate $\hat{V}_m^{(i+1)}$ by updating the critic weights $w_{vm}^{(i+1)}$ and $v_{vm}^{(i+1)}$;
(2.2) Train the actor network of agent m: Calculate $\hat{r}_m^{(i+1)}(k)$ by updating the actor weights $w_{rm}^{(i+1)}$ and $v_{rm}^{(i+1)}$;

End for;

3: If $\left\| \hat{V}_m^{(i+1)} - \hat{V}_m^{(i)} \right\| > \varepsilon$ for any agent m, then set $i = i + 1$ and go back to Step 2(2); Otherwise stop and obtain the approximate optimal control policies $\hat{r}_m^{(i+1)}$.

Remark 8.34 Since we try to find the optimal ϑ-tuple of control polices by using the value iteration approach (Wei et al. 2014; Mu et al. 2017), then the almost same learning framework as them, while with different iterative Bellman equation (8.133) and policy updating (8.134) for each agent, is adopted in Algorithms 8.5 and 8.6.

Remark 8.35 When implementing Step 2(1–3) in Algorithm 8.6. $p_j(k)$ can be generated by using two approaches, one is from the real operational processes online real time wherein $\hat{r}_j^{(i)}(k)$ act the processes, and the other is to estimate them using the NNs modeled in (8.138). To guarantee that the future policies be optimal, it is preferred to collect $p_m(k)$ online real time. If it is desired to avoid much interference with the real industrial processes, offline estimation of $p_m(k)$ can be used.

Fig. 8.13 Shaft furnace and grinding process

Remark 8.36 In Algorithm 8.6, the value iteration is combined with neural network approximation to learn the global Nash equilibrium solutions. If proper learning rates and probing noise are chosen, then the approximation errors between $\hat{V}_m^{(i+1)}$ and $V_m^{(i+1)}$ are bounded and as small as desired. In this sense, the approximate optimal operational indices can be derived for achieving the optimality of the production indices. Here it is emphasized that the proposed Algorithm 8.6 is general in the sense that it is valid for arbitrary unknown dynamics of agents even though they are nonlinear and heterogeneous.

Remark 8.37 Algorithm 8.6 is implemented by using only data from the plant unlike (Ding et al. 2016) where operators' experience is needed to repeatedly correct the operational indices when implementing performance evaluation.

8.4.4 Industrial Application in Iron Ore Processing

In this subsection, industrial emulations are conducted using data from a large iron ore processing plant in Gansu Province, China (see Fig. 8.13) to show (a) the effectiveness of the proposed Algorithm 8.6 when achieving the optimality of the plant-wide performance; (b) the significant improvement by comparing the proposed nonzero-sum game RL with the existing methods. First, the plant-wide performance optimization problem of mineral processing factory is formulated, and then the emulations using the proposed nonzero-sum game RL are implemented. Finally, comparisons with the existing methods are made to show the significant improvement of using the proposed nonzero-sum game RL method.

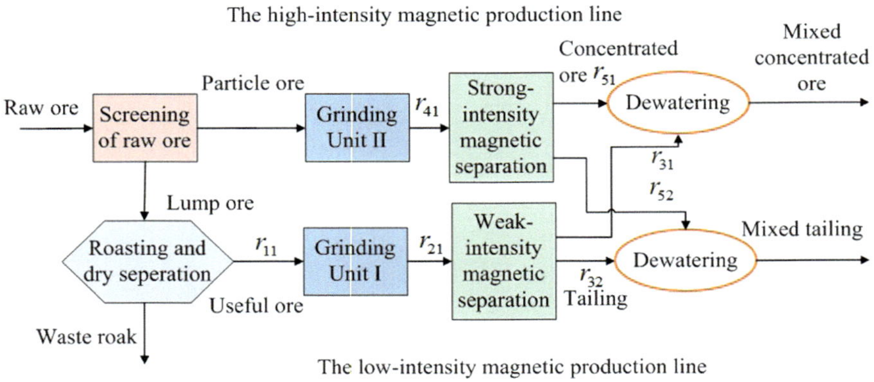

Fig. 8.14 Production line for mineral processing production plant

Table 8.3 Operational indices and their bounds of mineral processing

Units	Operational indices
Shaft furnace	r_{11}: Magnetic tube recovery rate (MTRR, %) $r_{11,\max} = 100$, $r_{11,\min} = 80$
Grinding unit I	r_{21}: Grinding particle size (PSWMPL, %) $r_{21,\max} = 95$, $r_{21,\max} = 60$
Separation (Weak)	r_{31}: Concentrate grade (CGWMPL, %) $r_{31,\max} = 100$, $r_{31,\min} = 45$ r_{32}: Tailing grade (TGWMPL, %) $r_{32,\max} = 20$, $r_{32,\min} = 0$
Grinding unit II	r_{41}: Grinding particle size (PSSMPL, %) $r_{41,\max} = 90$, $r_{41,\min} = 58$
Separation (Strong)	r_{51}: Concentrate grade (CGSMPL, %) $r_{51,\max} = 100$, $r_{51,\min} = 45$ r_{52}: Tailing grade (TGSMPL, %) $r_{52,\max} = 22$, $r_{52,\min} = 0$

8.4.4.1 Mineral Processing of Hematite Iron Ore Formulation

Figure 8.14 shows the production structure of the biggest iron ore processing plant in Gansu Province, China, which consists of some units: screening, shaft furnace roasting, grinding and low-intensity and high-intensity magnetic separation (Ding et al. 2016). In addition, there are two dewatering units used for producing the final mixed concentrated ore and tailings. The daily mixed concentrate grade p_1 and the daily yield of concentrated ore p_2 are the production indices focused in this paper and taken as the system outputs. The operational indices $r_{11}, r_{21}, r_{31}, r_{32}, r_{41}, r_{51}, r_{52}$ of all unit processes are the system inputs, and they need to satisfy the constraints in Table 8.3.

According to the real degree of correlation between these operational indices and production indices, the daily mixed concentrate grade p_1 is mainly affected by the operational index $r_1 = [r_{31}\ r_{32}\ r_{51}\ r_{52}]^T$, while the operational index

$r_2 = [r_{11}\ r_{21}\ r_{41}]^T$ is the prominent factor that causes the changing of the daily yield of concentrated ore p_2 [1]. Moreover, the two production indices p_1 and p_2 are coupled under the operational indices r_1 and r_2, and they are time-varying with the forms of $p_1(k+1) = f_1(p_1(k), r_1(k), a_2 r_2(k), \pi(k), q(k))$ and $p_2(k+1) = f_2(p_2(k), r_2(k), a_1 r_1(k), \pi(k), q(k))$, where $a_1 = \begin{bmatrix} I \\ 0 \end{bmatrix}$, $a_2 = \begin{bmatrix} 0 \\ I \end{bmatrix}$ and I being an identity matrix with appropriate dimension. Agent p_1 will try its best to follow the target p_1^* by designing an approximate operational index r_1, meanwhile, the changing of r_1 will influence agentp_2 as well. So does agent p_2. In this sense, agents p_1 and p_2 could be viewed as two nodes in a strong connected graph $\mathcal{G}$. The connectivity matrix $E = [e_{ij}]$ is a 2×2 matrix, where $e_{12} = 1$ and $e_{21} = 1$. This is indeed a nonzero-sum graphical game problem, in which all agents could reach the Nash equilibrium at last.

The target production indices are chosen as $[p_1^*, p_2^*] = \{[58.25,\ 346], [58,\ 318], [58,\ 1.342]\}$ for three cases of working conditions (average process throughput per hour $\pi(k)$, the process raw material quality $q(k)$). The control objective here is to find the proper operational indices r_1, r_2 within bounds at each working condition, under which the production indices p_1 and p_2 can respectively track their own targets p_1^* and p_2^*. This can be described in Problem 8.5. To achieve this objective, Algorithm 8.6 will be implemented, such that Problem 8.5 can be solved by the nonzero-sum graphical game RL approach, since Problem 8.6 is more easily solvable to handle Problem 8.5.

8.4.4.2 Learning the Optimal Operational Indices

The actual data of the production indices $p_m(kT)$ $(m = 1, 2)$ and the unit operational indices r_1, r_2 are collected after each production run k with running period $T = 1.2$h. The collected data are used for NNs modeling the dynamics f_1 and f_2 of the two production indices.

First, we run Step 1 in Algorithm 8.6. Let the NNs structures of both the daily mixed concentrate grade (agent 1) and the daily yield of concentrated ore (agent 2) be 14-15-1. Let the learning rates be 0.01 and the training errors be 0.001 for all these two NNs. For training the networks of production indices, 466 data of the total 532 data collected from the biggest iron ore processing plant in Gansu Province, China are used to estimate the weights of these two networks. The left 66 data are utilized to validate the trained models. Figure 8.15 shows the results of validation using the trained neural networks of the production indices.

Secondly, Step 2 in Algorithm 8.6 is implemented. Let the structures of critic NNs for agent 1 and agent 2 be 3-6-1, and let the structures of actor NNs for agent 1 and agent 2 be 3-14-4 and 3-14-3, respectively. Choose $Q_1 = Q_2 = 10$ and $\gamma = 0.5$. Let the learning rates be 0.01 and the training errors be 0.001 for all these NNs. The training or iteration results of the critic and actor networks for agent 1 and agent 2 at the first working condition are shown in Fig. 8.16a, b, c, d, respectively. Thus, the approximate optimal operational indices can be derived after 5th

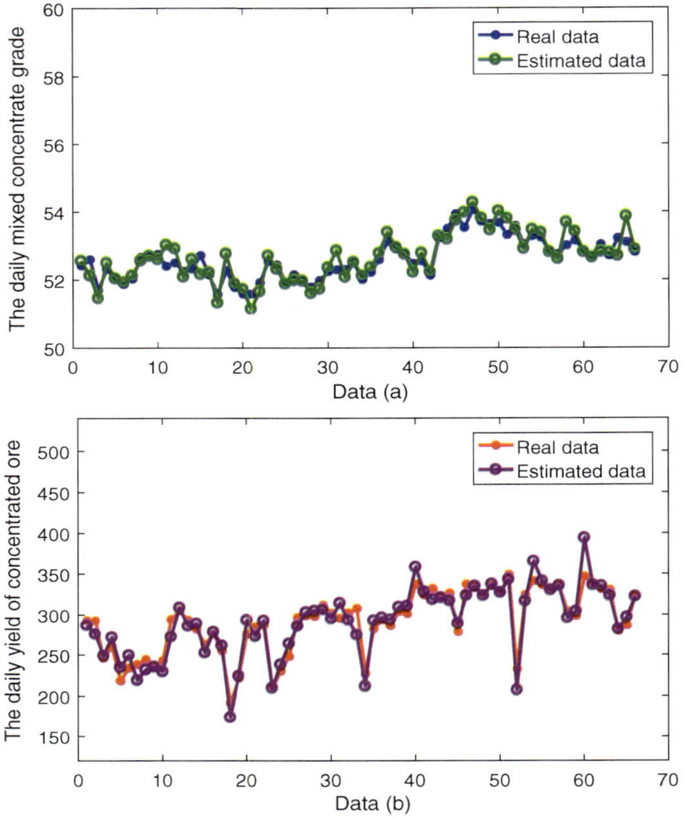

Fig. 8.15 Validation results using the trained weights of production indices

iteration. Figures 8.17 and 8.18 present the tracking results of the mixed concentrate grade p_1 and the yield of concentrated ore p_2 under the learned approximately optimal operational indices shown in Figs. 8.19 and 8.20 after implementing Algorithm 8.6 under the above three cases of working condition. For another target values $[p_1^*, p_2^*] = \{[58, 341], [58.25, 328], [58.1, 339]\}$ under another three cases of working conditions, Figs. 8.21 and 8.22 present the tracking results of the mixed concentrate grade p_1 and the yield of concentrated ore p_2 under the learned approximately optimal operational indices shown in Fig. 8.23.

8.4.4.3 Analysis and Comparisons with Existing Methods

From Figs. 8.17, 8.18, 8.21 and 8.22, one can find that tracking objective of the two production indices are achieved under the learned approximated optimal operational

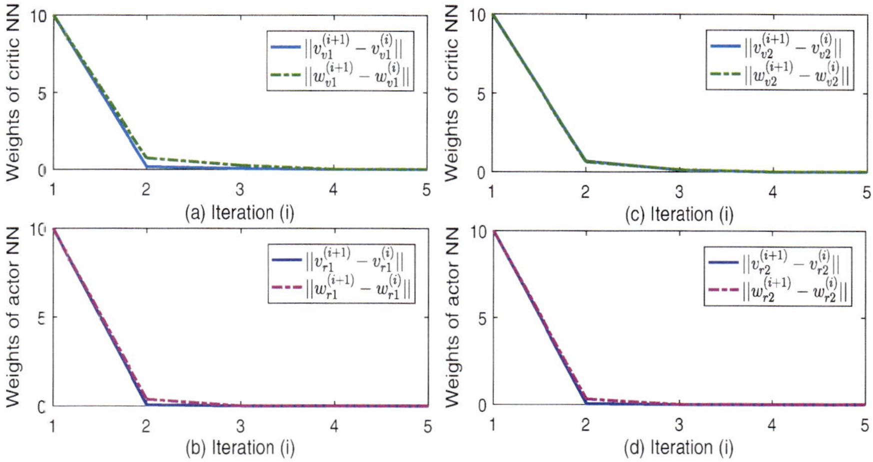

Fig. 8.16 Learning results of weights of critic and actor neural networks

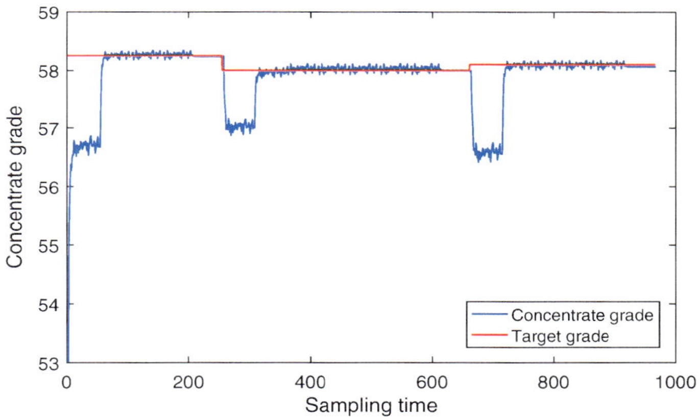

Fig. 8.17 The tracking results of the mixed concentrate grade

indices shown in Figs. 8.19, 8.20 and 8.23. Moreover, it is clear that all learned optimal operational indices using the proposed Algorithm 8.6 are within the ranges of their own constraints from Table 8.3 and Figs. 8.19, 8.20 and 8.23. To evaluate the performance, Table 8.5 lists the rewards calculated based on (8.101) under the learned optimal operational indices at each working condition.

According to the data collected from the on-site mineral processing of hematite iron ore, the actual mixed concentrate grade and the actual yield of concentrated ore are given in Table 8.4. These are achieved based on the operators' experience. Compared with the measured results from manual regulations, the mixed concentrate grade and the yield of concentrated ore for these six cases of working condition

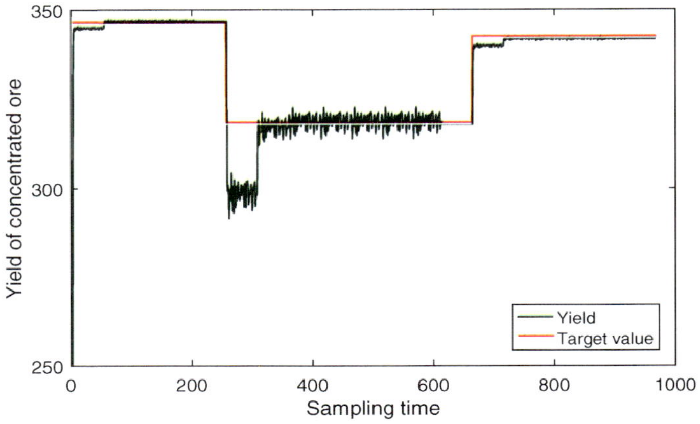

Fig. 8.18 The tracking results of the yield of concentrated ore

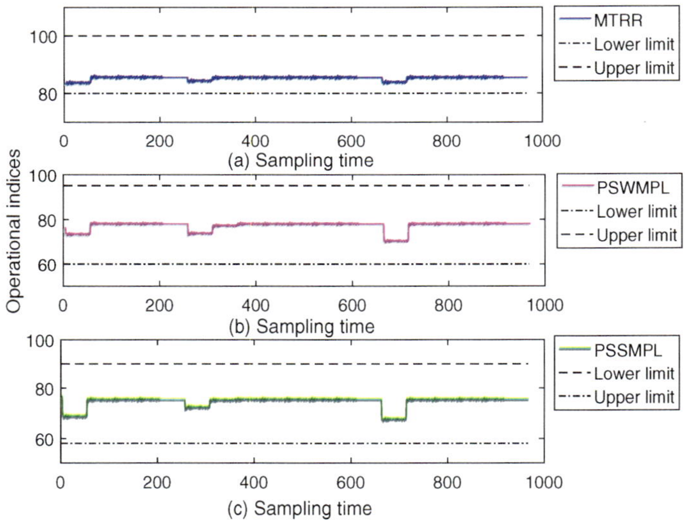

Fig. 8.19 The approximately optimal operational indices

are averagely improved 5.82% and 73.34t/h, respectively, by using the proposed Algorithm 8.6 in this paper.

To further show the efficiency of this proposed game RL algorithm, the computational complexity and the running time of the above-mentioned first three working conditions are compared using the proposed Algorithm 8.6 in this paper with the augmented RL way (Neural networks and RL are combined to learn value function and operational indices for augmented form (8.103) and (8.104) like (Zhang et al. 2009)) in Tables 8.6 and 8.7. If we use the augment way, then only one critic NN

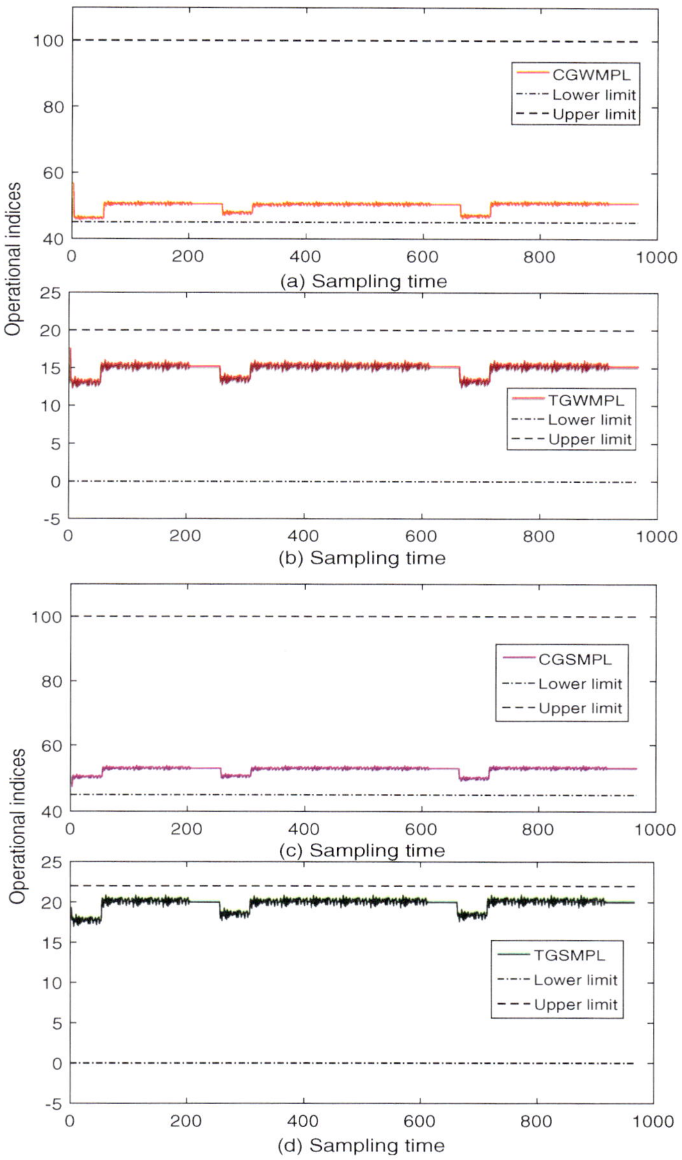

Fig. 8.20 The approximately optimal operational indices

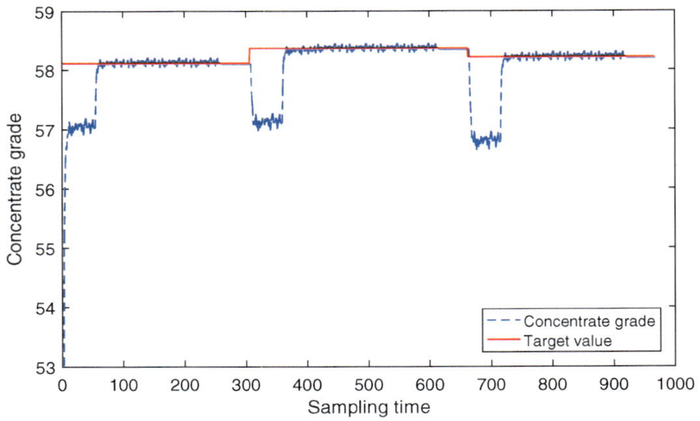

Fig. 8.21 The tracking results of the mixed concentrate grade

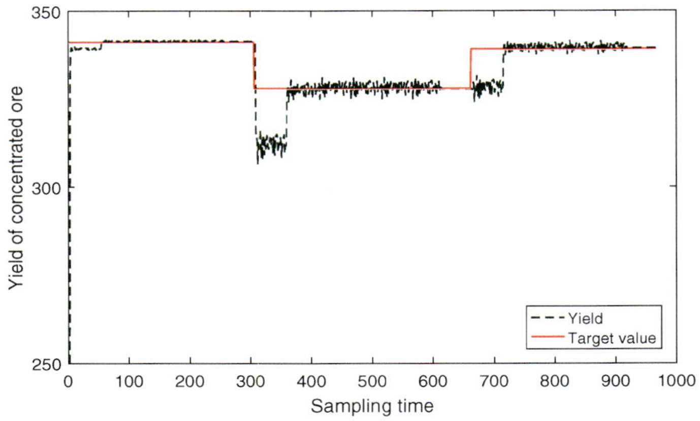

Fig. 8.22 The tracking results of the yield of concentrated ore

and one actor NN are needed to learn in Step 2(2) and Step 2(3) in Algorithm 8.6. The same number of hidden layer as the game RL approach here is used to make a comparison. From Table 8.6, one can see that more 57 and 72 unknown elements than those in agent 1 and agent 2, respectively, if training the weights of the iterative actors and iterative critics using the augmented RL way. More computational complexity and computer storage might produce long running time of simulations shown in Table 8.7. (In these simulations, CPU is Intel, Core^{TM}, I7-7700 and internal storage is 8GB).

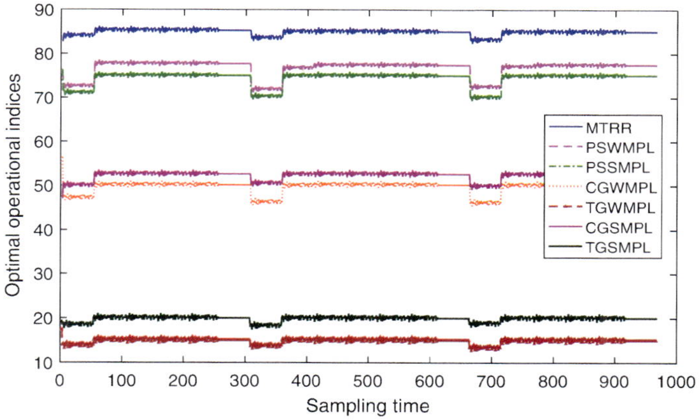

Fig. 8.23 The approximately optimal operational indices

Table 8.4 Manual regulation by operators' experience

	Real measured grade (%)	Measured yield (t)
Case 1	52.40	290.57
Case 2	52.11	229.50
Case 3	52.55	285.98
Case 4	52.27	286.67
Case 5	52.29	244.20
Case 6	52.48	237.05
Mean values	52.35	262.33

Table 8.5 Rewards calculated using Algorithm 8.2

	Mixed concentrate grade (%)	Yield (t)
Case 1	0.8960	675.11
Case 2	0.5279	20339
Case 3	0.3708	482.13
Case 4	0.6512	415.42
Case 5	0.4732	22388
Case 6	0.5062	6819

Table 8.6 The number of unknown variables

	Algorithm 8.2 (Game RL)	Augmented RL way
Mixed concentrate grade	Critic NN $w_{v1}^{(i+1)}$: 7; $v_{v1}^{(i+1)}$: 24 Actor NN $w_{r1}^{(i+1)}$: 60; $v_{r1}^{(i+1)}$: 64	Critic NN $w_v^{(i+1)}$: 7; $v_v^{(i+1)}$: 30
Yield	Critic NN $w_{v2}^{(i+1)}$: 7; $v_{v2}^{(i+1)}$: 24 Actor NN $w_{r2}^{(i+1)}$: 45; $v_{r2}^{(i+1)}$: 64	Actor NN $w_r^{(i+1)}$: 105; $v_r^{(i+1)}$: 70

Table 8.7 Running time

	Case 1	Case 1 and Case 2	Case 1, Case 2 and Case 3
Algorithm 8.2 (Game RL)	2.876s	3.328s	3.405s
Augmented RL way	23.703s	Out of memory	Out of memory

8.5 Conclusion

In this section, the off-policy RL algorithms are presented to solve the OOC problem and plant-wide performance optimization problem for industrial processes using only measured data. For the OOC problem, the zero-sum game method is employed for H_∞ tracking problem resulting in the DT GAREs to solve the optimal set-point and the worst disturbance. An off-policy Bellman equation is derived to evaluate the target set-point and the target disturbance to find improved target set-point. Operational process applies a behavior set-point to collect data which is used to learn the solution to the off-policy Bellman equation. This set-point is different from the target set-point which is not actually applied to the systems and is evaluated and learned to find the optimal set-point. The proposed approaches do not require any knowledge of the industrial operational process dynamics. The flotation process examples with single cell and multiple cells show the effectiveness of the proposed methods. Moreover, a nonzero-sum game RL-based approach combined with game theory is developed to achieve plant-wide performance optimization of large-scale industrial processes using only data and without requiring the dynamics of unit processes and the dynamics of production indices to be known a priori. A multi-agent game optimization problem is constructed by formulating the plant-wide performance optimization with strict constraints of operational indices, and its exact global Nash equilibrium solutions are derived by the nonzero-sum graphical game. The stability and the global Nash equilibrium are rigorously proven. Then the RL-based algorithm with critic–actor structure is proposed to learn the Nash equilibrium solution for reaching the global Nash equilibrium of production indices. The convergence of the proposed algorithm is analyzed, and the emulations are conducted to show the effectiveness of the proposed algorithm by using the data from a large mineral processing plant in Gansu Province, China.

References

Al-Tamimi A, Abu-Khalaf M, Lewis FL (2007) Adaptive critic designs for discrete-time zero-sum games with application to H_∞ control. IEEE Trans Syst Man Cybern Part B (Cybern) 37(1):240–247

Al-Tamimi A, Lewis FL, Abu-Khalaf M (2007) Model-free Q-learning designs for linear discrete-time zero-sum games with application to H-infinity control. Automatica 43(3):473–481

Başar T, Olsder GJ (1998) Dynamic noncooperative game theory, 2nd edn. SIAM, PA

Başar T, Bernhard P (2008) H-infinity optimal control and related minimax design problems: a dynamic game approach. Springer Science & Business Media

Chai T, Ding J, Wu F (2011) Hybrid intelligent control for optimal operation of shaft furnace roasting process. Control Eng Pract 19(3):264–275

Chai T, Zhao L, Qiu J, Liu F, Fan J (2012) Integrated network-based model predictive control for setpoints compensation in industrial processes. IEEE Trans Industr Inf 9(1):417–426

Chai T, Qin SJ, Wang H (2014) Optimal operational control for complex industrial processes. Ann Rev Control 38(1):81–92

Cheng Y, Shi D, Chen T, Shu Z (2015) Optimal data scaling for principal component pursuit: A lyapunov approach to convergence. IEEE Trans Autom Control 60(8):2057–2071

Dai W, Chai T, Yang SX (2014) Data-driven optimization control for safety operation of hematite grinding process. IEEE Trans Industr Electron 62(5):2930–2941

Desbiens A, Hodouin D, Najim K, Flament F (1994) Long-range predictive control of a rougher flotation unit. Miner Eng 7(1):21–37

Ding J, Modares H, Chai T, Lewis FL (2016) Data-based multiobjective plant-wide performance optimization of industrial processes under dynamic environments. IEEE Trans Industr Inf 12(2):454–465

Duchesne C, Bouajila A, Bartolacci G, Pelletier P, Breau Y, Fournier J, Girard D (2003) Application of multivariate image analysis (mia) to predict concentrate grade in froth flotation processes. In: Proceedings of the 35th annual meeting of the Canadian mineral processors, vol 511. CIM, Ottawa, Canada

Dünnebier G, Van Hessem D, Kadam JV, Klatt KU, Schlegel M (2005) Optimization and control of polymerization processes. Chem Eng Technol: Industr Chem Plant Equip Process Eng Biotechnol 28(5):575–580

Ellis M, Christofides PD (2014) Integrating dynamic economic optimization and model predictive control for optimal operation of nonlinear process systems. Control Eng Pract 22:242–251

Engell S (2007) Feedback control for optimal process operation. J Process Control 17(3):203–219

Fan J, Jiang Y, Chai T (2016) Operational feedback control of industrial processes in a wireless network environment. Acta Autom Sinica 42(8):1166–1174

Ferreira J, Loveday B (2000) An improved model for simulation of flotation circuits. Miner Eng 13(14–15):1441–1453

Gao W, Jiang Y, Davari M (2019) Data-driven cooperative output regulation of multi-agent systems via robust adaptive dynamic programming. IEEE Trans Circuits Syst II: Express Briefs 66(3):447–451

Hou ZS, Wang Z (2013) From model-based control to data-driven control: Survey, classification and perspective. Inf Sci 235:3–35

Huang X, Chu Y, Hu Y, Chai T (2005) Production process management system for production indices optimization of mineral processing. IFAC Proceed 38(1):178–183

Hulbert D (1995) Multivariable control of pulp levels in flotation circuits. IFAC Proceed 28(17):91–97

Jiang H, He H (2019) Data-driven distributed output consensus control for partially observable multiagent systems. IEEE Trans Cybern 49(3):848–858

Jiang Y, Jiang ZP (2012) Computational adaptive optimal control for continuous-time linear systems with completely unknown dynamics. Automatica 48(10):2699–2704

Jiao Q, Modares H, Xu S, Lewis FL, Vamvoudakis KG (2016) Multi-agent zero-sum differential graphical games for disturbance rejection in distributed control. Automatica 69:24–34
Johnson M, Kamalapurkar R, Bhasin S, Dixon WE (2014) Approximate N-player nonzero-sum game solution for an uncertain continuous nonlinear system. IEEE Trans Neural Networks Learn Syst 26(8):1645–1658
Kim JH, Lewis FL (2010) Model-free H_∞ control design for unknown linear discrete-time systems via Q-learning with LMI, vol 46. Elsevier
Kiumarsi B, Lewis FL (2015) Actor-critic-based optimal tracking for partially unknown nonlinear discrete-time systems. IEEE Trans Neural Networks Learn Syst 26(1):140–151
Kiumarsi B, Lewis FL, Modares H, Karimpour A, Naghibi-Sistani MB (2014) Reinforcement Q-learning for optimal tracking control of linear discrete-time systems with unknown dynamics. Automatica 50(4):1167–1175
Kiumarsi B, Lewis FL, Jiang Z (2017) H_∞ control of linear discrete-time systems: Off-policy reinforcement learning. Automatica 78:144–152
Lee JY, Park JB, Choi YH (2012) Integral Q-learning and explorized policy iteration for adaptive optimal control of continuous-time linear systems. Automatica 48(11):2850–2859
Lewis FL, Vrabie D, Syrmos VL (2012) Optimal control. John Wiley & Sons, New York, NY, USA
Lewis FL, Vrabie D, Vamvoudakis KG (2012) Reinforcement learning and feedback control: Using natural decision methods to design optimal adaptive controllers. IEEE Control Syst Mag 32(6):76–105
Li J, Modares H, Chai T, Lewis FL, Xie L (2017) Off-policy reinforcement learning for synchronization in multiagent graphical games. IEEE Trans Neural Networks Learn Syst 28(10):2434–2445
Li J, Chai T, Lewis FL, Ding Z, Jiang Y (2019) Off-policy interleaved Q-learning: Optimal control for affine nonlinear discrete-time systems. IEEE Trans Neural Networks Learn Syst 30(5):1308–1320
Li K, Chen R, Fu G, Yao X (2018) Two-archive evolutionary algorithm for constrained multiobjective optimization. IEEE Trans Evol Comput 23(2):303–315
Liu D, Xu Y, Wei Q, Liu X (2017) Residential energy scheduling for variable weather solar energy based on adaptive dynamic programming. IEEE/CAA J Autom Sinica 5(1):36–46
Liu F, Gao H, Qiu J, Yin S, Fan J, Chai T (2013) Networked multirate output feedback control for setpoints compensation and its application to rougher flotation process. IEEE Trans Industr Electron 61(1):460–468
Luo B, Wu HN (2013) Computationally efficient simultaneous policy update algorithm for nonlinear H_∞ state feedback control with Galerkin's method, vol 23. Wiley Online Library
Luo B, Liu D, Huang T, Wang D (2016) Model-free optimal tracking control via critic-only Q-learning. IEEE Trans Neural Networks Learn Syst 27(10):2134–2144
Lv Y, Ren X (2019) Approximate nash solutions for multiplayer mixed-zero-sum game with reinforcement learning. IEEE Trans Syst Man Cybern: Syst 49(12):2739–2750
Lv Y, Na J, Ren X (2019) Online H_∞ control for completely unknown nonlinear systems via an identifier-critic-based adp structure. Int J Control 92(1):100–111
Maldonado M, Sbarbaro D, Lizama E (2007) Optimal control of a rougher flotation process based on dynamic programming. Miner Eng 20(3):221–232
Mayne David Q (1970) Differential dynamic programming. American Elsevier Pub, Co
Modares H, Lewis FL (2014) Linear quadratic tracking control of partially-unknown continuous-time systems using reinforcement learning. IEEE Trans Autom Control 59(11):3051–3056
Modares H, Lewis FL, Jiang ZP (2015) H_∞ tracking control of completely unknown continuous-time systems via off-policy reinforcement learning. IEEE Trans Neural Networks Learn Syst 26(10):2550–2562
Modares H, Lewis FL, Kang W, Davoudi A (2018) Optimal synchronization of heterogeneous nonlinear systems with unknown dynamics. IEEE Trans Autom Control 63(1):117–131
Movric KH, Lewis FL (2014) Cooperative optimal control for multi-agent systems on directed graph topologies. IEEE Trans Autom Control 59(3):769–774

Mu C, Ni Z, Sun C, He H (2016) Data-driven tracking control with adaptive dynamic programming for a class of continuous-time nonlinear systems. IEEE Trans Cybern 47(6):1460–1470
Mu C, Wang D, He H (2017) Novel iterative neural dynamic programming for data-based approximate optimal control design. Automatica 81:240–252
Narayanan V, Jagannathan S (2017) Event-triggered distributed control of nonlinear interconnected systems using online reinforcement learning with exploration. IEEE Trans Cybern 48(9):2510–2519
Nash J (1951) Non-cooperative games. Ann Math 286–295
Ochoa S, Wozny G, Repke JU (2010) Plantwide optimizing control of a continuous bioethanol production process. J Process Control 20(9):983–998
Oliveira F, Gupta V, Hamacher S, Grossmann IE (2013) A lagrangean decomposition approach for oil supply chain investment planning under uncertainty with risk considerations. Comput Chem Eng 50:184–195
Pérez-Correa R, González G, Casali A, Cipriano A, Barrera R, Zavala E (1998) Dynamic modelling and advanced multivariable control of conventional flotation circuits. Miner Eng 11(4):333–346
Pomerleau A, Hodouin D, Desbiens A, Gagnon É (2000) A survey of grinding circuit control methods: from decentralized pid controllers to multivariable predictive controllers. Powder Technol 108(2–3):103–115
Pulkkinen K, Ylinen R, Jämsä-Jounela S, Järvensivu M, Woodcock J (1993) Integrated expert control system for grinding and flotation. In: XVIII IMPC, pp 325–334
Qin J, Li M, Shi Y, Ma Q, Zheng WX (2019) Optimal synchronization control of multiagent systems with input saturation via off-policy reinforcement learning. IEEE Trans Neural Networks Learn Syst 30(1):85–96
Qin SJ, Badgwell TA (2003) A survey of industrial model predictive control technology. Control Eng Pract 11(7):733–764
Rojas D, Cipriano A (2011) Model based predictive control of a rougher flotation circuit considering grade estimation in intermediate cells. Dyna 78(166):29–37
Thornton A (1991) Cautious adaptive control of an industrial flotation circuit. Miner Eng 4(12):1227–1242
Vamvoudakis KG, Lewis FL (2011) Multi-player non-zero-sum games: Online adaptive learning solution of coupled Hamilton-Jacobi equations. Automatica 47(8):1556–1569
Vamvoudakis KG, Lewis FL, Hudas GR (2012) Multi-agent differential graphical games: Online adaptive learning solution for synchronization with optimality. Automatica 48(8):1598–1611
Vega P, Revollar S, Francisco M, Martín J (2014) Integration of set point optimization techniques into nonlinear mpc for improving the operation of wwtps. Comput Chem Eng 68:78–95
Wang D, He H, Liu D (2017) Adaptive critic nonlinear robust control: A survey. IEEE Trans Cybern 47(10):3429–3451
Wei Q, Liu D, Shi G (2014) A novel dual iterative Q-learning method for optimal battery management in smart residential environments. IEEE Trans Industr Electron 62(4):2509–2518
Wei Q, Lewis FL, Sun Q, Yan P, Song R (2017) Discrete-time deterministic Q-learning: A novel convergence analysis. IEEE Trans Cybern 47(5):1224–1237
Wei Q, Liu D, Lin Q, Song R (2018) Adaptive dynamic programming for discrete-time zero-sum games. IEEE Trans Neural Networks Learn Syst 29(4):957–969
Wongsri M, Siraworakun C (2010) Dynamic performance-based design for plantwide control structure. Chem Eng Commun 197(5):753–774
Wu Z, Wu Y, Chai T, Sun J (2014) Data-driven abnormal condition identification and self-healing control system for fused magnesium furnace. IEEE Trans Industr Electron 62(3):1703–1715
Würth L, Hannemann R, Marquardt W (2011) A two-layer architecture for economically optimal process control and operation. J Process Control 21(3):311–321
Xie Y, Wu J, Xu D, Yang C, Gui W (2016) Reagent addition control for stibium rougher flotation based on sensitive froth image features. IEEE Trans Industr Electron 64(5):4199–4206
Xu C, Sand G, Harjunkoski I, Engell S (2012) A new heuristic for plant-wide schedule coordination problems: The intersection coordination heuristic. Comput Chem Eng 42:152–167

Yin S, Gao H, Qiu J, Kaynak O (2016) Fault detection for nonlinear process with deterministic disturbances: A just-in-time learning based data driven method. IEEE Trans Cybern 47(11):3649–3657

Yuan Q, Wang F, He D, Wang H, Liu T (2016) A new plant-wide optimization method and its application to hydrometallurgy process. Can J Chem Eng 94(2):273–280

Zaragoza R, Herbst J (1988) Model-based feedforward control scheme for flotation plants. Min Metall Explor 5(4):177–185

Zhang H, Luo Y, Liu D (2009) Neural-network-based near-optimal control for a class of discrete-time affine nonlinear systems with control constraints. IEEE Trans Neural Networks 20(9):1490–1503

Zhang H, Jiang H, Luo Y, Xiao G (2017) Data-driven optimal consensus control for discrete-time multi-agent systems with unknown dynamics using reinforcement learning method. IEEE Trans Industr Electron 64(5):4091–4100

Zhang J, Wang Z, Zhang H (2019) Data-based optimal control of multiagent systems: A reinforcement learning design approach. IEEE Trans Cybern 49(12):4441–4449

Zhang R, Tao J (2016) Data-driven modeling using improved multi-objective optimization based neural network for coke furnace system. IEEE Trans Industr Electron 64(4):3147–3155

Zuo S, Song Y, Lewis FL, Davoudi A (2017) Output containment control of linear heterogeneous multi-agent systems using internal model principle. IEEE Trans Cybern 47(8):2099–2109

Appendix A

A.1 Matrix Calculus

The Kronecker product is an operation between two matrices of arbitrary size, and that of two matrices $A = [a_{ij}] \in \mathbb{R}^{m \times n}$ and $B = [b_{ij}] \in \mathbb{R}^{p \times q}$, which is expressed as

$$A \otimes B = \begin{bmatrix} a_{11}B & \dots & a_{1n}B \\ \vdots & \ddots & \vdots \\ a_{m1}B & \dots & a_{mn}B \end{bmatrix} = \left[a_{ij}B\right] \in \mathbb{R}^{mp \times nq} \tag{A.1}$$

or

$$A \otimes B = \begin{bmatrix} Ab_{11} & \dots & Ab_{1q} \\ \vdots & \ddots & \vdots \\ Ab_{p1} & \dots & Ab_{pq} \end{bmatrix} = \left[Ab_{ij}\right] \in \mathbb{R}^{mp \times nq}. \tag{A.2}$$

Set $x \in \mathbb{R}^n = [x_1\ x_2\ \dots\ x_n]^T$ to be a vector, $s \in \mathbb{R}$ to be a scalar, and $f(x) \in \mathbb{R}^m$ to be an m-vector function with respect to x. The differential regarding x can be shown as

$$dx = \begin{bmatrix} dx_1 \\ dx_2 \\ \vdots \\ dx_n \end{bmatrix}. \tag{A.3}$$

The derivative of x in terms of s (which could be time) is

J. Li et al., *Reinforcement Learning*, Advances in Industrial Control,
https://doi.org/10.1007/978-3-031-28394-9

$$\frac{dx}{ds} = \begin{bmatrix} dx_1/ds \\ dx_2/ds \\ \vdots \\ dx_n/ds \end{bmatrix}. \tag{A.4}$$

If s is a function of x, then the gradient of s in terms of x, shown below, is a column vector.

$$s_x \triangleq \frac{\partial s}{\partial x} = \begin{bmatrix} \partial s/\partial x_1 \\ \partial s/\partial x_2 \\ \vdots \\ \partial s/\partial x_n \end{bmatrix}. \tag{A.5}$$

Then the total differential for s is

$$ds = \left(\frac{\partial s}{\partial x}\right)^T dx = \sum_{i=1}^{n} \frac{\partial s}{\partial x_i} dx_i. \tag{A.6}$$

If s is a function with respect to two vectors x and y, then one has

$$ds = \left(\frac{\partial s}{\partial x}\right)^T dx + \left(\frac{\partial s}{\partial y}\right)^T dy. \tag{A.7}$$

The second derivative can be expressed as

$$s_{xx} \triangleq \frac{\partial^2 s}{\partial x^2} = \left[\frac{\partial^2 s}{\partial x_i x_j}\right]. \tag{A.8}$$

The Jacobi matrix of f in terms of x is the following $m \times n$ matrix:

$$f_x \triangleq \frac{\partial f}{\partial x} = \left[\frac{\partial f}{\partial x_1}\ \frac{\partial f}{\partial x_2}\ \cdots\ \frac{\partial f}{\partial x_n}\right] \tag{A.9}$$

and

$$\frac{\partial f^T}{\partial x} \triangleq \left(\frac{\partial f}{\partial x}\right)^T \in \mathbb{R}^{n\times m}. \tag{A.10}$$

Then the total differential of f is

$$df = \frac{\partial f}{\partial x} dx = \sum_{i=1}^{n} \frac{\partial f}{\partial x_i} dx_i. \tag{A.11}$$

If y is a vector and A is a matrix, then one can get some statements for the gradient as follows:

$$\frac{\partial}{\partial x}(y^T x) = \frac{\partial}{\partial x}(x^T y) = y \tag{A.12}$$

$$\frac{\partial}{\partial x}(y^T Ax) = \frac{\partial}{\partial x}(x^T A^T y) = A^T y \tag{A.13}$$

$$\frac{\partial}{\partial x}(y^T f(x)) = \frac{\partial}{\partial x}(f^T(x)y) = f_x^T y \tag{A.14}$$

$$\frac{\partial}{\partial x}(x^T Ax) = Ax + A^T x \tag{A.15}$$

and if Q is a symmetric matrix, one has the following forms about the quadratic derivative:

$$\frac{\partial}{\partial x}(x^T Qx) = 2Qx \tag{A.16}$$

$$\frac{\partial}{\partial x}(x - y)^T Q(x - y) = 2Q(x - y). \tag{A.17}$$

Some useful second-order partial derivatives are

$$\frac{\partial^2 x^T Qx}{\partial x^2} = Q + Q^T \tag{A.18}$$

if Q is symmetric, it follows

$$\frac{\partial^2 x^T Qx}{\partial x^2} = 2Q \tag{A.19}$$

$$\frac{\partial^2}{\partial x^2}(x - y)^T Q(x - y) = 2Q. \tag{A.20}$$

For a matrix Q:

- if $x^T Qx > 0$ for all nonzero x, then $Q > 0$ is a positive definite matrix.
- if $x^T Qx \geq 0$ for all nonzero x, then $Q \geq 0$ is a positive semi-definite matrix.
- if $x^T Qx \leq 0$ for all nonzero x, then $Q \leq 0$ is a negative semi-definite matrix.
- if $x^T Qx < 0$ for all nonzero x, then $Q < 0$ is a negative definite matrix.

A.2 Some Notation

$\mathbb{R}$	The p dimensional Euclidean space.
$\mathbb{R}^{p\times q}$	The set of all real p by q matrices.
$Q \geq 0$	Q is a positive semi-definite matrix.

$R > 0$	R is a positive definite matrix.
A^T	Matrix A transpose.
$\mathrm{tr}(A)$	The trace of matrix A.
$\otimes$	Kronecker product.
$\mathrm{vec}(X)$	Turning any matrix X into a single column vector, that is, $\mathrm{vec}(X) = [x_1^T, x_2^T, \ldots, x_m^T]^T$ with $x_i \in \mathbb{R}^s$ the columns of matrix $X \in \mathbb{R}^{s\times m}$.
$\mathrm{vecs}(L)$	$\mathrm{vecs}(L) = \left[l_{11}, 2l_{12}, \ldots, 2l_{1s}, l_{22}, 2l_{23}, \ldots, 2l_{s-1,s}, l_{ss}\right]^T \in \mathbb{R}^{\frac{1}{2}s(s+1)}$, where $L \in \mathbb{R}^{s\times s}$ is a symmetric matrix.
$\|x\|$	The Euclidean norm of vector x.
$\mathrm{diag}A$	Diagonal matrix of A.
j	Policy iteration step: $j = 0, 1, \ldots$
$\lambda_{\max}(Q)$	The maximum eigenvalue of positive definite matrix Q.

In Sect. 8.1**:**

$\mathcal{X},\ \mathcal{X}_i,\ \mathcal{U}$	Compact sets, denote $\mathbb{D} \triangleq \left\{(\delta, u, \delta')\mid\delta,\ \delta' \in \mathcal{X},\ u \in \mathcal{U}\right\}$ and $\mathbb{D}_i \triangleq \left\{(\delta_i, u, \delta_i')\mid\delta_i,\ \delta_i' \in \mathcal{X}_i,\ u \in \mathcal{U}\right\}$.
$\left\langle S_1(\delta, u, \delta'),\ S_2(\delta, u, \delta')\right\rangle_{\mathbb{D}} \triangleq \int_{\mathbb{D}} S_1^T(\delta, u, \delta')S_2(\delta, u, \delta')d(\delta, u, \delta')$	The inner product of column vector functions S_1 and S_2.
$r(k)$	Economic benefit at time instant k, and k is some non-negative integer.
r^*	Desired operational index.
l	Cell l ($l = 1, 2, \ldots, s$), s is some positive integer.
$M_{lp}^i,\ M_{le}^i$	Pulp mass and froth mass.
i	Mineralogical class 1, 2.
$h_{lp},\ q_{la}$	Pulp level and feed flow rate.
$L_{lcg},\ L_{ltg}$	Concentrate grade and Tail grade.
g_a	Feed mineral grade.
$Q_{esp},\ Q_{col}$	Frother and collector.
$x_l = \left[M_{lp}^i\ M_{le}^i\right]^T$	State.
$u_l = \left[h_{lp}\ q_{la}\right]^T$	Control input.
$y_l = \left[L_{lcg}\ L_{ltg}\right]^T$	Control output.
n	Sampling time instant of the lower layer control loop, $n = 0, 1, \ldots$
$L_{lp},\ K_{lI}$	Proportional coefficient and integral coefficient controller gains of proportional-integral (PI) controller.
$e_l(n) = w_l(n) - y_l(n)$	Tracking errors, $w_l(n)$ is set-point.
$w_l^*,\ \hat{w}_l^*$	Optimal set-point and approximate optimal set-point.
$\frac{\partial f_l}{\partial x},\ \frac{\partial f_l}{\partial u}$	Partial derivatives with respect to x and u.
$\frac{\partial g_l}{\partial r},\ \frac{\partial g_l}{\partial y}$	Partial derivatives with respect to r and y.
$p_l',\ p_l''$	First-order and second-order derivatives, respectively.
$\frac{\partial^2}{\partial x}f_l,\ \frac{\partial^2}{\partial u}f_l$	Second-order partial derivatives with respect to x and u.
$\frac{\partial^2}{\partial r}g_l,\ \frac{\partial^2}{\partial y}g_l$	Second-order partial derivatives with respect to r and y.
$x_{le},\ u_{le},\ r_e,\ y_{le},\ \hat{y}_e$	Equilibrium points.

γ	Attenuation factor $\gamma > 0$.
β	discount factor $0 < \beta \leq 1$.

In Sect. 8.2**:**

r	Operational index.
r^*	Desired value of the operational index.
$w,\ w^*$	Set-point and Optimal set-point.
u	Control input.
y	Control output.
x	System state.
n	Sampling time instant of the lower-layer control loop, $n = 0, 1, \ldots$
K_p	Proportional coefficient of proportional-integral (PI) controller.
K_i	Integral coefficient of PI controller.
e	Tracking error: $e = w - y$.
N_0	Positive integer.
$E(n)$	Summation of the tracking errors: $E(n) = \sum_{i=1}^{n-1} e(i)$.
k	Updating time instant of the upper-layer operational control loop: $k = N_0 n$.
$f_x,\ f_u$	Partial derivatives with respect to x and u.
$g_r,\ g_y$	Partial derivatives with respect to r and y.
$p',\ p''$	First-order and second-order derivatives, respectively.
$\frac{\partial^2}{\partial x} f,\ \frac{\partial^2}{\partial u} f$	Second-order partial derivatives with respect to x and u.
$\frac{\partial^2}{\partial r} g,\ \frac{\partial^2}{\partial y} g$	Second-order partial derivatives with respect to r and y.
$x_e,\ u_e,\ r_e,\ y_e$	Equilibrium points.
γ	Specific attenuation factor: $\gamma > 0$.
β	Discount factor: $0 < \beta \leq 1$.
$H_w,\ H_\chi$	Partial derivatives with respect to w and χ, respectively.
$\chi(k)$	Residual error.

Index

A

Actor Neural Network (NN), 99–101, 103, 105–109, 167, 198, 199, 202, 203, 205, 206, 287, 288, 291, 296

Adaptive Dynamic Programming combined with Reinforcement Learning (ADPRL), 3, 5, 6, 65, 66, 139, 144, 150

Admissible control policies, 15, 25, 43, 44, 69, 70, 125, 160, 194, 196, 209, 244

Analysis of observer state-based virtual Smith predictor, 144

B

Bias of solution analysis, 162

C

Cooperative game algebraic Riccati equation, 72

Coordination equilibrium, 65, 66, 69, 72, 78, 79

Coupled HJB equations, 139, 189, 190

Critic Neural Network (NN), 98–100, 103, 105–109, 168, 170, 198, 199, 203, 205, 206, 287, 291, 294

D

Data-based off-policy RL for multi-agent game, 203

Dual-rate rougher flotation processes, 235, 256, 258, 260

Dynamic programming, 1, 3, 11, 91, 110, 119, 126, 143, 156, 157, 185, 186, 211, 234, 287

F

Feedback PI control, 239, 258

G

Game algebraic Riccati equation, 18, 238, 244, 245

Global Nash equilibrium, 185, 186, 189, 191, 193–195, 203, 274, 282, 289, 298

Graphical game for optimal synchronization, 185

H

Hamilton–Jacobi–Bellman (HJB) equations, 2, 3, 14, 17, 91, 92, 94–98, 103, 104, 157, 172, 190, 191, 193–195, 198, 203

H_∞ control, 11–15, 19, 22, 27, 30, 35, 41–43, 45, 46, 48, 49, 53–56, 60

H_∞ Tracking Control for Operational Processes, 240

I

Interleaved robust reinforcement learning, 91, 98

L

Lifting technique, 240, 242, 259, 260

J. Li et al., *Reinforcement Learning*, Advances in Industrial Control,
https://doi.org/10.1007/978-3-031-28394-9

M

Markov decision process, 1, 2, 6, 156
Multi-agent graphical games, 186, 188, 274, 278, 279

N

Nash equilibrium, 7, 13, 14, 17, 18, 26, 41–43, 45, 46, 49, 52, 53, 60, 66, 113, 139–143, 147–150, 185, 189, 191, 207–209, 211, 212, 218, 228, 238, 250, 268, 279, 280, 282, 285–287, 291, 298
Network control systems, 4–6
Neural network-based off-policy interleaved Q-learning, 166
No bias analysis of the solution, 79

O

Off-policy game Q-learning, 6, 7, 11–13, 21, 22, 26–29, 36, 37, 49, 60, 145, 147, 148, 150, 185, 208, 212, 215, 219, 221, 222, 228, 230
Off-policy game Q-learning algorithm design, 145
Off-policy Q-learning for single-player networked control systems, 113
Off-policy RL for multi-agent games, 197
Off-policy RL for optimal operational control, 250
On-policy game Q-learning, 13, 17–21, 27–29, 35, 48, 208, 212, 213, 215, 221, 222, 228, 230
On-policy Q-learning based on zero-sum game, 262
Optimal control, 1, 3–7, 18, 25, 45–47, 65–69, 81, 91–94, 113, 115, 125–127, 132, 137, 139, 143, 155–159, 161–166, 169, 170, 174, 176–178, 181, 186, 193–195, 198, 202–204, 206–208, 212, 218, 222, 225, 233–235, 237, 238, 245, 256, 259, 265, 282, 286–288
Optimal control for affine nonlinear systems, 155, 156
Optimal Operational Control (OOC) problem, 234, 235, 237, 238, 240, 242, 243, 258–260, 298
Optimal robust control problem, 67
Optimum set-point selector, 269
Output regulation, 5, 6, 66–68, 78, 81, 88

P

Plant-wide optimization control, 233

R

Reinforcement learning, 1, 3, 4
Robust controller design, 91, 97, 109
Rougher flotation processes, 233–235, 256, 258–261, 265, 270, 271

S

Saddle point solution, 14
Simplified HJB equations, 93, 97, 101, 110
Synchronization of multi-agent systems, 139, 185–187
Synchronization problem, 185–188, 230

T

Two-time-scale operational, 238

V

Virtual Smith predictor, 5, 140–142, 144, 145, 148, 150

Printed by Printforce, the Netherlands